# Our Place in the Universe - II

Sun Kwok

# Our Place in the Universe - II

## The Scientific Approach to Discovery

 Springer

Sun Kwok
Department of Earth, Ocean, and Atmospheric Sciences
University of British Columbia
Vancouver, BC, Canada

ISBN 978-3-030-80259-2     ISBN 978-3-030-80260-8   (eBook)
https://doi.org/10.1007/978-3-030-80260-8

Cover illustration: Abell 2151 cluster of galaxies. Image taken with the MegaCam camera on the Canada-France-Hawaii Telescope. Credits: CFHT/Coelum - J.-.C. Cuillandre & G. Anselmi

This Springer imprint is published by the registered company Springer Nature Switzerland AG.
The registered company address is: Gewerbestrasse 11, 6330 Cham, Switzerland

# Preface

This is the second volume of the book *Our Place in the Universe*. The first volume *Understanding Fundamental Astronomy from Ancient Discoveries* covers the development of astronomy from ancient times to Newton. The book uses the historical development of astronomy to illustrate the process of rational reasoning and its effect on philosophy, religion, and society. This volume follows this theme and discusses the development of astronomy after Newton, and the parallel evolution of ideas in geology and biology. While the effect of science on technology is well known, the effects of science on how we see ourselves and our world are much less appreciated. The aim of this book is to demonstrate how science motivated intellectual thought and had a major impact on the social development of humans throughout history. Specifically, we use the examples in the development of astronomy to illustrate the process of science, and the effects of evolution in science on our perception of the Universe and on ourselves.

In our educational system, science is often presented to our students as a series of facts. In reality, science is about the process of rational thinking and creativity. What we consider to be the truth is constantly evolving and has certainly changed greatly over the history of humankind. The essence of science is not so much about the current view of our world, but how we changed from one set of views to another. This book is not about the outcome but the process.

As an example, every student knows that the Earth revolves around the Sun. They accept the heliocentric theory as a fact because this was told to them by an authority. However, in my experience, almost no one could cite a single piece of direct evidence for the Earth going around the Sun. As we will see in Chaps. 2 and 3 of this book, direct confirmation of the heliocentric theory is not trivial and only came two hundred years after Copernicus. The fact that this is not emphasized in our teaching of science is indeed worrying. We are asking our students to accept certain facts of science without telling them the tortuous process by which we came to that conclusion. The goal of this book is to show how we know.

Science is often presented as "logical" and the development of science is taught in textbooks as one success after another. The version of scientific development presented is often a sanitized version, where only successes are mentioned. In fact, there have been many (now forgotten) failures and misconceptions that were very popular at the time. When the correct theories came along, they were often resisted, ridiculed, or ignored by the contemporary authorities. If we are unaware of such struggles, we are likely to repeat the same mistakes.

Some may ask: why teach theories that we know to be wrong? The fact is that many of those theories were held up as the truth at the time. Only by tracing the process of discovery can we understand how science works. Students will be able to see current scientific theories in a more critical light and be able to more objectively assess information given to us by the media or authorities today. For scientists, if they are not aware of mistakes made in the past, they may find themselves making similar mistakes in their research now.

This book is based on a course designed for the Common Core Program of the University of Hong Kong (HKU). The HKU Common Core courses are not based on a specific discipline and are designed to help students develop broader perspectives and abilities to critically assess complex issues. I developed this course and taught it from 2010 to 2018. Every year, the class contained about 120 students from all faculties of the university, including architecture, arts, business and economics, dentistry, education, engineering, law, medicine, science, and social sciences. Because of the students' diverse background, no mathematical derivations or calculations were used. The students were, however, expected to understand qualitative concepts, develop geometric visualizations, and perform logical deductions.

At the end of each chapter are some discussion topics that can be used in tutorial sessions or be assigned as essay topics. These questions are designed to motivate students to think beyond the class materials and explore implications of the topics covered. Often these questions have no right or wrong answers and are open-ended to encourage creative thinking.

Jargons are great obstacles to learning. In this book, I try to minimize the use of jargons as much as possible and some technical terms are replaced by simple words with similar meaning. Some concepts have precise definitions, and the use of technical terms is unavoidable. All definitions are presented in the Glossary.

For more technical readers, I have added some optional mathematics and physics in this book, with additional materials presented in the Appendices. Non-mathematical readers can skip these parts. To focus on the evolution of concepts, I have deliberately omitted certain details. Readers who wish a more in-depth understanding of certain topics should consult the respective textbooks.

Every year, students ask me whether they will be handicapped by their lack of previous knowledge of physics and astronomy. In fact, the reverse is true. Students in science are told all the modern notions but usually have often never learned how we arrived at those conclusions. In this book, we try to retrace historical steps to find out how we got to these conclusions.

In addition to lectures, we had weekly tutorials, quizzes, assignments, computer laboratory exercises, a planetarium show, and exams. For the first half of the course, a planetarium show was developed to illustrate the celestial motions observed in different parts of the world and at different times in history. The laboratory exercises were based on computer software, allowing students to have first-hand experience viewing and recording data from simulated observations. The assessments were designed to test whether the students had understood the course materials, could connect material from different parts of the course, had achieved some degree of synthesis, and could apply the acquired knowledge to new situations.

My hope is to help students develop their sense of curiosity and acquire the confidence to ask questions and challenge assumptions. Modern university students should be knowledgeable about our world and aware of how Nature works. From the historical development of science, I hope students will learn to think analytically and quantitatively, keep an open mind, and remain independent from public opinion. They should be able to make rational judgments on the complicated issues facing society today and rise above the ignorance and prejudice that are prevalent in society. These ideas are reflected in my article "Science Education in the 21st Century" published in *Nature Astronomy*, Vol. **2**, p. 530–533 (2018).

For instructors, the two books *Our Place in the Universe: Understanding Fundamental Astronomy from Ancient Discoveries* and *Our Place in the Universe: The Scientific Approach to Discovery* can serve as textbooks for two one-semester general education courses. The first course can be taught independently, but it is advisable that students take the first course before taking the second course.

In the later chapters of this book, I have included some recent research discoveries in astrobiology. I am privileged to have contributed to research in this field during these exciting times. I am also grateful for the opportunity to work with and learn from many contemporary pioneers in this field. I would like to express my sincere appreciation to them, although many of their names are not specifically mentioned in this book.

Over the eight years that I taught the course while serving as Dean of Science, I was ably assisted by lecturers/instructors Tim Wotherspoon, Jason Pun, Anisia Tang, and Sze Leung Cheung. I thank Gray Kochhar-Lindgren, Director of the HKU Common Core Program, for his unyielding support.

Bruce Hrivnak and Anisia Tang provided helpful comments on an earlier draft. I thank Ramon Khanna, my editor at Springer, for encouraging me to publish this book. Clara Wang skillfully drew many of the illustrations in this book.

I first became interested in the subject of the history of science during my second year of undergraduate study at McMaster University, where Prof. Bertram Brockhouse (Nobel Prize in Physics, 1994) introduced me to the subject. His teaching made me realize that physics is more than just mechanical calculations; it is a subject with philosophical and social implications as well.

Vancouver, BC, Canada                                                        Sun Kwok

# Prologue

天地玄黃, 宇宙洪荒。 日月盈昃, 辰宿列張。
寒來暑往, 秋收冬藏。 閏餘成歲, 律呂調陽。

千字文 周興嗣

In the beginning, there was the black heaven and the yellow earth. The Universe was vast and without limit. The Sun rises and sets, the Moon goes through phases, and the stars spread over distinct constellations in the sky. The warm and cold seasons come and go, while we harvest in the fall and store our grains for the winter. A year is composed of an uneven number of months, and harmony of music governs the cosmos.

First eight verses from the *Thousand Character Essay* by Zhou Xing Si (470–521 A.D.), translated from Chinese.

Zhou, an official in the Court of the Liang Dynasty, was asked by the Emperor Wu 梁武帝 (reigned 502–549 A.D.) to arrange a set of 1,000 characters into an essay for the education of the young princes. He composed a rhymed essay of 250 four-character verses where each character was used only once. From the sixth century to the early twentieth century, this essay was commonly used as a primary text to teach young children the Chinese characters.

The essay begins with eight verses that express humans' desire to understand the Universe and their appreciation for the celestial objects' orderly movements. As Zhou describes it, people also recognize that observations of the Sun, Moon, and stars have led to the development of calendars and that the structure of the Universe can be understood by theoretical models.

These verses exemplify the yearning for knowledge of our place in the Universe, which is shared by all ancient cultures. Through tireless observations, our ancestors on different continents observed the behavior of the Sun, Moon, planets, and the stars. They were aware that these patterns were regular, but by no means simple. Although the data collected were similar across cultures, the interpretations of the celestial patterns differed. These interpretations were incorporated into social, religious, and philosophical structures. Throughout history, the evolution of our models of the Universe led to changes in these structures. This book is an attempt to tell the story of the evolution of astronomical development over two millennia and its effect on our society.

# Contents

# About the Author

**Sun Kwok** is a professional astronomer and author, specializing in astrochemistry and stellar evolution. He is best known for his theory on the origin of planetary nebulae and the death of Sun-like stars. His recent research covered the topic of the synthesis of complex organic compounds in the late stages of stellar evolution. His recent books include *The Origin and Evolution of Planetary Nebulae* (Cambridge, 2000), *Cosmic Butterflies* (Cambridge, 2001), *Physics and Chemistry of the Interstellar Medium* (University Science Books, 2007), *Organic Matter in the Universe* (Wiley, 2012), *Stardust: The Cosmic Seeds of Life* (Springer, 2013), and *Our Place in the Universe: Understanding Fundamental Astronomy from Ancient Discoveries* (Springer 2017). He has lectured extensively at major universities, research institutes, and public forums all over the world. He has been a guest observer on many space missions, including the *Hubble Space Telescope* and the *Infrared Space Observatory*.

He has performed extensive service in international organizations, including as the President of Astrobiology Commission of the International Astronomical Union (IAU), President of IAU Interstellar Matter Commission, chairman of IAU Planetary Nebulae Working Group, and an organizing committee member of IAU Astrochemistry Working Group. His academic affiliations include Dean of Science and Chair Professor of Space Science at the University of Hong Kong, Director of the Institute of Astronomy and Astrophysics, Academia Sinica in Taiwan, Killiam Fellow of the Canada Council for the Arts, and Professor of Astronomy at the University of Calgary in Canada. He currently works at the University of British Columbia in Canada.

# List of Figures

# List of Tables

# Chapter 1
# From Copernicus to Enlightenment

As soon as humans developed a sense of awareness of their surroundings, they realized that their lives are governed by daily cycles of day and night and annual cycles of seasons. They also made the association that these cycles are related to the Sun. For practical and religious reasons, they became keen observers of the heavens. Our ancestors were in awe not only of the existence of celestial objects – the Sun, the Moon, and the stars – but also of the fact that they all move. Not only do they move, but they also do so in regular patterns. The Sun, the Moon, and the stars rise day after day without fail. After the Sun sets and the Earth is covered in darkness, stars can be seen to rise from the eastern horizon, move across the sky, and set in the opposite horizon. The existence of these repeated patterns was believed to be a sign of divine governance and a source of messages from the gods. Motivated by the desire to decipher the hidden meaning of these patterns, our ancestors observed the movement of celestial objects with great diligence.

They soon learned that these patterns, although regular, were complicated. The Sun rises at different times and rises to different heights each day. The length of daylight varies through the year, with longer days in summer and shorter days in winter. When the Sun is highest in the sky, a stick planted vertically on the ground always casts its shadow in the same direction, which is referred to as "north."[1] The existence of this special direction allowed our ancestors to designate four directions: south being opposite to north, and east and west perpendicular to the line of north and south. Although the Sun rises in the general eastern direction, the exact location of sunrise varies each day. The Sun rises in more northerly directions in the summer and in more southerly directions in the winter. As early as 3000 years ago, ancient civilizations realized that there are four special days in a year: summer solstice when the Sun rises in the northernmost direction, winter solstice when the Sun rises in the

---

[1] In order to simplify the narrative, our references to seasons and other observed events are for the northern hemisphere. This statement is true for an observer in mid-northern latitudes.

© Springer Nature Switzerland AG 2021
S. Kwok, *Our Place in the Universe - II*,
https://doi.org/10.1007/978-3-030-80260-8_1

southernmost direction, and vernal and autumnal equinoxes when the Sun rises in the exact east.

Both the Sun and the stars appear to revolve around the Earth each day, but the period of revolution of the stars is slightly shorter (by about 4 minutes) than that of the Sun. This is manifested by stars rising slightly earlier each night, and new stars appearing on the eastern horizon the following night. This results in different constellations of stars being seen at different times of the year: for example, the constellation of Orion is seen in the winter and the constellation of Sagittarius in the summer.

Like the Sun, the Moon also rises in the east and sets in the west. However, unlike the Sun, the Moon changes its shape daily, going through phases. The Moon is sometimes seen at night and sometimes during the day, but its time of appearance is related to its phases. The full Moon rises in the early evening and is seen only during the night, whereas the new Moon rises in the early morning. Although the new Moon is in the sky throughout the day, it is hard to see as it is close to the Sun.

The Sun's and stars' asynchronous movements suggest that in addition to the daily motion of the Sun, the Sun is also simultaneously slowly moving relative to the stars over the period of a year. The daily motion of the Sun is from east to west, but its yearly motion through the stars is from west to east. Like the Sun, the Moon also moves relative to the stars, but much more quickly. Instead of a year, the Moon goes through the stars over a period of 27.3 days. This period is close to, but not exactly the same, as the period of the phases of the Moon, which is 29.5 days.

The Sun and the Moon are not the only objects that move relative to the stars. Our ancestors also knew of five bright celestial objects that move through the stars. For this reason, they are called planets, which originated from the Greek word for wandering stars. The planets Mercury, Venus, Mars, Jupiter, and Saturn all move relative to the stars along the same path of the Sun (called the ecliptic), but with different periods. Like the Sun, they usually travel from west to east relative to the stars. But most strangely, they appear to reverse directions from time to time.

The regular patterns of motions of celestial objects provided humans with the first motivation to understand the working of the heavens. This cultivated the seeds for rational thinking. The first example of abstract thinking is the development of a model of the Universe. The fact that the Sun and the stars appear to revolve around the Earth in circular paths led to the first cosmological model of a flat Earth and a spherical sky. The celestial objects move from horizon to horizon on the surface of a sphere, called the celestial sphere. The axis upon which the stars and the Sun turns is assumed to be the pole that holds up the canopy of the sky (Fig. 1.1).

## 1.1   A Spherical Earth

Although the Earth appears obviously flat from our everyday experience, there are some disconcerting observations that caused ancient astronomers to have doubts about this perception. If we live on a flat Earth, we would see the same stars no

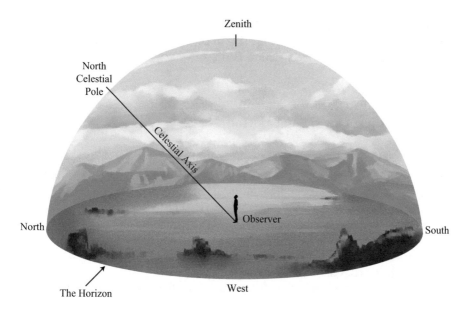

**Fig. 1.1** The first cosmological model of a flat Earth and a spherical sky. The celestial axis around which the stars revolve is inclined with respect to the horizon (shown in green). The north celestial pole is the only point on the celestial sphere that does not turn. This figure corresponds to an observer located in the mid-northern latitude

matter where we are. However, some stars can only be seen from certain locations on Earth: Canopus, the second brightest star in the sky, can be seen from Alexandria but not from Athens. Contrary to intuitive expectations, this polar axis around which the stars turn is not planted perpendicularly but is inclined with respect to the horizon. This inclination angle varies from place to place: at Athens, the polar axis is inclined 38° relative to the horizon; in Alexandria, Egypt, it is 31°.

Around 500 B.C., Greek astronomers deduced that the Earth is round from the shape of its shadow during lunar eclipses. Eclipses occur at a precise time, but the hour of occurrence (as measured from the time at noon) as recorded by observers in the east is always later than that recorded by those in the west. The evidence for a spherical Earth was summarized by Claudius Ptolemy (100–175) in his book *Almagest*.

Using a simple stick planted on the ground and measuring the length and angle of its shadow over the course of a day, ancient observers could trace the exact daily paths of the Sun. Over the course of a year, these paths vary from one day to the next. Observers at different locations on Earth also see different apparent paths of the Sun across the sky (Chap. 5, Vol. 1). An observer in Athens never sees the Sun directly overhead but an observer in southern Egypt does. These results are not compatible with the idea of a flat Earth. By measuring the angle of the Sun at Alexandria on the longest day of the year (summer solstice) when the Sun shines vertically onto the city of Syene in southern Egypt, Eratosthenes was able to determine the size of the spherical Earth (Sect. 12.1, Vol. 1). Through the observations of the Sun, the Moon, and the stars, ancient astronomers were able to determine not only that the Earth was round, but also the size of the spherical Earth.

## 1.2  The First Cosmological Models

After our ancestors realized that the Earth is round and not flat, they were able to make sense of many peculiarities of the motion of celestial objects. The different inclination of the polar axis at different locations is because observers at different latitudes have different horizons (Fig. 1.2). The change in orientation of the horizon accounts for the reason that different stars are visible by observers at different locations. While an observer will see stars rise and set, a star that is too far south of the celestial equator is never visible to an observer in the northern hemisphere because it never rises above his horizon (Fig. 1.2).

All the apparent complexities of the motion of the Sun and the stars can be explained by a two-sphere universe model where a spherical Earth is situated at the center of a celestial sphere upon whose surface the stars lie. In this model, the celestial sphere revolves around the Earth from east to west each day. Because of the rotation of the celestial sphere, stars rise above the horizon in the east and set below the horizon in the west. Some stars can be seen throughout the night but some stars cannot be seen at all (Fig. 1.2).

The Sun also revolves around the Earth, but its motion is slightly slower than that of the stars by four minutes each day. Over the course of a year, the Sun travels across the celestial sphere relative to the stars from west to east along the plane of the ecliptic, which is inclined 23.5° with respect to the equator of the spherical Earth (Fig. 1.3). Because of this inclination angle, the Sun goes through a north-south motion over the course of a year.

The intersection points between the ecliptic and the celestial equator are the vernal and autumnal equinoxes. When the Sun crosses these two points, sunlight shines directly on the equator and the lengths of day and night are equal. On these two days, the Sun rises exactly in the east and sets exactly in the west. The northernmost and southernmost points of the ecliptic are the summer and winter solstices. When the Sun is at these two points, the difference in lengths of day and night is most extreme. Sunrise and sunset will be at their northernmost (on summer solstice) and southernmost (on winter solstice) directions of the year. The two-sphere model, developed by the Greeks over 2000 years ago, can successfully explain the changing seasons and all the known observational facts of the apparent motions of the Sun and the stars.

A practical device called the armillary sphere (Fig. 1.4) based on the two-sphere universe model can be used to predict the time and direction of sunrise/sunset from any place on Earth. It can also determine the length of each day of the year for anywhere on Earth. The two-sphere Universe model was extremely successful.

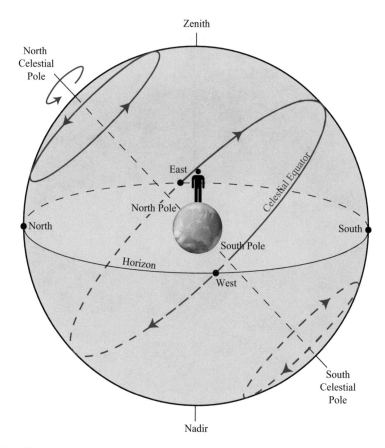

**Fig. 1.2** Changing horizon on a spherical Earth. The sphere of stars revolves around an axis, whose angle of inclination with respect to the horizon (the green plane) is dependent on the location of the observer on the spherical Earth. This diagram illustrates the example of an observer at approximately 45° latitude north. The blue circles are examples of trails of three stars at different declinations over one sidereal day. Parts of the trails that are below the horizon (shown as dashed lines) cannot be seen by the observer

## 1.3   Uneven Movements of the Sun and the Planets

While the two-sphere Universe model represented a tremendous success in explaining the movement of celestial objects, unfortunately, there are other complications. Although the seasons return in regular annual intervals, the lengths of each season are unequal. This anomaly was known as early as 330 B.C. by Callippus (370–330 B.C.), and the lengths of the seasons were measured by Hipparchus (185–120 B.C.) to be 94½, 92½, 88⅛, and 90⅛ days for spring, summer, autumn, and winter, respectively (in the northern hemisphere). Although the planets generally revolve around the Earth from west to east, as does the Sun, they appear to reverse direction (retrograde motion) from time to time. The sphere of fixed stars is not

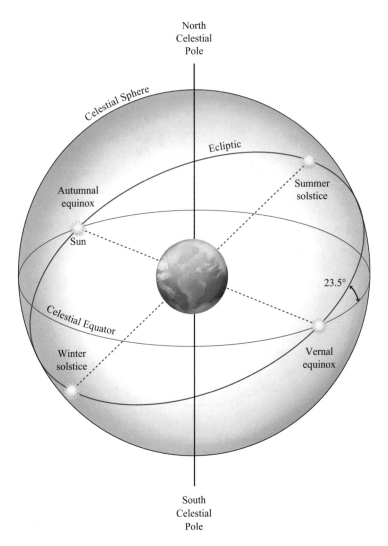

**Fig. 1.3** The two-sphere Universe model. The model consists of an inner sphere (the Earth) and an outer sphere (the celestial sphere) and can successfully explain the apparent motion of the Sun at all locations on Earth. The daily (east to west) motion of the Sun is explained by the rotation of the celestial sphere around the north-south axis of the Earth. Perpendicular to this axis is the celestial equator (in red), which is an extension of the Earth's equator to the celestial sphere. The annual (north-south) motion of the Sun is explained by its movement along the ecliptic (in blue), which is inclined with respect to the celestial equator by 23.5°

oriented permanently with respect to the ecliptic, the annual path of the Sun. The intersection point between the celestial equator and the ecliptic shifts about 1° every 100 years. This phenomenon is known as the precession of the equinox.

**Fig. 1.4** A model of an armillary sphere. The spherical Earth is at the center. Depending on the latitude of the observer, the north-south axis of the Earth is inclined with respect to the horizon (the thick outside plate). Five rings perpendicular to the north-south axis are the Arctic Circle, Tropic of Cancer, the Equator, Tropic of Capricorn, and Antarctic Circle. The ecliptic (a thick band) is inclined 23.5° with respect to the equator (Chap. 7, Vol. 1). Image created by the Technology-Enriched Learning Initiative of the University of Hong Kong

In order to explain these anomalies, Ptolemy in the first century A.D. introduced additional elements into the two-sphere universe model. Epicycles (cycles upon cycles) were introduced to explain the retrograde motions of the planets. A planet moves at uniform speed around an epicycle whose center revolves along a large circle called the deferent. When the planet is inside of the deferent, it is in retrograde motion (Fig. 1.5).

The concepts of eccentric and equant were also needed to explain the unequal lengths of the seasons, the uneven motion of the planets, and the precession of the equinox. The center of the deferent is separated from the position of the Earth by the eccentric. The center of the epicycle revolves uniformly relative to the equant, not the center of the deferent. All three mathematical constructs are necessary to successfully explain planetary motions.

Ptolemy's model of the Universe consists of a stationary Earth at the center, with a celestial sphere of stars revolving around it. Between the Earth and the celestial sphere are the Sun, the Moon, and the planets all revolving around the Earth near the plane of the ecliptic. The order of celestial bodies from near to far is the Moon, Mercury, Venus, the Sun, Mars, Jupiter, and Saturn (Fig. 1.6). To Ptolemy, the epicycles are not just mathematical entities but carry real physical meaning. The orbits and epicycles of the Sun, the Moon, and the planets fill the interior of the celestial sphere without any empty space.

By noting that the horizon of an observer always bisects the whole celestial sphere into two halves no matter where the observer is located on the spherical Earth, Ptolemy concluded that the Earth must be very much smaller than the celestial sphere. In the model of a two-sphere Universe, the Earth is just a point in comparison to the size of the celestial sphere.

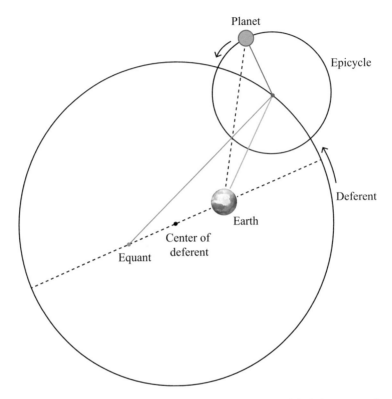

**Fig. 1.5** The use of eccentric, epicycle, and equant in Ptolemy's model of planetary motion. The Earth and the equant are placed at equal distances on opposite sides of the center of the deferent. The planet moves uniformly around the center of the epicycle which revolves around the deferent centered on a point eccentric from the Earth. The angular speed of the center of the epicycle is uniform relative to the equant point

Given the complexity of the motions of the Sun, the Moon, and the planets, Ptolemy faced a formidable task. The high degree of accuracy of his model's predictions – which can quantitatively forecast the positions of the Sun, the Moon, and the planets hundreds of years into the future – is a testament to the model's success. Ptolemy's book *Almagest* represents the pinnacle of Greek science and stood for 1500 years before being challenged.

## 1.4    The Copernican Model

Aristarchus (310–230 B.C.) was the first person to suggest that the Sun, not the Earth, is the center of the Universe. Using his measurement that the Sun is 20 times farther away from the Earth than the Moon is, he estimated that the Sun is seven times larger than the Earth (Sect. 12.2, Vol. 1). He argued that it was unreasonable to

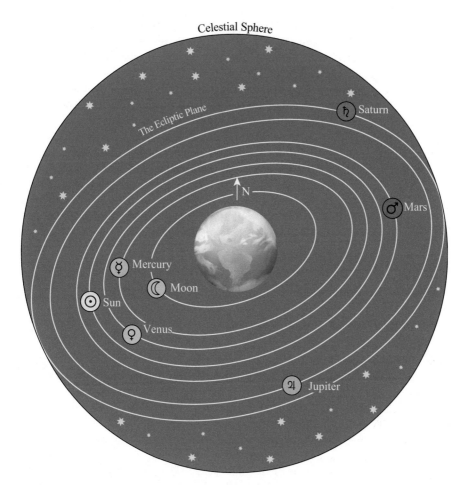

**Fig. 1.6** Schematic diagram of Ptolemy's cosmological model. The celestial sphere upon which all stars lie revolves around a stationary Earth once a day. The Sun, the Moon, and the five planets revolve around the Earth on the plane of the ecliptic also once a day. In addition, these seven celestial objects also move relative to the stars with different periods. The plane of the ecliptic is inclined with respect to the equator of the Earth. The epicycles are not shown in this diagram. The sizes of the planetary orbits and the celestial sphere are not to scale. The Earth should be just a point in comparison to the size of the celestial sphere

assume that a larger body (the Sun) should revolve around a smaller body (the Earth). The heliocentric model was taken up by Nicolaus Copernicus (1473–1543). Copernicus considered the concept of equant in Ptolemy's model as violating the spirit of uniform circular motion and proceeded to seek an alternative without the use of the equant. Copernicus placed the Sun near the center of the Universe, with the Earth and the other five planets revolving around it. The only object that revolved around the Earth is the Moon.

Although Copernicus's model was not more accurate and had more epicycles than the model of Ptolemy, he did remove some unexpected coincidences in Ptolemy's model such as the alignment between the lines connecting the planets to the centers of their epicycles being always in parallel with the line connecting the Earth and the Sun, and the planets revolving around the centers of their epicycles at the same rate as the Sun revolves around the Earth (Sect. 15.3, Vol. 1). Copernicus's model also has the esthetic advantage of providing a natural explanation for the retrograde motions of the planets. From the observed synodic periods (time from conjunction to conjunction or from opposition to opposition) of the planets, he was able to derive the true periods of revolution (sidereal periods) of the planets (Sect. 16.1, Vol. 1). The heliocentric model also naturally explains why the superior planets (Mars, Jupiter, and Saturn) are brightest during opposition, without assuming artificial periods for their epicycles (Sect. 15.3, Vol. 1). Most significantly, Copernicus can predict the relative distances between the planets and the Sun, therefore resolving the ambiguity in the order of Mercury and Venus in Ptolemy's model (Sect. 16.2, Vol. 1).

## 1.5   Immutability of the Heavens

One of the essences of Aristotelian cosmology is the immutability of the heavens. This was further entrenched into the Christian doctrine that the fixed stars came into existence at the time of Creation and were everlasting and unchanging. In 1572, Tycho Brahe (1546–1601) noted the appearance of a new star in the constellation Cassiopeia. Tycho carefully measured the separations of the new star from nearby stars as the stars move from horizon to near the zenith and to the opposite horizon. He found no change in the separations, suggesting that the new star was located far above the Moon and was part of the fixed stars. The appearance of this new star represented a clear violation of the principle of the immutability of the heavens.

## 1.6   A Physical Universe

Like Ptolemy's model, Copernicus's model relies purely on geometry to describe the motions of celestial objects. All objects move in perfect circles and at uniform rates. By removing both assumptions, Johannes Kepler (1571–1630) greatly simplified Copernicus's model. From Tycho's accurate measurements of Mars's apparent motion in the sky, Kepler found that its orbit around the Sun is best described by an ellipse, with the Sun located in one of the two foci of the ellipse (his first law of planetary motion). He also found that the revolution speed of a planet varied during its orbit, and did so in relation to its distance to the Sun (faster when they are near the Sun and slower when they are far from the Sun). The varying speeds of Mars's and the Earth's motions around the Sun can be summarized by his second law which states that planets sweep through equal areas in their orbit over equal time. By removing the requirement of uniform circular motion, Kepler completely removed the need for eccentrics and epicycles. Not only did Kepler discover new rules that govern the motion of individual planets (his first and second laws), he also found a new rule that governs the orbital motion of all planets: the orbital periods of planets are proportional to 3/2 power of their distances to the Sun (his third law).

Ptolemy's geocentric model centers on a physical object, the Earth. Copernicus's model is centered at an imaginary point in space near the Sun, and the Sun plays no role in the motion of the planets. Kepler was the first to suggest that the movements of the planets are driven by the Sun, giving the planetary motions a physical origin. This is a departure from the Aristotelian doctrine, in which celestial bodies just move naturally without cause.

Aristotle maintained that there are two kinds of terrestrial motions: one is the natural tendency for objects to fall, and the other is violent motion when an object is pushed or thrown. The need to understand the relationship between the two became more urgent after the invention of cannons for military use. A better understanding of the trajectory of a cannonball was needed to aim the cannon in order to accurately predict its position of fall. Galileo Galilei (1564–1642) was the first person to reconcile the two. He suggested that a cannonball will continue to move at its initial speed while at the same time it is subjected to the pull of the Earth. The combination of the two motions yields a trajectory in the form of a parabola. This physical explanation of the motion of terrestrial objects opened the possibility of also giving the motion of celestial objects a physical interpretation.

This physical model was mathematically formulated by Isaac Newton (1643–1727), through the development of the theory of gravitation and extension of the laws of motion of Galileo. The planets are attracted to the Sun through a central gravitational force which decreases as the inverse square of the distance. An initial tangential velocity allows the planets to remain in a stable elliptical orbit around the Sun, which is located at one of the two foci of the ellipse (Chap. 21, Vol. 1).

The Universe, therefore, runs like clockwork machinery. By introducing the hypothesis of an action-at-a-distance gravitational force, Newton could derive all three laws of Kepler's planetary motion. The gravitational force acts instantaneously between two objects without the need for physical contact and does not require a medium to transmit. The magnitude of the force is proportional to the gravitational masses of the two objects, which are intrinsic physical properties of the objects. The same force acts on all physical objects that have mass, whether it is a planet or a cannonball. While ancient astronomers recognized that the Sun, the Moon, and the Earth have different sizes, Newton introduced the idea that they also have different masses. By assigning the same physical property (mass) to both terrestrial and celestial objects, and the same rules that govern their motions, Newton cast doubt on the distinction between heaven and Earth.

After Newton, scientists came to the belief that the gravitational force is universal and acts across all spatial and temporal scales. It acts on the smallest entities on Earth and applies to all objects in the Universe, past and future.

## 1.7    Social Implications of the Post-Renaissance Model
of the Universe

After the reintroduction of Greek knowledge to Europe from the Muslim world in the twelfth century, the cosmology of Ptolemy was integrated with the metaphysical philosophy of Aristotle to form a new worldview. This view was further reconciled with the Christian faith by placing Hell in the center of the Earth and Heaven outside of the celestial sphere. Humans occupy the central place in the Universe and their souls can descend to Hell or ascend to Heaven. The perpetual revolution of the planets and the stars are driven by the hand of God. Ptolemy, Aristotle, and Christianity have become the three pillars of knowledge. It was impossible to challenge one aspect of the system (e.g., the geocentric model of planetary motion) without affecting the other two pillars. One cannot affirm the heliocentric model of Copernicus without also resolving the contradiction between the motion of the Earth and Aristotle's laws of motion. The degradation of Earth to one of the six planets also raised theological questions about the privileged position of men.

If the celestial sphere does not rotate (according to Copernicus) and stars are no longer required to lie on the surface of the same sphere, then stars can be at any distance from Earth. This opens the possibility that there are an infinite number of stars, as suggested by Thomas Digges (1546–1595). If the Universe is infinite in size, where is Heaven? If the Earth revolves around the Sun and is not at the center of the Universe, where do we place Hell? The possibilities of stars being distant suns and the existence of planets around distant stars hypothesized by Giordano Bruno (1548–1600) also raised questions about the uniqueness of humans.

The scientific revolution began in the sixteenth century with Copernicus's publication of the heliocentric theory, continued with the refinement of the laws of planetary motions by Kepler, the theory of motion of terrestrial bodies by Galileo, and cumulated with Newton's theory of gravitation and unification of terrestrial and celestial motions near the end of the seventeenth century. This sequence of scientific progress eroded the theological basis of cosmology, and totally dismantled the integrated worldview of Aristotle, Ptolemy, and Christianity built up over the previous 1000 years. While scientists held on to the view that God created the Universe, the operation of the Universe was no longer believed to be driven by direct divine intervention but was instead controlled by a mechanical system governed by physical laws. It is through the process of science that God's design can be understood.

The scientific revolution also raised doubts in the reality of prevalent religious doctrines, and by implication, the divine right of kings. The consequences of the idea of Copernicus introduced skepticism into the minds of the people. Instead of blindly following the words and dictum of kings, nobles, the Pope and bishops, people began to question whether they could make their own decisions based on facts and reason. At the end of the seventeenth century, philosophers began to promote the view that reason, not faith, was the path to truth. From the ignorance and superstition common in the Dark Ages, human society was elevated to self-awareness and rational thinking.

The impact of the scientific revolution has extended beyond our attempt to understand the natural world, to the realm of society. People no longer accepted the hierarchical social structures imposed on them by the monarchy and hereditary aristocracy and began to demand rights for individuals. The concepts of liberty and equality that we now regard as obvious and self-evident, were totally foreign and revolutionary in the eighteenth century. This social movement is called The Enlightenment.

The most significant political events that emerged from the Enlightenment are the French and American revolutions. Although there had been peasant uprisings before, such uprisings only replaced one absolute ruler with another. The French and American revolutions instead overthrew entire existing social structures and opened new models of society and governance. The power of the Church was eroded and its control over people's lives diminished. People in the low classes of society were given new rights, treated as (more or less) equals, and gained access to opportunities and education.

For comparison, we may examine the case of China which developed almost entirely independently from the west. Unlike the Greeks, the Chinese had little knowledge of the true size of the world and over the centuries considered themselves the "Middle Kingdom" – in which they possessed all the knowledge that needed to be learned and wealth that needed to be gathered. This perception minimized their desire to explore the world, even though they had the technical capability to do so. For over two thousand years, the social structure of China was dominated by Confucian philosophy. The fact that the scientific revolution never took place in China could be one of the reasons its social structure remained intact for so long. Chinese society did not begin to change until western science and social ideas were introduced.

One could argue whether the Enlightenment would have occurred eventually on its own without the scientific revolution. Without the demonstration of the mechanical explanation of celestial motions and the validity of the scientific method, it would be near impossible to shake the foundation of the existing social structures of the Church and State. With all these things in mind, the scientific revolution can therefore be considered the most important achievement of modern human society.

## Questions to Think About

1. The Greeks knew that the Earth is round as early as 500 B.C. But 1000 years later, Roman scholars Lactantius (250–325 A.D.) and Kosmas (sixth century A.D.) argued that the Earth is flat. How can the progress of science go backward?
2. If the scientific revolution from Copernicus to Newton were delayed by 300 years, what kind of society do you believe we would be living in today?
3. How did the concept of "mass" originate? What are the philosophical implications of the introduction of this concept?
4. How are the concepts of liberty and equality of the Enlightenment traceable to the scientific revolution?

5. Both the Catholic and Protestant churches were initially against the heliocentric theory and regarded it as in conflict with the Bible. What eventually changed their minds?
6. Even today, there are authoritarian regimes that resist the free enquiry of science. Cite some examples from recent history to illustrate this point.
7. Like European monarchs, Chinese emperors also declare that they inherit their power from Heaven and emperors of China are called "son of Heaven." There was no scientific revolution in China and the imperial system lasted until 1911. Was the lack of scientific progress a reason for the delay in social change in China?
8. Will humans undergo another Age of Enlightenment sometime in the future that overthrows our existing beliefs?

# Chapter 2
# Empirical Evidence for the Heliocentric Model

Copernicus published his heliocentric theory in 1543, but acceptance was by no means immediate. His theory was not simpler than Ptolemy's (Copernicus's model has more epicycles) or more accurate, and it was contradictory to common sense. Copernicus replaced the hypothesis of the rotation of a celestial sphere with the hypothesis of a rotating Earth, and the apparent annual motion of the Sun by the Earth revolving around the Sun. Both of these hypotheses can be tested observationally. The lack of evidence for the Earth's rotation and revolution was the reason most scholars were skeptical about the heliocentric theory.

Some of the scientific arguments against the heliocentric theory were presented in the book *Mathematical Disquisitions Concerning Astronomical Controversies and Novelties* published by German astronomer Johann Georg Locher (1580–1630) in 1614. The most obvious problem of the heliocentric theory was the failure to detect the expected shift in apparent positions of nearby stars as the Earth revolves around the Sun. With his precise instruments, Tycho Brahe was able to set an upper limit of 1 arc minute for the shift in star positions as the result of the orbital motion of the Earth. This was the main reason Tycho was unwilling to accept the heliocentric theory. Instead, he chose to develop his own compromise version (the Tychonic system) where the planets revolve around the Sun, but the Sun revolves around the Earth (Fig. 2.1). This geo-heliocentric model preserves some of the advantages of the Copernican model: natural explanation of the retrograde motions of the planets, derivation of the relative sizes of the planetary orbits, and the order of the planets while avoiding the problems associated with a moving Earth (Sect. 19.1, Vol. 1).

The invention of the telescope and its use in the observation of celestial bodies provided new observational data to test the heliocentric theory. The discovery of the phases of Venus leaves no doubt that Venus revolves around the Sun, but this fact is still compatible with the Tychonic system. The discovery of four moons of Jupiter demonstrates that not every celestial object revolves around the Earth, but this fact could also be interpreted as evidence for the Ptolemaic model as the moons are in epicycles around Jupiter.

© Springer Nature Switzerland AG 2021
S. Kwok, *Our Place in the Universe - II*,
https://doi.org/10.1007/978-3-030-80260-8_2

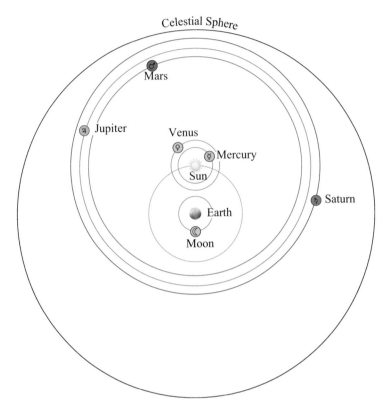

**Fig. 2.1** A schematic drawing of the Tychonic system. While the five planets revolve around the Sun, the Sun and the stars revolve around the Earth, which remains at the center of the Universe. The orbits are not to scale. Adapted from original drawing in *De mundi aetherei recentioribus phaenomenis* by Tycho Brahe (1588)

If the Earth revolves around the Sun and the apparent shift in stellar positions as the result of orbital motion of the Earth is not observed, stars must be very far away. However, in naked-eye observations of stars, they appear to have finite sizes. Tycho Brahe estimated the apparent size of a 3rd magnitude star to be 1 arc minute. This is probably the result of human vision which tends to make us perceive bright stars to be bigger. If stars are indeed very far away, then their observed apparent sizes imply that they must have enormous physical sizes, far beyond what is deemed reasonable. This erroneous perception persisted even after the use of telescopes. The finite size of stars that one sees in a telescope are not the actual sizes of stars, but diffraction patterns introduced by the telescope's lens. This was finally explained by English astronomer George Airy (1801–1891) in 1835 and the "Airy disk" diffraction pattern is named after him. For astronomers at the time, the apparent finite size of stars was a very convincing argument against the heliocentric theory.

Although the Copernican heliocentric theory was promoted by Galileo, who supposedly uttered the phrase "E pur si muove" (and yet it moves) during his trial at the Vatican, he actually had no proof that the Earth was moving. The only "proof" that he came up with was his explanation of ocean tides, which turned out to be wrong. The actual direct observational evidence for the Earth's rotation upon its own axis and its revolution around the Sun came much later, more than 200 years after the death of Copernicus.

## 2.1   Empirical Evidence for the Rotation of the Earth

Copernicus interprets the daily (diurnal) motion of the stars to be the result of the rotation of the Earth. If this is the case, there should be physical experiments that can be done to show that it is indeed the Earth that is rotating, not the stars. If the Earth is rotating with a uniform angular speed, the physical speed of the Earth's rotation varies with latitude. The rotation velocity is highest at the equator, decreases with increasing latitude, and is zero at the poles. A projectile launched from the North Pole due south along a single longitude will land west of its target because the earth is rotating eastward (Fig. 2.2). According to the dynamics of Galileo, a projectile launched from the equator to the north has an eastward component to its rotation. Because this eastward motion is larger than the rotation speed at a more northern latitude, the projectile will land to the east. The deflection is always to the right from the point of view of the thrower. This apparent deflection is now called the Coriolis effect (Sect. 13.2), named after the French mathematician Gaspard-Gustave de Coriolis (1792–1843) who mathematically formulate this effect in 1835.

The possibility that a rotating Earth could have an effect on the trajectory of cannonballs was considered by the Italian Giovanni Battista Riccioli (1598–1671) in 1651. Over short distances, the deflection is small, and the non-detection of this

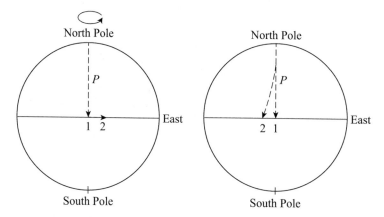

**Fig. 2.2** Illustration of the Coriolis force. For an outside observer viewing a rotating Earth (left panel), a projectile launched from the North Pole will land directly to the South. However, an observer located at the North Pole of the Earth (right panel) will see the projectile landing to the right

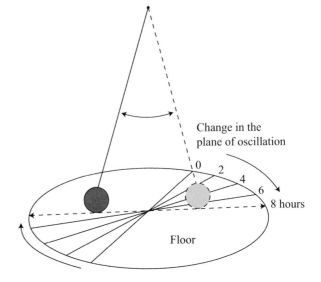

**Fig. 2.3** Foucault's
pendulum. As the pendulum
swings back and forth, the
plane of oscillation changes
with time as the Earth
rotates under it

effect was used by Riccioli against the heliocentric hypothesis. The same effect is
also expected for a ball dropped from a high tower. Because a ball at the top of a
tower will have a larger eastward rotational velocity due to the rotation of the Earth,
the position of a dropped ball at the foot of the tower is expected to be shifted to the
east. In 1668 Italian mathematician Giovanni Borelli (1608–1679) estimated the
amount of shift and concluded that it was too small to be measured, due to other
factors such as wind.

In 1803, Pierre Simon Laplace (1749–1827) and Carl Friedrich Gauss
(1777–1855) both independently derived a formula for the expected eastward
deflection of free-falling objects. This deflection was first successfully measured
by German physicist Ferdinand Reich (1799–1882), who performed experiments in
1831 in a mine 158.5 m deep in Freiberg, Saxony. A mine was used to minimize
other effects. From the results of 106 experiments, he found an average eastward
deflection of 2.8 cm, in agreement with the theoretical values calculated by Laplace
and Gauss. This experiment can be considered as the first experimental confirmation
of the rotation of the Earth, 288 years after the publication of the heliocentric theory
by Copernicus.

The Coriolis effect is also observable in the large-scale movement of air in the
atmosphere or water in the ocean. The explanation of the atmospheric circulation
pattern by the Coriolis effect was first proposed by American meteorologist William
Ferrel (1817–1891) in 1856.

In 1851, the French physicist Bernard Foucault (1819–1868) suspended a 28-kg
ball with a 67-m-long wire attached to the ceiling of the Pantheon in Paris. As the
ball oscillated back and forth like a pendulum, the plane of oscillation did not appear
to stay fixed but rotated from east to west (Fig. 2.3). This showed that the floor was
turning under the pendulum.

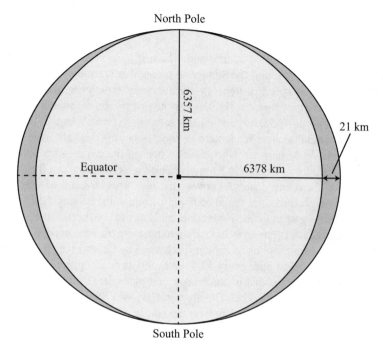

**Fig. 2.4** Oblateness of the Earth. The radius of the Earth at the equator is larger than the radius at the poles as the result of a rotating Earth

The direction of the plane's rotation is opposite to that of the Earth's rotation in the northern hemisphere. This effect varies with latitude. On the equator, there is no rotation of the Earth under the pendulum, and there is no change in the plane of oscillation. The effect is largest at the North Pole where the plane will rotate westward by a complete revolution once every sidereal day. At intermediate latitudes, a pendulum rotates with a period of 24 hours/sin $\phi$, where $\phi$ is the latitude. At $\phi = 50°$, sin $\phi = 0.77$, and the period is 31 hours. The period of the rotation of the Earth can be determined by measuring the rotation rate of the oscillation plane.

The third proof of the rotation of the Earth came with the observation of the oblateness of the Earth. A particle of mass $m$ at latitude $\phi$ on Earth experiences two forces: a gravitational force $GM/r^2$, toward the center of the Earth and a centripetal acceleration (Sect. 13.2) outward of a magnitude $\omega^2 (R_E \cos \phi)$, where $\omega$ is the angular rotation rate of the Earth. The effective gravity is therefore largest at the poles and smallest at the equator. Because the Earth is not totally rigid, this results in the Earth taking on a slightly oblate shape. The modern measurement of the radius of the Earth at the equator is 21 km larger than the radius at the poles (Fig. 2.4). This degree of oblateness, about 1/300, is an indication that the Earth is rotating.

The Coriolis effect, Foucault's pendulum, and the oblateness of the Earth are the three pieces of physical evidence for the rotation of the Earth.

## 2.2   Finite Speed of Light

Scientific discoveries are often intertwined. The first observational evidence for the revolution of the Earth around the Sun relied on another seemingly unrelated topic: the finite propagation speed of light. In our everyday experience, the reception of distant light seems instantaneous. Galileo was the first person reported to have tested whether this is true. In 1638, he had an assistant standing with a lantern on a distant hill stop, who would uncover his lantern as soon as he saw the light from Galileo's lantern. The time-lapse between when Galileo opened the cover of his own lantern and the time when he saw the light from his assistant would represent the time light needed to travel the return distance between the two. This lapsed time was so short that Galileo concluded that the speed of light, if finite, must be very fast.

The small time-lapse in Galileo's experiment is due to the short distance between the two. If the distance is larger, is it possible to measure the time needed for light to travel? Among the four moons of Jupiter discovered by Galileo (Chap. 20, Vol. 1), the moon Io orbits Jupiter once every 42.5 hours and its orbital plane is very close to that of Jupiter. This means that Io can be eclipsed by Jupiter from time to time.

In 1676, a Danish astronomer Ole Rømer (1644–1710) noticed that the time between eclipses of Io is not constant over the course of a year. He found that when the Earth and Jupiter are far from each other (near conjunction) the eclipse intervals are longer than when the Earth and Jupiter are close to each other (near opposition). This difference in time was found to be longer than 10 minutes. Assuming that this variation is not due to a change of Io's orbital period but instead is the result of time delay due to the difference in lengths that light needs to travel across the Earth's orbital diameter (Fig. 2.5), Rømer was able to calculate the speed of light. Because the value of the astronomical unit (Sect. 4.4) was not accurately known at the time, the speed of light that Rømer derived was lower than the modern value. Nevertheless, he was the first person to make a quantitative determination of the speed of light.

Rømer's results were not accepted by the Director of the Paris Observatory Giovanni Cassini (1625–1712), who cited various practical problems in the observations. The results were better received in England, including by Newton and Edmond Halley (1656–1742). The acceptance of the finite speed of light finally came after the discovery of the aberration of starlight by Bradley in 1729 (Sect. 2.3).

If the speed of light is finite and the distances to celestial objects are large, then the images that we see of celestial objects must have been formed sometime in the past, not that of the present. The Sun that we see was the Sun 8 minutes and 19 seconds ago. The time for light from Jupiter to reach us is from 35 minutes (during opposition) to 52 minutes (during conjunction). The times for light from the nearest stars to travel to Earth are measured in years. This illustrates the vastness of the Solar System and Universe.

The determination of the speed of light was improved by an experiment by French physicist Hippolyte Fizeau (1819–1896). In 1849, he allowed a beam of light reflected from a mirror 8 km away to pass through a rotating wheel with gaps in

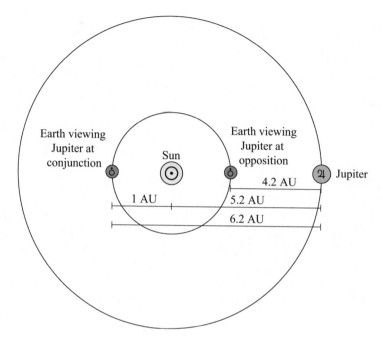

**Fig. 2.5** A schematic diagram illustrating the method used by Rømer to measure the speed of light. At opposition, Jupiter is 4.2 AU from Earth while in conjunction, the distance to Earth is 6.2 AU. The time for light to reach the Earth from Jupiter in conjunction is greater than the time from Jupiter at opposition by 16.6 minutes, using the modern value of the astronomical unit. Although Jupiter cannot be seen in conjunction because it is too close to the Sun, it can be seen near conjunction. This diagram is not to scale

it. By spinning the wheel and watching for the reflected light to come through the next gap in the wheel, he was able to determine the speed of light to within 5% of the modern value.

Because of the large distances between stars, the light-year has become a convenient unit of distance in astronomy. One light-year is defined as the distance that light travels in one year and has a value of 63,240 AU. The nearest star, Alpha Centauri, is 4.2 light-years away, and the brightest star in the sky Sirius is 8.6 light-years away. The discovery of the finite speed of light was crucial in the first detection of the orbital motion of the Earth around the Sun.

## 2.3   Search for Empirical Proof that the Earth Revolves Around the Sun

The variation of the apparent positions of stars as the Earth revolves around the Sun is a core prediction of the heliocentric theory. Because stars can be at different distances, the position of a nearby star relative to a background of distant stars is

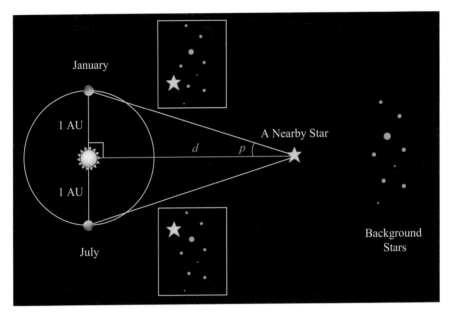

**Fig. 2.6** A schematic diagram illustrating the measurement of stellar parallax. As the Earth revolves around the Sun, the apparent directions of nearby stars will shift against the background of more distant stars. This shift is the maximum between two dates separated by half a year (upper and lower boxes). The angle *p* is defined as the parallax and can be translated into distance *d* using the value of the astronomical unit (AU). This diagram is not to scale

expected to shift with the seasons as the Earth moves to a different place in its orbit. This is illustrated in the schematic diagram in Fig. 2.6. As the Earth goes from January to July, the angular position of a nearby star can shift by $2p$. This angle $p$ is called the parallax and is given by the ratio of the astronomical unit to the distance to the star. The failure to find stellar parallax was one of the reasons that the heliocentric theory was not immediately accepted.

With the construction of larger telescopes and the improvement of the ability to measure small angular separations between stars, there was an all-out effort to search for the elusive stellar parallax.

James Bradley (1693–1762) installed a "zenith telescope" pointed straight up to observe the position of a single star to look for annual shift pattern in position (parallax). The star Gamma Draconis was chosen because it passes through the zenith at London's latitude and therefore suffers fewer effects introduced by the Earth's atmosphere. Bradley did detect a small shift of stellar position, but this shift cannot be attributed to parallax. Instead, he discovered a new phenomenon of aberration of starlight in 1729.

We can find an analogy for the aberration of starlight in our experience of running in falling rain. When we stand still, rain seems to fall vertically, and we need to hold the umbrella vertically. But once we start running, we must tilt the umbrella forward at an angle as the rain now seems to fall at an angle. If the speed of light is finite (as in

**Fig. 2.7** A schematic diagram illustrating the aberration of starlight. Left: light from a star at the ecliptic pole reaches the Earth vertically at speed $c$. Right: As the Earth orbits around the Sun with velocity $v$, the star would appear to shift in position by an angle of $\sim v/c$

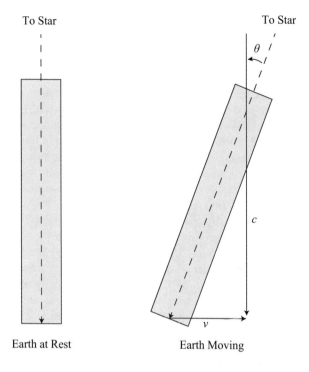

To Star

To Star

Earth at Rest

Earth Moving

the case of the speed of falling rain), then the direction of starlight falling on Earth will seem to tilt as the Earth revolves around the Sun. Because the direction of motion of the Earth at one side of the orbit around the Sun is opposite of its direction on the other side of the orbit, the direction of the tilt of the star is different. A measurement of this angular tilt will give us the orbital speed of the Earth around the Sun. For a star near the ecliptic pole (Fig. 2.7), the tilt is observed to be 20.5 arc seconds. Assuming the modern value for the speed of light ($c$), this translates to an orbital velocity for Earth of $v_e = c \tan (20.5'') = 29.8$ km/s, or over 100,000 km/hour

The discovery of the aberration of starlight was the first empirical proof of the Earth revolving around the Sun. Evidence for the heliocentric theory has gone beyond symmetry and beauty as outlined by Copernicus (Chap. 17, Vol. 1), and can be demonstrated observationally.

The second implication of Bradley's discovery is that it supports Rømer's theory of the finite speed of light, as this hypothesis explains both the changes in intervals between eclipses of Io during the course of the year (Sect. 2.2) and the phenomenon of the aberration of starlight.

## 2.4   Shift of Stellar Position as the Result of Orbital Motion of the Earth

The measurement of the accurate position of stars was not easy. Astronomers had to take into account the effects of precession, aberration, nutation, and atmospheric and instrumental effects. By the early nineteenth century, improvements in precisions set an upper limit of 1 arc second for the angular shift of nearby stars over the period of one Earth's orbit around the Sun. To get an idea of how difficult this is, the ability to measure an angle of 1 arc second is equivalent to trying to measure the size of a 1-cm coin at a distance of 2 km away. This put an upper limit for the parallax of 0.5 arc second. A parallax of less than 0.5 arc second implies that stars are at least AU/$p$, or 400,000 AU away.

Success finally came in 1838. Friedrich Wilhelm Bessel (1784–1846) had chosen the star 61 Cygni as a candidate for his search for parallax based on its very high proper motion (Sect. 5.2), meaning it was therefore likely to be a nearby star. He searched for evidence of periodic shifts over the course of a year in angular separations between 61 Cygni and two apparently nearby (but much farther from Earth) stars. After having made hundreds of measurements and painstakingly considered all sources of possible errors, Bessel found a parallax of 0.3 arc second to the star 61 Cygni. This puts 61 Cygni at a distance of 660,000 AU, or 10 light-years. Bessel's measurement is within 10% of the modern parallax value of 0.287 arc second.

This first success was soon followed by the determination of the parallaxes of Vega by Friedrich Georg Wilhelm von Struve (1793–1864) and of Alpha Centauri by Thomas Henderson (1798–1844). Vega is a bright star in the northern sky and can be conveniently observed by telescopes in the northern hemisphere. Using one of the largest (24-cm) refractor telescopes in the world at that time at Dorpat Observatory, Struve reported a parallax of 0.26 arc second for Vega in 1839. The modern value measured by the *Hipparcos* Satellite for the parallax of Vega is 0.129 arc second, implying a distance of 25 light-years.

Alpha Centauri is the third brightest star in the sky (after Sirius and Canopus), but it was not familiar to ancient northern civilizations in Europe or Asia because of its southern location (–60° in declination). Ancient Egyptians could observe it briefly over the southern horizon and this bright star was of great interest to them. Henderson began to observe the parallax of Alpha Centauri in 1833 at the Royal Observatory at the Cape of Good Hope in South Africa. Two months after the announcement of Bessel, Henderson reported a parallax of ~1 arc second for Alpha Centauri, corresponding to a distance of 200,000 AU (3.3 light-years). The modern value for the parallax is 0.75 arc seconds, implying a distance of 4.4 light-years.

Efforts to measure the parallax of Canopus, the second brightest star in the sky, turned out to be much more difficult. The upper limits set for the parallax of Canopus mean that it has to be very far away. Its apparent brightness is therefore not because of its proximity, but its intrinsic brightness. This raised the possibility that stars are not exactly like the Sun and can have very different intrinsic brightness (Sect. 7.3).

## 2.5   The Third Proof of the Earth's Revolution Around the Sun

In 1842, Austrian physicist Christian Doppler (1803–1853) noticed that the pitch of sound becomes higher or lower as the carrier of the sound approaches or recedes from us. This is now known as the Doppler effect. In 1848, French physicist Hippolyte Fizeau (1819–1896) found that this effect also applies to light: when an emitter of light is coming toward us, the wavelength of light becomes shorter (toward a bluer color), and when it is moving away, the wavelength becomes longer (redder in color). When the light carrier has a spectral line (either bright or dark), it will have a specific wavelength, and the speed of the object can then be measured very precisely. Because stars have dark lines in their spectra (Sect. 9.2), application of the Doppler effect will allow us to measure how fast the star is moving along our line of sight. A blue shift in wavelength will mean the star is approaching, and a red shift, receding.

After the invention of astronomical spectroscopy where stellar spectral lines can be measured (Chap. 9), the motion of the Earth around the Sun could be detected through the Doppler shift of stellar lines. The Doppler effect states that the wavelength $\lambda$ of light will change by $\Delta\lambda = \lambda(v/c)$, where $v$ is the line of sight velocity of the moving object. As the Earth orbits around the Sun, a star on the ecliptic plane will experience the maximum amount of Doppler shift. The effect decreases with ecliptic latitude and vanishes for stars on the ecliptic pole. A star on the ecliptic will appear to move toward or away from the Earth as the Earth orbits the Sun. At the visual wavelength of $\lambda = 500$ nm, the Earth's orbital speed will give a shift of 0.05 nm. In 1872, William Huggins (1824–1910) measured the shifts in wavelength of atomic lines for a number of stars and attributed part of the shift to the orbital motion of the Earth, and the remaining to the velocities of the stars themselves. His observations not only confirmed the orbital motion of the Earth, but also showed that stars are not stationary in space (Sect. 5.2).

## 2.6   A Long Journey from Theory to Confirmation

Copernicus published his heliocentric theory in 1543. Although the discovery of the phases of Venus and of moons around Jupiter gave support to the heliocentric theory, the evidence was not conclusive, as Tycho's model could also explain the phases of Venus while keeping the Earth stationary (Sect. 19.1, Vol. 1). It was not until almost 200 years later, in 1729, that Bradley discovered the phenomenon of aberration of starlight, giving the first observational proof that the Earth is indeed moving relative to the stars. The speed of Earth's revolution around the Sun as derived from the aberration of starlight is over 100,000 km/hour. This is very close to the expected value of the orbital speed of the Earth

$$v_E = \frac{2\pi(\text{AU})}{\text{year}}$$

using the modern value of the astronomical unit. The expected high orbital speed of the Earth that troubled Copernicus's contemporaries so much was finally confirmed.

Final conclusive support of the heliocentric observations came with the detection of stellar parallax in 1838, 300 years after Copernicus. The reason it took so long is that stars are very far away, making the angle of the expected parallax very small (less than 1 arc second) and difficult to measure.

Contrary to popular belief that the heliocentric theory is obvious and self-evident, its acceptance was not immediate. Ptolemy's geocentric theory was not overturned overnight, and the complete triumph of Copernicus's heliocentric theory took over 200 years.

**Question to Think About**
1. Copernicus replaced the daily rotation of the celestial sphere with the hypothesis of the Earth rotating on its own axis. This is part of his heliocentric model to better explain the motion of the planets. If the Earth were covered by such a thick atmosphere such that the study of celestial objects was impossible, how would we discover that the Earth is round or is rotating?

# Chapter 3
# Resolution of the Theoretical Objections to the Heliocentric Theory

When Copernicus's *Six Books on the Revolutions of the Heavenly Spheres* was published in 1543, he faced a number of theoretical objections to his theory. The hypotheses of rotation of the Earth and the Earth's revolution around the Sun are central to the heliocentric theory but they are counter-intuitive. If the Earth is rotating eastward, then any object in the sky, such as flying birds and clouds, would be expected to drift in the opposite direction according to our common sense. In the sixteenth century, the understanding of the motion of objects was based on Aristotle's distinction between "natural" and "violent" motions. If the rotation of the Earth is a natural motion (as it was interpreted by Copernicus) then a falling object would be left behind by the turning Earth. The fact that this is not observed was used as a strong argument against the rotation of the Earth.

The effects of a rotating Earth were discussed as early as the fourteenth century by the French bishop Nicole Oresme (1325–1382). He noted that the daily rotation of the celestial sphere can be explained by a rotating Earth, and argued that our failure to observe flying objects falling behind is because air and water rotate together with the Earth. If this is the case, he concluded that "we could make no observation that would establish that the heavens make a daily rotation and that the earth does not."

Our common sense also tells us that if the Earth is rotating and going around the Sun at high speed, we should feel it. If the Earth is rotating on its own axis, the speed of rotation at the equator is $2\pi R_E/24$ hours $= 1670$ km per hour. Because the radius of the Earth ($R_E$) has been relatively well determined since the time of Eratosthenes, the rotation speed of the Earth could be accurately estimated at the time of Copernicus. The speed of the Earth revolving around the Sun is $2\pi$ AU/year, where AU stands for the astronomical unit, defined as the mean value of the distance between the Earth and the Sun. Using the modern value for AU, the speed of the revolution of the Earth is 107,229 km per hour. Both the rotation and revolution speeds of the Earth are very high compared to speeds we experience in our daily lives, yet we do not feel their effects. Copernicus was unable to answer these objections.

© Springer Nature Switzerland AG 2021
S. Kwok, *Our Place in the Universe - II*,
https://doi.org/10.1007/978-3-030-80260-8_3

## 3.1   The Concept of Inertia

We all live on the surface of the Earth. Our common sense tells us that the Earth is stationary. We do not feel the Earth moving under us. Aristotle explained objects falling onto the ground as the result of natural motion and a stone thrown into the air or birds flying as examples of violent motions. If the Earth is turning under us, falling stones and flying birds should be left behind by the rotating Earth.

The resolution of this dilemma began with the introduction of the concept of inertia by Galileo. In his book *Dialogue on the Great World Systems* published in 1632, Galileo described an experiment of dropping a stone from the top of a moving ship's mast. The stone will fall to the bottom of the mast but not behind. Although this is empirically true, it is not intuitively obvious. If this question is posed to the general public, most would answer that the stone would fall behind the mast. The recognition of inertia of motion was one of the greatest achievements of Galileo.

Galileo further argued the same is true for the rotating Earth: a stone thrown vertically up in the air will inherit the eastward rotation of the Earth and continues to move along with the rotating Earth. It will fall back directly to the point from which it was thrown, not west of it. A bird flying in the air already shares the eastward rotation of the Earth and will not fall behind because the Earth below is rotating. The same is true for clouds and the atmosphere. The concept of inertia was further expanded into a general study of the physics of moving objects in his book *Dialogues Concerning Two New Sciences* published in 1638, in which he stated that objects which start with a certain motion will continue with that motion. Galileo's study of dynamics laid the foundation for Newton's more comprehensive study of the laws of motion.

## 3.2   Why We Do Not Feel We Are Moving Around the Sun?

One of the strongest arguments against the heliocentric theory was because it was contrary to our common sense. The question "if we are moving at high speed around the Sun, how come we do not feel it?" was extremely difficult to answer and it was not until 140 years later that this was resolved by Newton.

Newton extended the concept of inertia introduced by Galileo and integrated the motions of terrestrial and heavenly objects under the same set of rules (Newton's laws of motion and the theory of gravitation). According to Newton (his first law), an object under no force will maintain at rest or move uniformly in a straight line. When a force is applied, the object's motion ceases to be uniform and/or remain on a straight line, and the acceleration of the object is proportional to the force acting on it (Newton's second law). In the example of the gravitational force acting by the Sun on the planets, this inward central force bends the initially straight, uniform motion of a planet into a circle, ellipse, or parabola (Chap. 21, Vol. 1).

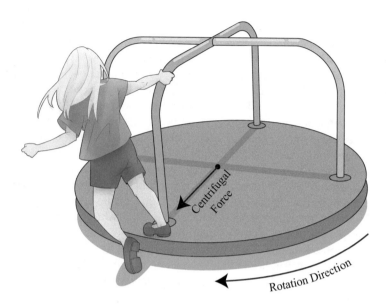

**Fig. 3.1** A merry-go-round. Everyone has the experience of feeling an outward force when on a merry-go-round. We know that we are rotating even if we close our eyes

The presence of a force distinguishes two kinds of motions: (1) those which are at rest or moving uniformly in a straight line, and (2) those which are accelerating or rotating. However, for the concept of motion to be meaningful, it must be measured against a standard of rest. A person located in an object undergoing the first kind of motion relative to this standard of rest is considered to be in an inertial frame of reference, and a person located in an object undergoing the second kind of motion is considered to be in a non-inertial frame of reference.

A person's experience in these two frames of reference is very different. For example, when we are in a smoothly moving car and close our eyes, we have a perception that we are at rest. We may also have experienced the situation when we are in a stationary train at the station and a train goes by on the adjacent track. We have a feeling that we are moving in the opposite direction, but actually, we are not. In other words, it is difficult for us to tell whether we are at rest or in a state of uniform motion in a straight line.

But we can clearly tell that we are moving when a car we are in sudden brakes, as the car's deceleration will cause our body to fall forward. This force is very real because we can feel it. When we are standing on a rotating merry-go-round, we can feel an unmistakable force outward from the center of the rotation platform (Fig. 3.1). We can tell that we are turning even if we close our eyes. There is a fundamental difference in personal perception whether one is in a uniform moving situation or in an accelerating/rotating situation.

A person standing near the outer edge of a merry-go-round with a radius of 1 m and a rotation period of 5 seconds will have a velocity of 4.5 km per hour and the person can easily feel the rotation. In comparison, the velocity of the Earth revolving around the Sun is $2\pi$(AU)/year, or over 100,000 km per hour. The revolution speed is far greater than that of the merry-go-round and yet we do not feel it.

There are a lot of misconceptions about this question of why we do not feel the motion of the Earth. The most common answer you will find given today (e.g., on the internet) is that it is like flying on an airplane. While the plane is going at high speed, we do not feel the movement. This is incorrect because in a plane flying at constant speed in a straight line, we are in an inertial frame of reference. As soon as the plane turns or decelerates, we will immediately feel the effects. While we cannot distinguish between uniform movement and being at rest, we can tell without looking when the vehicle that we are in is accelerating or rotating.

Newton resolved this problem by introducing a fictitious force called the inertial force in a non-inertial frame of reference (Chap. 13). In the case of the Earth revolving around the Sun, the inertial force is the centrifugal force. Because the Earth is in a state of free fall, this centrifugal force is exactly balanced by the gravitational force from the Sun. The net balance of the two forces is zero. This is the reason we do not feel the Earth's revolution around the Sun. While traveling around the Sun, our everyday experience is still dominated by the gravitational force acting on us by the Earth.

## 3.3  Why We Do Not Feel the Rotation of the Earth?

Another argument commonly raised against a rotating Earth was that bodies not attached to the Earth will fly off into space. It was well known in our everyday experience that if we swing around a stone attached to a string, the stone will fly off if we let go of the string. To this question, Galileo had no answer as he admitted in *Dialogue on the Great World Systems*.

The resolution of this problem was provided by Newton after he developed the theory of gravitation. Objects on the surface of the Earth are bound to the Earth by gravity. We do not fly off into space because the gravitational attraction of the Earth keeps us on the ground. An object with mass $m$ on the surface of the Earth is subjected to an attractive force of $mg$, where $g$ is the gravitational acceleration on the surface of the Earth with a value of 9.8 m sec$^{-2}$. The object is also subjected to a centrifugal force due to the rotating Earth. For an object at the equator, the centrifugal acceleration is $\omega^2 R_E$, where $\omega$ is the angular speed of the Earth and $R_E$ is the radius of the Earth. Because the Earth rotates once every 24 hours, $\omega$ is $7.3 \cdot 10^{-5}$ radians sec$^{-1}$, and the centrifugal acceleration has a value of 0.03 m sec$^{-2}$. This centrifugal acceleration is about 0.3% of gravity. The gravitational attraction is the dominant force we feel. Because we are used to being bound to the ground, we do not feel the rotation of the Earth. However, if the Earth were to rotate much faster, the effective gravity would be lower, and we would feel like we are walking on the Moon. If the Earth rotates even faster, we would be floating in the air.

## 3.4 Evolution to a True Heliocentric Model

Another fundamental conceptual difficulty of Copernicus's model is that it is not a truly heliocentric model. Figure 3.2 shows a schematic diagram of Copernicus' model of planetary motion. The Earth revolves around a point that is offset (eccentric) from the Sun. A planet revolves around an epicycle whose center ($E$) revolves around a point that is eccentric from the center of the Earth's orbit. By employing both the techniques of eccentric and epicycles, Copernicus was able to simulate the non-circular and non-uniform movement of a planet. Although the Sun is stationary in Copernicus's model, it is strictly speaking not a Sun-centered (heliocentric) model.

In Ptolemy's model, the Universe has a physical center: The Earth. In Copernicus's model, there is no such center and each planet's orbit has a different

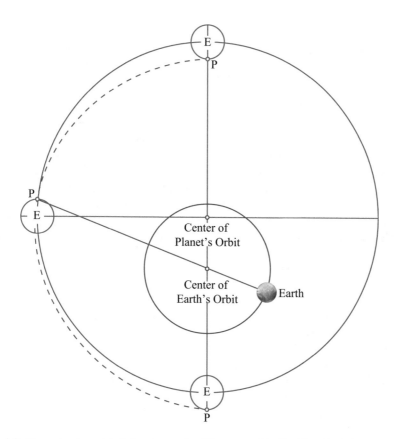

**Fig. 3.2** Copernicus's model of planetary motion. The center of the Earth's orbit (in blue), which is offset from the Sun (not shown). A planet ($P$) revolves around in an epicycle (in green) whose center ($E$) revolves around (in red) a point that is eccentric from the center of Earth's orbit. The final trajectory of the planet is shown by the dashed line (in purple)

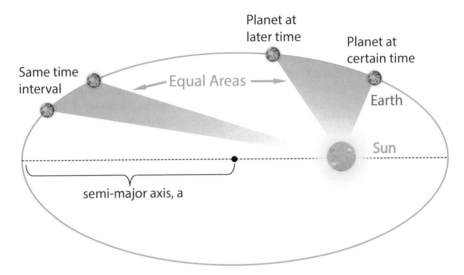

**Fig. 3.3** Kepler's law of planetary motion. Kepler's discoveries that planetary orbits are ellipses with the Sun at one of the foci (Kepler's first law) and that planets sweep through equal areas in equal time (Kepler's second law) were able to greatly simplify Copernicus's heliocentric model

center. The center of Earth's orbit is not the Sun (Fig. 3.2). The center of Jupiter's orbital is near the sphere of Mercury and the center of Saturn's orbit is outside of the sphere of Venus. If these orbits are real, would they not interfere with each other? It was not clear how the different centers of planetary orbits relate to the Sun.

One of the motivations for Copernicus to develop a new cosmological theory is to remove the construct of equant, which he considered to be against the Aristotelian spirit of uniform motion. In order to explain the unevenness of the seasons and the precession of the equinox, Copernicus had to make the Earth undergo all kinds of circular motions. While these motions work mathematically, it was difficult to regard them as real. His contemporary scholars admired Copernicus for his mathematical rigor, but they also considered his model too complicated and arbitrary to be real. In spite of his claim that his model is a Sun-centered model, the role of the Sun is not obvious in his geometric construction.

Kepler's discovery of the elliptical nature of planetary orbits finally gave real significance to the Sun. The Sun is located at one of the two foci of the ellipses and all planets truly revolve around the Sun. Kepler also conceived the idea that planetary motions are driven by a force emanating from the Sun. With Kepler's second law, the non-uniform motions of the planets are accurately described (Fig. 3.3). The first and second laws of Kepler, therefore, eliminated all the complications of eccentrics and epicycles that Copernicus found necessary.

## 3.5 Artificial Satellites and Interplanetary Travel

The greatest achievement of Newton was his unification of terrestrial and planetary motions. Using his theory of gravity and laws of motion, he could explain the dynamics of both terrestrial and celestial bodies. The rules that govern the trajectory of cannonballs are the same as those that govern how the planets move. While the planets revolve around the Sun, the Moon revolves around the Earth under the same rules.

The unity of motion of cannonballs and the Moon can be illustrated in this thought experiment. When a cannonball is fired horizontally from a high mountain, it will travel for a distance before dropping onto the ground due to gravity (trajectory A, Fig. 3.4). With a larger initial horizontal velocity, the projectile can travel a larger distance (trajectory B). With enough initial velocity, the projectile can go into a permanent circular orbit around the Earth without falling and ends with an orbit similar to that of the Moon (trajectory C). If we can put a projectile into orbit in this way, the object will become an artificial satellite of the Earth. With even higher initial speeds, the projectile can go into an elliptical orbit (trajectory D) and even an unbound parabolic orbit (trajectory E).

A human passenger inside a spacecraft undergoing orbital motions (trajectory C and D) will experience weightlessness. A common misconception about

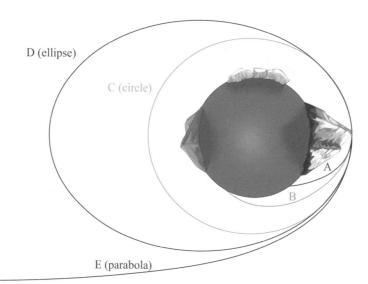

**Fig. 3.4** Artificial satellites. If an object is propelled tangentially from the top of a high mountain, it can land at farther distances with higher initial velocities (A and B). With even higher velocities, it can go into circular (C) or elliptical (D) orbits around the Earth. With sufficient velocity, it can even escape the gravity of the Earth and go into a parabolic orbit (E). This diagram is an extrapolated version of the sketch by Newton in his book *De mundi systemate* (*On the System of the World*), published in 1687

weightlessness in space is that it is the result of lower gravity in a spacecraft far from Earth. In fact, the difference in gravitational force for an astronaut in a space station on a 400 km orbit and the surface gravity on Earth is minimal (about 11%). The correct explanation is that an astronaut in an Earth-orbiting space station is in a state of free fall where the effect of gravity is exactly balanced by the centrifugal force— hence the weightlessness.

Once projected into orbit, an artificial satellite will revolve around the Earth perpetually, just like the Moon revolves around the Earth. For artificial satellites in low Earth orbit, atmospheric drag will eventually slow down the satellite and cause it to fall back to Earth.

The same principle can be used for interplanetary travel. Contrary to popular belief, spacecraft do not require fuel for interplanetary travel. Fuel is required to launch the spacecraft from the rocket with sufficient speed to enter the proper Earth's orbit, but after that, the spacecraft travels on its own under free fall, and no further fuel is required until it gets near the target. For a spacecraft to go from Earth to another planet, it would never fly directly from the Earth to the planet because this would require an incredible amount of fuel. Instead, we inject the necessary horizontal speed to put the spacecraft into an elliptical orbit with a trajectory that will intersect with the orbit of the planet. When the spacecraft gets close to the target planet, it does need to burn fuel to slow down in order to enter the orbit of the planet. During most of the journey, the spacecraft is in a state of free fall and occupants will be under a state of weightlessness.

In order to enter the desired elliptical orbit, a certain energy is required to provide the correct initial horizontal velocity of the spacecraft. For example, to go to Mars with minimum energy, we would design a spacecraft trajectory such that the launch point from Earth would be at the perihelion of the orbit and at a time such that the spacecraft would arrive at Mars at the aphelion (Fig. 3.5).

## 3.6   Final Vindication of Copernicus

The objection to a moving Earth on both scientific and religious grounds persisted well into the eighteenth century. As we have seen in this chapter, the resolution of this issue is by no means trivial. Far from being immediately accepted, the helio-centric model took over 200 years to be firmly established both on theoretical and observational grounds.

While Copernicus believed that he had a good idea, he was also fully aware of the shortcomings of his model. Because he was trying to fit the same data that Ptolemy used and using the same mathematical techniques as Ptolemy, his model became complicated and a large number of epicycles was required. He also realized that his model seemed artificial with all the arbitrary eccentrics for the planets. Still under the influence of Aristotle, Copernicus was unable to explain why we did not feel the movement of the Earth. It took the work of Kepler, Galileo, and Newton to finally settle all these issues.

**Fig. 3.5** Space flight to Mars. A schematic drawing for the trajectory of least energy to Mars. The blue line indicates the orbit of Earth around the Sun, the red line the orbit of Mars around the Sun. The black line is the trajectory of the spacecraft from Earth to Mars. Graph not to scale

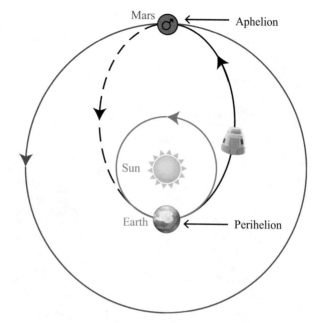

Newton put the heliocentric model into a firm physical basis, with the Sun driving the movement of the planets through a gravitational force. After Newton, almost all scientists have come to accept the heliocentric theory, although at that time there was still no empirical evidence for the revolution or the rotation of the Earth. Newton's theory of planetary motion was so comprehensive and elegant, it was difficult to perceive any other explanation. Faith in the theory of Newton was so great that when the long-sought-after stellar parallax was finally detected in 1838, no one was surprised.

## 3.7 A Turning Point in Our View of the Universe

For over 2000 years until the sixteenth century, our perception of the cosmos consisted of a stationary Earth orbited by the Moon, the Sun, and five planets, all revolving around the Earth on the plane of the ecliptic. Beyond the outermost planet, Saturn, is the celestial sphere of fixed stars. After the integration of Ptolemaic cosmology and the Christian doctrine, a finite celestial sphere takes on a religious meaning as the lower boundary of Heaven (Fig. 3.6). A stationary Earth is significant because its center toward which all objects naturally fall, is where Hell is located. The removal of Earth as the center of the Universe requires a new religious interpretation and the structure of the Universe, as well as a revision to Aristotle's concepts of motion.

**Fig. 3.6** "Empedocles Breaks through the Crystal Spheres" is a wood engraving by an unknown artist. It first appeared in 1888 in a book by Camille Flammarion. The art depicts a man peeking through the junction between the Earth and the celestial sphere to view Heaven, which was believed to be located beyond the celestial sphere. The colorization of this woodcut is by Clara Wang

After Copernicus, the Sun is the center of the Universe, with six planets revolving around it. The stars are no longer on the surface of a sphere but are distributed over different distances. It was not known whether there was an outer boundary beyond the most distant stars. By the early nineteenth century, the Solar System had expanded to include eight planets, plus occasional intruders in the form of comets. The absolute physical size of the Solar System was known through the determination of the astronomical unit. After the determination of distances to nearby stars by the measurement of parallax, astronomers learned that the stars are very far away (the nearest ones more than 100,000 times the distance from Earth to the Sun) and the size of the Universe is much larger than previously thought.

**Questions to Think About**
1. How much should we rely on our common sense? Does our common sense deceive us sometimes?
2. A social experiment: ask your friends why they do not feel the effect of the Earth traveling at high speed and analyze the statistics of the responses.

3. The notions of violent and natural motion of Aristotle may seem arbitrary to modern readers. If so, how can you explain the overwhelming prominence of these ideas over many centuries?

# Chapter 4
# Size of the Solar System

The first attempt to determine the size of the Solar System was by Ptolemy, who assumed that the Solar System was filled by the epicycles of the planets and there is no empty space in the Universe. From the sizes of the planetary epicycles need to fit the observed planetary orbits, he was able to estimate the size of the Solar System (Sect. 15.1, Vol. 1). In his book *Planetary Hypothesis*, Ptolemy listed the maximum distance of Saturn from Earth as 19,865 Earth radii, and this was used as the maximum extent of the Solar System. In the heliocentric model, Copernicus was able to determine the relative sizes of the planetary orbits (Sect. 16.2, Vol. 1). Because the physical sizes of the planetary orbits are scaled to the distance between the Earth and the Sun, this distance represents the key parameter for determining the size of the Solar System.

The first attempt to measure the Earth-Sun distance was by Aristarchus in the third century B.C. He measured the angle between the Earth, quarter-Moon, and the Sun to be 87° (Sect. 12.2, Vol. 1), giving an Earth-Sun distance of $1/cos\,(87°)$ times the Earth-Moon distance. Using an Earth-Moon distance of 70 times the radius of the Earth, Aristarchus estimated an Earth-Sun distance of 1400 times the size of the Earth. Similar values were assumed by Ptolemy, Copernicus and Tycho. Using data from Tycho and his own model for the orbit of Mars, Kepler revised the Earth-Sun distance to 3500 times the radius of the Earth.

After the introduction of the heliocentric model, astronomers expected a shift of the celestial sphere as the Earth goes around the Sun. Because this is not detected, the celestial sphere must be much larger than the orbit of the Earth around the Sun. Assuming Aristarchus's value of an Earth-Sun distance of 1400 Earth radii, the minimum size of the celestial sphere must be 800,000 Earth radii (Sect. 18.1, Vol. 1).

One of the major successes of Copernicus's model was that he could derive the relative sizes of planetary orbits (Chap. 16, Vol. 1). Using Tycho's observational data and the assumption that planetary orbits are elliptical, Kepler improved upon Copernicus's values. For example, the orbit of Venus was changed from 0.719 AU to 0.724 AU (the modern value is 0.723 AU), and Saturn's orbit from 0.92 AU to 0.95 AU (modern value is 0.954 AU). Although these represent major improvements

© Springer Nature Switzerland AG 2021
S. Kwok, *Our Place in the Universe - II*,
https://doi.org/10.1007/978-3-030-80260-8_4

over Ptolemy, the absolute size of the planetary distances from the Sun remained uncertain due to the imprecise knowledge of the value of AU.

## 4.1   How Fast Is Heaven Turning?

By observing the daily turning of the stars, we can determine that the celestial sphere revolves around the north-south axis once every day with an angular period of 23 hours and 56 minutes. In the geocentric model, the rotation of this large sphere posed problems. The physical speed of the celestial sphere is proportional to the size of the celestial sphere. From the sizes of the epicycles, Ptolemy estimated the size of the celestial sphere to be 20,000 times radius of the Earth (Sect. 15.1, Vol. 1). Because the radius of the Earth had been determined by Eratosthenes in 250 B.C., we can use the period of the celestial sphere to calculate the turning speed of the celestial sphere to be 10,000 km/s.

The very fast rotation rate of the celestial sphere as implied by the size of the celestial sphere was philosophically unsatisfying. In the view of Aristotle, all celestial objects move naturally and there is no need to explain how the celestial sphere moves. After the integration of Ptolemaic cosmology, Aristotelian metaphysics, and the religious doctrine of Christianity, angels were believed to be the agents guiding the daily turning of the celestial sphere.

In the heliocentric model of Copernicus, the daily rotation of the stars is explained by the rotation of the Earth, and there is no need for the celestial sphere to rotate. The structure of the Universe has become a static one, with stars all fixed in the sky, and only the Earth, the Moon, and the planets moving at a fast rate.

## 4.2   No More Points Of Light

Among the celestial objects known in ancient times – the Sun, the Moon, planets, and stars – only the Sun and the Moon have an extended finite angular size in the sky observable by the naked eye. The five planets and the stars were points of light, and they were believed to be just geometric points with no physical sizes. As the resolving power of telescopes improved in the seventeenth century, astronomers found that planets were also extended objects.

As early as the fifth century B.C., Greek astronomers were already aware that the Moon shines from reflected sunlight (Chap. 8, Vol. 1). A similar conclusion was arrived at by Chinese astronomers in the first century B.C. The possibility that planets also shine from reflected sunlight was suggested by Chinese astronomer Jing Fang (京房, 78–37 B.C.), This view was echoed by Zhang Heng (張衡; 78–139 A.D.), although neither provided detailed reasoning about how they arrived at this conclusion. The variation of the brightness of planets was well known by the Greeks. Because Ptolemy had a good model for the motions of the Sun and the planets, he

**Table 4.1** Minimum and maximum angular sizes of planets

| Planet | Minimum angular size (arc sec) | Maximum angular size (arc sec) |
|---|---|---|
| Mercury | 4.5 | 13.0 |
| Venus | 9.7 | 76 |
| Mars | 3.5 | 25.1 |
| Jupiter | 29.8 | 50.1 |
| Saturn | 14.5 | 20.1 |

should have been able to predict the variation of the brightness of planets based on their changing distances from the Earth as they move along their epicycles. It is therefore likely that Ptolemy could distinguish whether planets shine by intrinsic light or through reflection of sunlight.

Due to the relative motions of the planets and the Earth around the Sun, the distances between planets and the Earth also vary with time. They are closer to Earth in opposition than in conjunction (for the superior planets Mars, Jupiter, and Saturn), and closer to Earth in inferior conjunction than in superior conjunction (for the inferior planets Mercury and Venus) (Figure 16.3, Vol. 1). Their apparent angular sizes also vary accordingly (Table 4.1).

Jupiter, with an angular size of 50 arc seconds at opposition, was the first planet to be spatially resolved. When Mercury and Venus move across the surface of the Sun (planetary transits, Sect. 4.4), their angular sizes can be compared directly to the angular size of the Sun, which is 31.5 arc minutes (or 0.507 degrees). The angular size of Venus was found by English astronomer Jeremiah Horrocks (1618–1641) to be 76 arc seconds during transit of Venus in 1639. Because Venus is in inferior conjunction during a transit, it is at its closest approach to Earth and this is Venus's maximum angular size in the sky as viewed from Earth. Similarly, Mercury undergoes a transit across the Sun during its inferior conjunction, and its angular size was measured by the French priest Pierre Gassendi (1592–1655) in 1631.

## 4.3 How Far Are the Planets?

From our everyday experience, we know that when we place an object farther and farther away, it appears smaller and smaller. The apparent (angular) size of the object is inversely proportional to its distance (Fig. 4.1). So, if we assume a planet to have a certain physical size ($R_p$), then we can estimate its distance ($D_p$) by measuring its apparent angular size ($\theta_p$) in units of radians:

$$D_p = \frac{R_p}{\theta_p}$$

In 1659, the Dutch astronomer Christian Huygens (1729–1695) assumed that the planet Mars has a size 60% of that of the Earth and proceeded to measure that

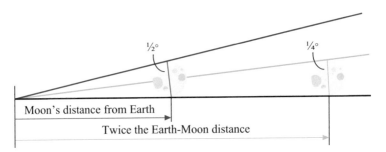

**Fig. 4.1** Relationship between distance and angular size. If the Moon is placed at a distance twice as far away as its present distance, its angular size will change from ½ a degree to ¼ of a degree

angular size of Mars during opposition when it was closest to Earth. Mars has a synodic period of 780 days (Table 16.1, Vol. 1), so Mars will be in opposition about every 26 months. With his 7-cm aperture telescope equipped with an eye piece fitted with a micrometer, he found the angular diameter of Mars to be ~20 arc seconds. The distance to Mars (at its closest approach at opposition) is therefore 0.6 $R_E$/ 10 arc sec = 12400 $R_E$, or about 79 million km. From the work of Copernicus, the radius of Mars's orbit is 1.52 AU (Table 16.2, Vol. 1), so the Mars-Earth distance at opposition is 0.52 AU. The value of AU therefore can be determined to be 23,800 Earth radii (or 152 million km), 17 times the value found by Aristarchus.

However, the actual situation is slightly more complicated because the orbit of Mars around the Sun is not a circle. Kepler realized that Mars had an elliptical orbit, and its closest approach to Earth occurs during perihelion and the distance to Earth can be as short as 0.38 AU if Mars's perihelion also happens to be in opposition (Fig. 4.2). Every 15 or 17 years, Mars opposition occurs within a few weeks of Mars's perihelion.

Because Huygens's assumption of the physical size of Mars is nothing but a guess, his estimate of the value of AU could therefore be wrong by a factor of two or more. However, the significance of this work is that his value for AU was much higher than previous values.

Humans are equipped with a pair of eyes to provide a sense of depth for distant objects. Our two eyes, separated by a few centimeters, will perceive a distant object at slightly different angles. These two separate pieces of information allow our brain to infer the distance of the object observed. The same method can be applied to determine the distance of planets. If a planet is observed simultaneously by two observers separated by a large distance, then the apparent positions of the planet against the background stars would be slightly different. If we know the separation of the two observers, the distance to the planet can be calculated. This method is known as triangulation or parallax (Fig. 4.3). The best time to make such a measurement is during the opposition of Mars when it is closest to Earth.

In 1672, Giovanni Cassini (1625–1712), together with other French astronomers Ole Rømer, Jean Richer, and Jean Picard arranged to have the planet Mars at opposition simultaneously observed from Paris and French Guiana in South

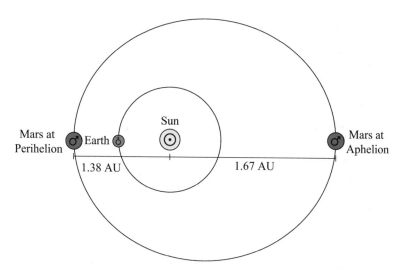

**Fig. 4.2** Huygens's method to determine the Earth-Sun distance. During opposition (left), Mars is closest to Earth and its angular size is the largest. The shortest distance between the Earth and Mars happens when Mars is also at perihelion. This diagram illustrates the closest approach of Mars when opposition coincides with perihelion. By measuring the angular size of Mars and assuming a physical size, the Mars-Earth distance can be obtained. Because the perihelion distance to Mars was known by Kepler to be ~0.38 AU, the value of the astronomical unit can be derived

America. Light rays from the stars, being very distant, will be essentially parallel and appear the same to both observers. However, Mars at opposition is relatively close to Earth and the two different lines of sight to Mars will make the angular position of Mars appear different relative to the stars. Cassini was able to determine an absolute distance to Mars which is within 10% of modern values. From this distance, he derived a value of AU of 21,600 Earth radii. However, due to the smallness of the angles, uncertainties in distances between observers, effects of atmospheric refraction, and synchronization of the clocks, other analyses yielded different answers.

## 4.4   How Far Away Is the Sun?

Because all the planets revolve around the Sun with orbits close to the ecliptic plane, there will be times that the inner planets will pass in front of the Sun when viewed from Earth (Fig. 4.4). This occurrence is called planetary transit and occurs during an inferior conjunction (Figure 16.3, Vol. 1). Because the orbital planes of the inner planets and the Earth are not exactly aligned (Mercury and Venus are inclined 7.0° and 3.4° relative to the ecliptic, respectively), transits require special alignments of the Earth, planet, and the Sun and therefore are rare in occurrence. The English astronomer Edmond Halley became very keen about planetary transits after he

**Fig. 4.3** The method of
triangulation. Through
observations from two
separate locations on Earth,
the apparent position of a
nearby object (e.g., Mars)
relative to the fixed stars
should be different (see
bottom panels). The shift of
the angular position of Mars
(see bottom right box) is
twice the parallax. A
measurement of the parallax
yields the distance to the
object. This diagram is not
to scale

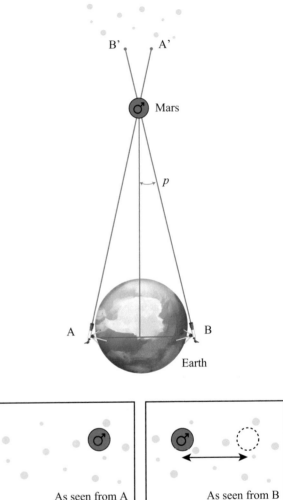

As seen from A

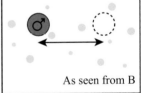
As seen from B

observed the transit of Mercury on October 28, 1677 from the South Atlantic island
of St. Helena, the same island where Napoleon was sent to exile in 1815.

In 1664, James Gregory (1638–1675) suggested that the distance to the planet can
be determined by accurately timing the planet's entrance and departure from the
solar disc by observers at different locations on the Earth. Because of the orbit
orientation of Venus, transits of Venus occur in paired events once every 120 years.
Edmond Halley realized that while he missed the transits of Venus in 1631 and 1639,
the upcoming transits of 1761 and 1769 can be used to determine the distance to
Venus. Halley did not live to do the measurements himself, but he outlined the
procedures for future observers for the upcoming transits.

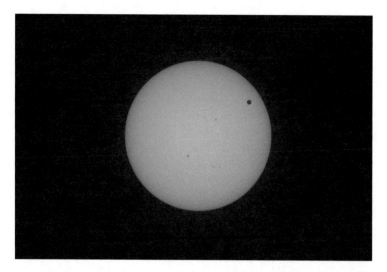

**Fig. 4.4** Transit of Venus on June 6, 2012. Venus is seen as a black dot on the disc of the Sun. Picture credit: Hong Kong Space Museum

By 1761, the astronomical community had become very well organized. The transit of Venus on June 6 was observed by 120 scientists in 62 different locations including Canada, South Africa, Siberia, and India. However, uncertainties in the longitudes and latitudes of some of the observing stations, as well as the uncertain timing of the beginning and ending of the transit introduced errors. The June 3, 1769 transit of Venus was even better organized, with 250 observers making measurements in 130 locations. The most well-known was the expedition of Captain James Cook (1728–1779) who took the British ship HMS Endeavour to Tahiti. These observations resulted in a determination of the astronomical unit to an accuracy within 1% of the modern value.

## 4.5   A Changing Perception of the Size of the Solar System

Figure 4.5 illustrates the evolution of our estimate of the Earth-Sun distance between 900 A.D. and 2000 A.D. Because the size of the planetary orbits is all scaled by the Earth-Sun distance, our perception of the size of the Solar System was greatly expanded after the seventeenth century. Because the size of the Sun scales directly with its distance, it was recognized that the Sun was much larger (109 times) than the Earth, instead of just seven times larger as it was first estimated by Aristarchus (Chap. 12.2, Vol. 1). Because the mass of an object scales by the 3rd power of size (assuming similar densities), the Sun is therefore much, much more massive than the Earth. With an accurate value of the astronomical unit and Newton's law of

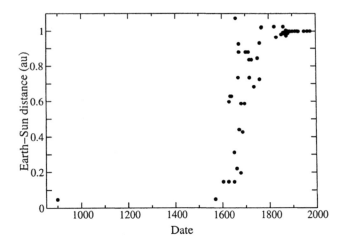

**Fig. 4.5** The change of Earth-Sun distance over history (900–2000 A.D.). The convergence to the correct value happened after the measurements from the transits of Venus in 1769. Figure from Hughes, D.W. 2001, Six stages in the history of the astronomical unit, *J. Astron. History and Heritage*, **4**, 15–28

gravitation, the mass of the Sun can be determined to be 333,000 times mass of the Earth. The drastic differences in mass and size of the Sun compared to the planets suggest that the Sun is in a different class from the planets, and not all the seven luminaries are equal. Heavenly objects can therefore be loosely classified into three types: The Sun (the most dominant object in the Universe known at the time), the planets and their moons, and the stars. At the end of the eighteenth century, astronomers could determine the size of the Sun and the planets, but no such values could be assigned to the stars. They remained distant points of light.

**Question to Think About**
1. Robert Grant (1814–1892) said "the determination of the distance from the Sun to any of the planets revolving around him, is one of the most important problems of astronomical science." Why is it so important?

# Chapter 5
# Celestial Navigation and Exploration of the Heavens

Much of the ancient knowledge of the world was derived from astronomical observations. By observing the shadow of the Earth during a lunar eclipse, Greek astronomers knew that the Earth was spherical in shape. By comparing the maximum altitudes of the Sun at noon on the longest day of the year (summer solstice) at Syene and Alexandria in Egypt and measuring the distance between Syene and Alexandria, Eratosthenes obtained a good estimate of the size of the Earth in the third century B.C. (Chap. 12, Vol. 1). By noting that the star Canopus could be observed just at the horizon at the island of Rhodes but rises to a maximum altitude of 1/48 of a full circle (7.5°) at Alexandria, Posidonius (135–51 B.C.) determined the circumference of the Earth by estimating the distance between these two locations. By the first century B.C., not only the spherical shape of the Earth was known, its size had also been measured to a reasonable degree of accuracy. Based on this knowledge, Greek geographer Strabo (64 B.C.–24 A.D.) wrote in his book *Geography* about the possibility of traveling to Asia by sailing west from Europe, 1500 years before Christopher Columbus.

In order to mark locations on the surface of a spherical Earth, the concepts of longitudes and latitudes were introduced as geographical coordinates. Armed with the knowledge that the Earth is a sphere, ancient astronomers could determine the latitude of a location by measuring the altitude of the pole star above the horizon. As a result, the Greeks knew with relative precision the geographical locations of various cities around the Mediterranean (mostly southern Europe and North Africa). Unlike the latitude, where the equator is the natural choice to be defined as the zero line, there is no preferred beginning line for longitudes (the prime meridian), as all longitudinal lines are equivalent on a sphere. Ptolemy chose the Canary Islands, the most western location known to Europeans, as the prime meridian (Sect. 5.4, Vol. 1).

In his world Atlas *Geographi*, Ptolemy listed the longitudes and latitudes of a number of city locations. However, knowledge of the rest of the world remained vague. From word-of-mouth information gathered from travelers, the Romans were aware of a large civilization (China) far in the east and knew that the land mass of

© Springer Nature Switzerland AG 2021
S. Kwok, *Our Place in the Universe - II*,
https://doi.org/10.1007/978-3-030-80260-8_5

**Fig. 5.1** World map based on the work of Ptolemy. A map made by Nicolaus Germanus in 1467 based on descriptions given in Ptolemy's *Geographi*. Also shown are longitudinal and latitude lines. The two thick latitude lines are the equator and the Tropic of Cancer. The longitude lines range from 0° (centered on the Canary Islands) to 180°, covering about half of the spherical Earth

Africa extends to the south well beyond Egypt. However, no quantitative information on the locations of these regions was available.

In the Middle Ages, the European view of the world consisted of three continents: Europe, Asia, and Africa. A fifteenth-century version of Ptolemy's world map is shown in Fig. 5.1. Besides the Mediterranean, there were two large oceans, the western (Atlantic) and eastern (Indian) oceans. The large island below India is Ceylon (or today Sri Lanka) and the sea east of the large Southeast Asian peninsula is the South China Sea. China is in the far right. The range of longitudes depicted is from 0° (centered on the Canary Islands) to 180°. The map therefore covers about half of the spherical Earth. The world east of China and west of the Atlantic Ocean was unknown.

Because of religious beliefs, map makers at the time put Jerusalem at the middle of the world, with Asia to the east, Europe to the west, and Africa to the south. The shapes and sizes of the European, Asian, and African continents were poorly known, and they were believed to cover most of the surface of the Earth. There were also speculations that there might be an unknown continent somewhere in the southern hemisphere. Beyond the eastern and western land edges was the great ocean which separates Europe from Asia. It was not until the age of exploration between the

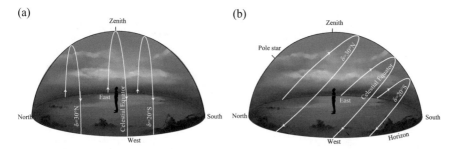

**Fig. 5.2** Paths of stars at different latitudes. (**a**) Left: Paths of three stars at declinations $\delta = 30°N$, 0, and 20°S as seen by an observer at the equator. (**b**) Right: paths of the same 3 stars observed at latitude 45°N. The Pole star is at an altitude of 45° for this observer

sixteenth and nineteenth century that an accurate map of the world finally took shape.

The exploration of the world beyond the coastlines of continents required knowledge of navigation. Once beyond the coast, sailors relied on the Sun and the stars for navigation. All ancient civilizations knew how to use stars as compasses. Although all stars move through the night, there is a star that does not move – the pole star Polaris. An observer at mid-northern latitude will see Polaris at a certain altitude above the horizon. A line projected vertically from Polaris to the horizon gives the direction of north (Fig. 1.1). An observer near the equator will see Polaris near the horizon, which directly indicates the direction north. This same observer will see all other stars rise near vertically in the eastern horizon, trace semi-circular paths across the sky, and set vertically in the western horizon. A star at the celestial equator (declination zero) will rise exactly in the east, will go through the zenith, and then set exactly in the direction west. A star north of the celestial equator will rise in the north-eastern direction, and rises to a maximum altitude when it crosses the great circle (the meridian) connecting the north, zenith, and south before setting in the northwest direction (Fig. 5.2a).

An observer at mid-northern latitudes will see stars rise at an angle with respect to the eastern horizon, rise to a maximum altitude when they cross the meridian and set in the western horizon (Fig. 5.2b).

For the purpose of navigation, a sailor near the equator can use Polaris as a pointer to the direction north, and the path of a star that rises in the east to define the east-west line (Fig. 5.3). However, for a sailor in mid-northern latitude, a star that rises in the east will shift to the south during the course of the night and will be in its most southern direction when it crosses the meridian, before setting in the west. Such rudimentary knowledge of stellar motions was adequate for sailors to navigate in the Mediterranean Sea and the Polynesians to sail from island to island across the Pacific.

In the mid-fifteenth century, Portuguese explorers began to sail south along the west African coast. Contrary to belief at the time that the equatorial region was too hot for human survival, the Portuguese explorer Lopo Gonçalve successfully crossed

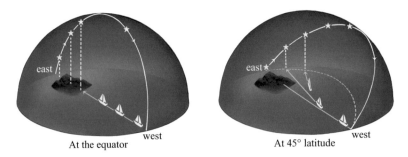

**Fig. 5.3** Celestial navigation. The Sun and the stars were the primary guides of navigation for early sailors. Near the equator, a star that rises in the east will remain in the direct east-west directions and therefore can serve as a good direction indicator. However, in mid-latitudes, the same star will shift gradually to the south after rising. Figure adapted from Aveni, A. 1999, *Stairways to the Stars*, Wiley

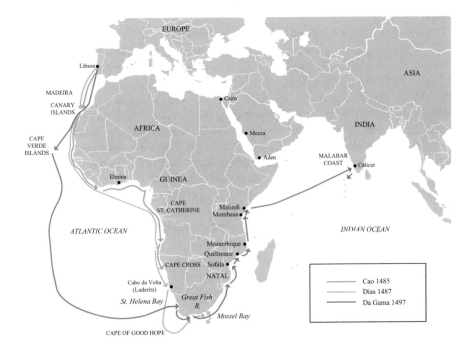

**Fig. 5.4** Portuguese explorations

the equator in 1473. Once sailors crossed the equator, they could no longer see the pole star Polaris and they learned to use the Southern Cross as a guide (Sect. 4.1, Vol. 1). In 1488, Portuguese explorer Bartolomeu Dias (1450–1500) became the first European to sail around the southernmost point of Africa, later named as the Cape of Good Hope by King John II of Portugal (Fig. 5.4). In 1492, Christopher Columbus (1451–1506) commanded a Spanish fleet crossed the Atlantic to reach the West

Indies, which he mistakenly believed to be Asia. In 1497, Giovanni Caboto (1450–1500) led an English expedition from Bristol, England to Newfoundland across the northern Atlantic. In 1498, the Portuguese explorer Vasco da Gama (1460–1524) sailed around the Cape of Good Hope, across the Indian Ocean, to reach India. In 1501, Amerigo Vespucci (1454–1512) showed that the land that Columbus landed on was not Asia, but a new large continent (now named America after him) in its own right. In 1519, Ferdinand Magellan (1480–1521) led a Spanish fleet to sail through a strait near the southern tip of America and into the vast Pacific Ocean which separates the Americas from Asia. The southern continent Australia was finally colonized by England in 1788 after explorations by Dutch and British expeditions. The period between fifteenth and nineteenth centuries was the age of exploration, when the entire Earth was thoroughly explored and mapped.

## 5.1 The Longitude Problem

In open ocean voyages beyond the sighting of coastlines, navigation became a serious problem. While the Sun and Polaris can be used as indicators of latitude, there was no reliable way to determine longitude at sea. As intercontinental commerce and trade increased, the need for accurate longitude determination became paramount. Valuable goods and lives were frequently lost due to inadequate information on the locations of sailing ships. All sea-faring nations in Europe devoted huge amounts of resources to seeking solutions to this problem. In 1666, King Louis XIV of France established the Paris Observatory for the purpose of longitude determination. The Royal Observatory at Greenwich was established by King Charles II of England in 1675 to produce an accurate catalog of star positions so that the lunar distance method, discussed below, could be applied to measure longitudes at sea. To provide further incentives to find a solution, the English parliament offered a prize of £20,000 (equivalent to millions of pounds today) in 1714 to the person who could determine longitude at sea to the accuracy of half a degree.

In theory, the determination of longitude is easy. The Earth rotates once every 24 hours. This means that every hour, the Earth rotates $360°/24 = 15°$, so at every $15°$ west in longitude, the Sun rises one hour later, and noon occurs one hour later. To know one's longitude, one only needs to compare the local time to the time at a location for which we know the longitude. If the local noon occurs 4 hours later than London, we are located at $60°$ west in longitude from London. All a ship needs to do is to carry a clock that keeps good London time.

The problem was that there was no mechanical clock that could run accurately at sea. The alternative was to find an astronomical phenomenon that is known to occur at precise times. By the sixteenth century, the occurrence of solar and lunar eclipses could be predicted relatively accurately. A sailing ship only needs to keep a table of timing of future solar or lunar eclipses at London and compare the time that they are observed to occur at sea. However, eclipses occur very infrequently, so this is not a practical solution.

Astronomers also knew the Moon changes its position in the sky with respect to the stars over the cycle of a sidereal month (Sect. 8.2, Vol. 1). If we have a time table of the Moon's position for every day in the calendar relative to the stars at a location (again, say London), one can observe the angular separation between the Moon and the star at sea, then one can estimate the time difference between the ship and London. This lunar distance method was proposed by German astronomer Johannes Werner (1468–1522) in 1514. However, this method required accurate star charts that were not yet available and models of lunar motion beyond the theoretical capability at the time. Furthermore, making angular astronomical measurements at sea on a rolling ship was difficult.

When Galileo observed Jupiter with his telescope in 1610, he found that Jupiter has four moons (Chap. 20, Vol. 1). At periodic intervals, these moons will move behind Jupiter and cannot be seen. Because the timings of the disappearances and re-appearances of the moons can be observed, one can produce a timetable of the expected eclipses for a certain location and give such a table to a sea captain for comparison with his observations at sea. The disadvantage of this method is that Jupiter cannot be seen in the daytime and can only be observed at night during part of a year. The observation of the Jupiter moons requires clear weather which is not always available.

With improved star charts (Sect. 5.2) and sophisticated celestial mechanics to calculate the lunar positions, the lunar distance method showed promise; all that was needed was a device to make astronomical measurements at sea. In 1731, the sextant (Fig. 5.5) was invented which allowed easy measurements of the altitude of celestial objects above the horizon. Measurements of the altitude of the Sun at noon or the

**Fig. 5.5** A sextant is used to measure the altitude of celestial objects above the horizon. By employing double reflecting mirrors, the relative angular separation between a celestial object with the horizon can be made, even under unsteady conditions at sea. This English sextant was made around 1770. Image courtesy of Adler Planetarium, Chicago, Illinois

altitude of Polaris at night will easily yield the latitude of the observer. It can also be used to measure the angular separation between the Moon and a star. By comparing the results with lunar tables for known locations, the longitude of the observer can be determined. In 1768, the French astronomer Pierre Antoine Véron (1736–1770), on board the French frigate *Boudeuse* commanded by Louis Antonine de Bougainville (1729–1811), used astronomical methods to determine their longitude in the western Pacific, thereby providing the first accurate estimate of the extent of the Pacific Ocean.

Just when the lunar distance method was getting close to being practical, an accurate mechanical time piece was built by John Harrison (1693–1776). Through a series of models built between 1737 and 1770, Harrison showed that his mechanical clocks and watches were more accurate and easier to use than any other astronomical methods. Subsequent versions of these time pieces were widely adopted by national navies and merchant marines of the world for longitude determination at sea. The longitude problem was finally solved.

## 5.2   Stars Are Moving

The thousands of stars in the sky represent a major part of our cosmos observed by ancient civilizations. Although distributed randomly in the sky, they seem to remain fixed relative to each other. Stars rotate together every day with the period of a sidereal day (23 hours and 56 minutes, Chap. 9, Vol. 1). This led to the concept of a celestial sphere upon whose surface all stars lie. After acceptance of the Copernican idea of a heliocentric Universe, the concept of the daily rotation of the celestial sphere was replaced by the rotation of the Earth around the north-south axis. In the Copernican model, stars are fixed and motionless in space. Furthermore, there is no longer a need for stars to lie on the surface of a celestial sphere, and they can be located at any distance from the Earth (Sect. 18.2, Vol. 1). Instead of a two-dimensional celestial sphere, stars can be anywhere in three-dimensional space.

Chinese began compiling star charts as early as the fourth century B.C. In the second century B.C., Hipparchus (185–120 B.C.) cataloged 850 stars from his observatory in Rhodes. Ptolemy listed 48 constellations and the coordinates of 1020 stars in Almagest (Sect. 13.1, Vol. 1). Because the sky is a sphere, the most logical presentation is to create a sky map on a sphere and Ptolemy gave detailed instructions for building such a celestial globe in *Almagest*. The oldest surviving star chart is the Chinese Dunhuang star map made in the Tang Dynasty (618–907 A.D.), which shows 1350 stars in 13 flat two-dimensional sections. Using the most sophisticated instruments available for naked eye observations, Tycho Brahe determined accurate positions for ~1000 stars at the observatory he built in 1581 in Denmark. The invention of the telescope led to the discovery of faint stars which cannot be seen by the naked eye. It became clear that there is not a fixed number of stars in the sky and the number of observable stars increases with the sensitivity of our

**Fig. 5.6** The Big Dipper as
seen in 50,000 B.C. (left)
and 2020 A.D. (right). The
change in the shape of the
constellation is due to the
proper motion of stars. The
directions of proper motion
for the seven stars are
indicated by arrows in the
right panel. Star charts
generated by *Starry Night
Pro. Version 7.6.6*
© *copyright Imaginova
Corp. All rights reserved.*
www.StarryNight.com

 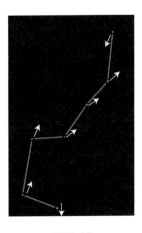

50 000 BC                                    2020 AD

observing instruments. If the Universe is infinite in size, there could be an infinite
number of stars, a possibility first raised by Thomas Digges (1546–1595).

The seventeenth century saw the beginning of a systematic telescopic exploration
of the heavens and the cataloging of positions of stars both old and new. In part
motivated by the need for accurate stellar positions for use by the lunar distance
method for longitude determination, English astronomer John Flamsteed
(1646–1719) carried out a survey over the course of 40 years at the Royal Observatory at Greenwich. When the catalog *Historia Coelestis Britannica* was finally
published in 1725 after his death, it contained accurate positions of ~3000 stars.

Because the Earth is a sphere, there are stars that could not be observed from the
northern hemisphere (Chap. 6, Vol. 1). During the age of exploration, astronomers
came into contact with the southern sky, and began to catalog the positions of
southern stars. In 1676, Edmond Halley measured the positions of 341 southern
stars from St. Helena (latitude 16° south). From the Cape of Good Hope at the
southern tip of the African continent, French astronomer Nicolas Louis de Lacaille
(1713–1762) observed 10,035 stars. Lacaille also introduced 14 new southern
constellations in his southern catalog *Coelum Australe Stelliferum* published
in 1763.

Although the notion of a fixed number of stars was discarded, stars were still
believed to be fixed in their relative positions in the sky. In 1718, Halley studied
positions of stars listed in Almagest and compared their positions obtained by
contemporary observations. He was surprised to find that some stars seemed to
have moved, and this is not just due to precession of the equinox. The bright stars
Arcturus and Sirius had changed positions in the celestial sphere by a degree and half
a degree, respectively, in 1500 years. These results imply that stars are not truly fixed
in their relative positions in the sky, and the shape of the constellation can in fact
change over time (Fig. 5.6). With the instruments available in the eighteenth century,
the apparent motions of stars in the sky could be measured. This apparent motion of

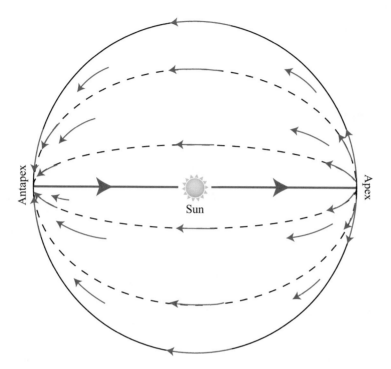

**Fig. 5.7** The effect of solar motion on the proper motions of stars. The apparent motions of nearby stars as the result of the movement of the Sun are indicated by arrows. The mean motions of stars only cancel out at the apex and antapex. An analysis of the proper motion of nearby stars, therefore, allows us to determine the direction of the Sun's motion

stars relative to the Sun[1] is called proper motion and is expressed in units of arc seconds per year.

## 5.3   The Solar System Is Moving

If stars are moving, can the Sun be moving too? This can be tested by observing the proper motion of stars in a small patch of sky. If the Sun is not moving and the stars are moving randomly, then the proper motion of this group of stars should average to zero. If the Sun is moving, this average will not be zero, except for a group of stars in the direction that the Sun is moving to (called the solar apex) or moving away from (the solar antapex) (Fig. 5.7). The first attempt to answer this question was made by William Herschel, who in 1783 suggested that Sun is traveling toward the

---

[1] To obtain the correct value of proper motion, the effects of apparent stellar position shifts arising from the Earth's motion have to be first accounted for.

constellation of Hercules. This result was confirmed in 1837 when proper motions of 390 stars were used. Not only the Earth moves, but the Sun and the whole Solar System also moves.

## 5.4  Unexpected Dividends

Science progresses in different ways. Contrary to the popular perception that science advances linearly and logically, discoveries were often made as the result of the pursuit of totally unrelated goals. The age of exploration was motivated by commerce and trade, and voyages across the oceans required accurate navigation, which was mostly supplied by observations of celestial objects. Because of the practical need to determine longitude at sea, a large amount of resources was devoted to astronomical work during the age of exploration. Although the astronomical methods eventually lost out to a mechanical clock, they did yield unexpected dividends. Due to the efforts invested in the measurement of eclipses of the moons of Jupiter, Rømer noticed in 1676 that when Jupiter is close to Earth (during opposition), the eclipses would occur ahead of schedule. When Jupiter is far from the Earth (at conjunction), the eclipses would occur behind schedule. This led Rømer to discover and measure the finite speed of light (Sect. 2.2).

The lunar distance method for longitude determination requires accurate models of the motion of the Moon with respect to the stars. As a result, a great deal of effort was invested in the theoretical work of celestial mechanics. Orbits of Solar System objects were calculated using the laws of motion by Newton, while also taking into account the gravitational perturbations of other nearby planetary bodies. This interest in celestial mechanics at the Paris Observatory laid the groundwork for Le Verrier's prediction of the existence of Neptune (Sect. 6.3).

## 5.5  Cosmological Implications

In the two-sphere universe cosmology (Chap. 7, Vol. 1), the Sun undergoes two types of motions: a daily (diurnal) motion and an annual motion. After Copernicus, the Sun became stationary, situated at the center of the Universe of a sphere of stars. As the celestial sphere is no longer rotating, there is no need for the stars to all be located on the surface of the celestial sphere. Stars can be at different distances from the Sun but they were still believed to remain fixed relative to each other in angular separation and orientation. Although the apparent distribution of stars in the sky is random, their positions relative to each other seem fixed.

The concerted effort in producing accurate star chart for the lunar distance method led to the discovery of stellar proper motion and the realization that stars are not fixed in the sky. Instead of remaining in fixed orientations with respect to each other, stars seem to move in a random manner across the sky. The Sun is no exception. The fact that the Sun also moves raises questions about its central role in the Universe. Is the Sun really special? Is the Sun, and by implication us, at the center of the Universe? This is the question that we will address in a later chapter.

**Questions to Think About**

1. The Earth undergoes a daily rotation and an annual revolution around the Sun, therefore allowing an observer at the equator to see all the stars in the celestial sphere. If the Earth does not rotate, what does it take for an observer to be aware of all the stars in the celestial sphere?
2. What is the philosophical significance of the fact that the Sun is not stationary in the Universe but moves with respect to the stars?
3. Astronomy was a very practical subject in the seventeenth century. What are the practical applications of astronomy today? Does modern astronomical research have economical implications?

# Chapter 6
# New Members of the Solar System

For thousands of years, humans on Earth across all cultures knew of seven luminaries: The Sun; the Moon; and five planets, Mercury, Venus, Mars, Jupiter, and Saturn, all of which move relative to the fixed stars. The number seven was therefore considered special and sacred (Chap. 10, Vol. 1) and the possibility that there could be more than five planets had never entered anyone's mind. After Copernicus proposed the heliocentric model, the Solar System was centered around the Sun, which had six planets (including the Earth) revolving around it. The Earth, therefore, became a part of the cosmic structure. Although the positions of the Solar System objects were re-arranged, the total number of celestial objects remained the same. When Galileo began his telescopic observations, he found that Jupiter has four moons similar to the Earth's Moon (Chap. 20, Vol. 1). So, the Solar System consisted of a total of 11 objects: The Sun, six planets, and five moons. The special significance given to the number seven was therefore no longer secure.

Motivated by Galileo's discovery of the moons of Jupiter, Dutch astronomer Christiaan Huygens discovered a moon around Saturn in 1655. This new satellite was later named Titan by John Herschel (1792–1871). Under the mistaken belief that there could not be more satellites than planets (six were then known), Huygens stopped searching for further moons. Such metaphysical thinking was not uncommon in the seventeenth century. Fortunately, this did not deter Cassini, who went on to discover more satellites around Saturn: Iapetus in 1671, Rhea in 1672, and Tethys and Dione in 1684.

Could there be more objects in the Solar System besides the Sun, the Moon, planets, and planetary satellites? Ancient sky watchers had long been aware that there are celestial events such as comets and meteors, but these transient phenomena were not considered to be in the same class as the permanent and everlasting seven luminaries. Encouraged by the discovery of moons around Jupiter, astronomers were keen to use the increased capabilities offered by telescopes to search for new objects in the Solar System.

© Springer Nature Switzerland AG 2021
S. Kwok, *Our Place in the Universe - II*,
https://doi.org/10.1007/978-3-030-80260-8_6

**Fig. 6.1**   An artist's depiction of the comet of 1577 seen over Prague. Engraving by Jiri Daschitzky

## 6.1   The Origin of Comets

Comets are magnificent sights that seemingly appear in the sky at random times. Aristotle believed that comets were an atmospheric phenomenon and they resided in the sublunary sphere. When Tycho observed the comet of 1577 (Fig. 6.1), he was unable to detect any changes in its apparent positions from different locations, as should occur for a nearby object. This forced him to conclude that this comet was a distant object located far beyond the orbit of the Moon (Chap. 19, Vol. 1). In 1578, German astronomer Michael Mästlin (1550–1631) determined distances for the comet ranging from 155 Earth radii to 1495 Earth radii, therefore putting the comet within the orbits of Mercury and Venus, according to distance scales published in Ptolemy's *Planetary Hypothesis*[1]. If this is the case, then comets belong with the planets as part of the celestial world. Because the orbits of comets cut through the orbits of planets, they also imply that planetary epicycles are not physical entities, as envisioned by Ptolemy (Chap. 15, Vol. 1).

However, the nature of comets was by no means certain. In 1618, Galileo expressed the opinion that comets were not physical objects but optical illusions.

---

[1] In *Planetary Hypothesis*, Ptolemy listed the minimum and maximum distances to Mercury as 64–166 Earth radii, and to Venus as 166–1079 Earth radii.

**Fig. 6.2** Conic sections.
When a plane is cut through
a cone at various angles, the
cross-sections can be in the
forms of a circle, ellipse,
parabola, or hyperbola

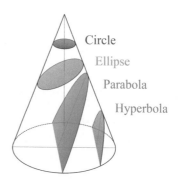

One of the reasons he gave was that comets did not have circular orbits like the planets.

Planetary orbits were found to be elliptical by Kepler. Newton showed the elliptical orbits are the natural results of a central force between the planets and the Sun, and the magnitude of the force varies with the inverse distance squared. However, ellipses are not the only possible orbit for Sun-orbiting objects. Newton's theory of planetary motions also predicts that the trajectory of an object revolving around the Sun can, in general, be any of the four conic sections: circle, ellipse, parabola and hyperbola (Fig. 6.2).

Edmond Halley (1656–1742) thought that comets could also be objects in the Solar System revolving around the Sun, governed by Newton's law of gravitation. If this is the case, their orbits can be elliptical, parabolic, or hyperbolic. A comet with a highly elliptical orbit will have large variations in its distance from the Sun. If comets also shine from reflected sunlight like planets, then their brightness will be dependent on their distance to the Sun. When observed from Earth, a comet will appear to get brighter as it moves close to the Sun, and will appear fainter and eventually fade from view as it recedes from the Sun. After passing the aphelion (the farthest point from the Sun), an elliptical comet will return and eventually comes into view again.

Halley studied orbits of the comets that appeared in 1531, 1607, and 1682, and concluded in 1705 that they were reappearances of the same object. Based on the seemingly fixed time interval of 75 years between appearances, Halley predicted that the same comet would return in 1758. Unfortunately, Halley died before witnessing the fulfillment of his own forecast. His predicted return of the comet was confirmed in December 1758, and then in 1835, 1910, and 1986. This returning comet is now named Halley's Comet in honor of Halley's contribution.

Based on the observed period of Halley's comet, certain historical sightings of comets can be associated with those of Halley's comet. Halley's comet's appearance in 1066 in Europe was described as being as bright as the quarter Moon, and it was depicted in the Bayeux Tapestry (Fig. 6.3). Halley's comet's appearances were also identified in various Asian records. Because of the comet's perceived astrological significance, the Chinese kept extensive records of comet sightings and approximately 30 separate sightings can be associated with Halley's comet, the earliest of which was at 240 B.C.

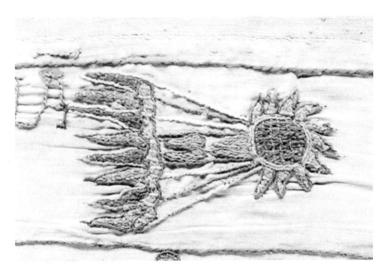

**Fig. 6.3** Halley's Comet in 1066 depicted in the Bayeux Tapestry. The scene on the tapestry suggests that the comet foretells the demise of King Harold, who was defeated by William the Conqueror in the battle of Hastings on October 14, 1066

Halley's explanation of comets conclusively showed that comets are not an atmospheric phenomenon as believed by Aristotle, but are members of the Solar System. The observational confirmation of Hally's prediction was a good example of the scientific method at work (Chap. 8). The identification of the nature of comets as objects in the Solar System also cast further doubts on the concept of an immutable heaven. The number of celestial objects is not fixed. The possibility of comets having parabolic and hyperbolic orbits suggests that there could be newcomers to the Solar System – objects that come into our view of the heavens only once and then leave forever. The invention of the telescope showed that the number of stars in the sky is more than we previously knew. These developments then opened the possibility that the number of planets might also be larger than previously thought.

## 6.2  Discovery of Uranus

If there is nothing special about the number of planets, is it possible that more planets could be discovered through new telescopic observations? On March 13, 1781, William Herschel (1738–1822), a German musician and self-taught astronomer residing in England, found a fuzzy object which resembled a comet. In subsequent observations, he found this object to be moving against the background of fixed stars, but unlike comets, it did not grow a tail or increase in brightness with time. Based on the observed positions, the Finnish-Swedish mathematician Anders Johan Lexell (1740–1784) calculated the orbit of this object and found it to be nearly circular. This made it unlikely to be a comet, but it could possibly be a new planet.

Because it moves across the sky more slowly than Saturn, the new planet must be located beyond the orbit of Saturn. The first new planet since ancient times had been discovered.

After a long debate, the name of this new planet was eventually settled as Uranus, a Greek god of the sky, father of Saturn (Cronus in Greek mythology), and grandfather of Jupiter (Zeus in Greek mythology). In recognition of his discovery, Herschel received a grant from King George III of England in 1787 to build the world's largest telescope with a mirror 48-inches (120 cm) in diameter, and a tube 40 feet (12 m) long, weighing a ton. During his visit to this new observing facility, The King told the Archbishop of Canterbury: "Come, my Lord Bishop, I will show you the way to Heaven."

As it turns out, Herschel was not the first one to have seen this new planet. With a mean magnitude of 5.7, Uranus is just at the limit of naked-eye visibility. Many astronomers have observed this object but did not realize its significance. These earlier observations of Uranus were later compiled to provide a longer time base span to determine its orbit. Uranus' distance from the Sun is about 19.2 AU, curiously close to the prediction of the empirical formula for planetary distances known as Bode's law. In 1772, German astronomer Johann Elert Bode (1747–1826) proposed a mathematical formula that could successfully predict the relative distances of the known planets[2]. According to this empirical formula, the distance to the next planet beyond Saturn was 19.6 AU, reasonably close to the actual observed value.

The discovery of Uranus generated great interest in Bode's law[3]. The law predicts that there should be a planet between the orbits of Mars and Jupiter, and Bode advocated for a search for this missing planet. When Ceres was found by Italian priest Giuseppe Piazzi (1746–1826) in 1801 at a distance of 2.8 AU from the Sun, as predicted by Bode's law, it was seen as a triumph of the law. Ceres was declared as the eighth planet and the fifth planet from the Sun. Ceres' planetary status lasted half a century. As more and more objects were discovered with approximately the same orbital radius of Ceres[4], it was then realized that Ceres is just one of many small Solar System bodies between Mars and Jupiter. These objects were later named minor planets and this region called the asteroid belt.

By the late eighteenth century, mathematical techniques had been developed to apply Newtonian mechanics to the celestial motion of the planets, considering not only the gravitational attraction of the planets by the Sun but also the mutual gravitational attractions among themselves. Although the Sun exerts the largest

---

[2]The Bode's law can be expressed in a simple arithmetic form: $a = 0.4 + 0.3 \times 2^n$, where $a$ is the orbital radius of a planet in units of AU, and $n = -\infty, 0, 1, 2, 3, 4, 5, 6, 7$ for Mercury, Venus, Earth, Mars, Ceres, Jupiter, Saturn, Uranus, and Neptune, respectively.

[3]To the laymen, the term "law" suggests "authority", but in science "law" means an empirical relationship which may or may not be real. Bode's law is an example of the latter.

[4]Pallas was discovered in 1802, Juno in 1804, Vesta in 1807, followed by more smaller objects.

force, the forces of other planets, especially the most massive planet Jupiter, are not negligible. This branch of theoretical astronomy is called celestial mechanics.

When the mathematical techniques of celestial mechanics were applied to fit the observed orbit of Uranus, they failed. Even after considering the gravitational effects of the other planets, the observed orbit of Uranus was significantly different from the theoretical model. There could be several reasons for this failure. First, Because Uranus moves slowly (its sidereal period is 84 years), it may not have been observed long enough to calculate its orbit accurately. Second, the mathematical techniques may have been imperfect, resulting in calculation errors. The third possibility was that the Newtonian theory of gravitation was wrong.

## 6.3   Prediction of Neptune

The task of solving the mystery of Uranus' orbit fell to a young French astronomer Urbain Jean Joseph Le Verrier (1811–1877) of the Paris Observatory. Le Verrier recognized that if none of the above three sources of error were true, the only other possibility was that the orbit of Uranus is being disturbed by another unseen object with an orbit beyond Uranus. Working through the equations of Newtonian dynamics of multiple sources of gravitational pull, Le Verrier predicted, in the summer of 1846, a hypothetical new planet located 36 AU from the Sun, beyond the 19 AU orbit of Uranus. However, his call for observations of this hypothetical planet at its predicted positions fell on deaf ears among his French colleagues. In desperation, he wrote to Johann Gottfried Galle, a young German astronomer working in Berlin, appealing to him to make an observational search. On September 23, 1846, Galle found an unknown 8th-magnitude object within 1° of the position predicted by Le Verrier. Upon further observations that this object indeed moves through the stars, it was identified as a new planet. Later, the new planet was named Neptune, for the Roman god of the sea (Poseidon in Greek mythology).

At almost the same time as Le Verrier was performing his calculations, a young English astronomer John Couch Adams (1819–1892) was also trying to predict the location of this unseen planet perturbing the orbit of Uranus. However, he was unable to convince the Astronomer Royal George Biddell Airy (1801–1892) to conduct an observational search. It was only after the success of Le Verrier that the English astronomical community campaigned to give the credit of discovery to Adams.

The discovery of Neptune was major news in Europe. A new planet had been found, not serendipitously as in the case of Uranus, but as the result of a theoretical prediction. This was a very powerful endorsement of Newtonian mechanics.

Many years later, after the orbit of Neptune was sufficiently well determined by observations, it was realized that the calculations of both Le Verrier and Adams were inaccurate. Both Le Verrier and Adams relied on a distance estimate based on Bode's law, which turned out to be too large. Instead of 38.8 AU as predicted by Bode's law, Neptune was only 30.1 AU from the Sun. In order to compensate for the

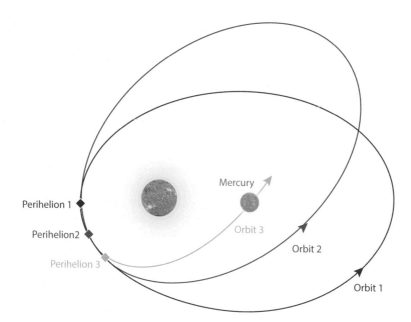

**Fig. 6.4** A schematic drawing illustrating the advance of the perihelion of Mercury. The scale of the diagram is exaggerated for clarity

larger distances between Neptune and Uranus and therefore the weaker gravitational pull between the two planets, Le Verrier and Adams had to assume a mass for Neptune that was too large. It was only luck that at the time of the search by Galle, the predicted position of the planet was close to the actual one. Had the search been done during any other parts of the orbit, the model errors would have been so great that the search would not have yielded anything. The discovery of the new planet Neptune would have probably been delayed for years, until a better theoretical calculation was done.

## 6.4   Search for Vulcan

After the success of predicting Neptune, Le Verrier turned his attention to the peculiar motion of Mercury. It was well known at that time that Mercury's orbit around the Sun is not stable. Its elliptical orbit does not remain fixed relative to the Sun but shifts with time (Fig. 6.4). This shift can be measured by the movement of Mercury's perihelion, which advances 565 arc seconds every century. The calculations of Le Verrier were sophisticated enough that he could attribute most of these erratic behaviors to gravitational disturbances from the other planets – Venus, Jupiter, and the Earth (in order of degree of influence). However, he was unable to account for all the observed effects, with 38 arc seconds per century remaining unexplained.

Drawing a parallel with Uranus, the most logical explanation, if Newton's law is correct, was to assume that there was an unseen planet between Mercury and the Sun that affects the motion of Mercury. By making some reasonable assumptions, Le Verrier was able to guess the approximate mass and orbital size of this unseen planet. Following the tradition of Uranus and Neptune, this mysterious planet was named Vulcan, for the Roman god of fire, also the son of Jupiter, brother of Mars, and husband of Venus.

Because the orbit of this predicted planet is interior to that of Mercury, it is always close to the apparent position of the Sun in the sky. For Mercury, the maximum angular separation from the Sun (elongation, see Chap. 10, Vol. 1) after sunset (eastern elongation), and before sunrise (western elongation) is between 18° and 28°.[5] The maximum elongation expected for Vulcan is 8° and it would be very difficult to spot even in the early morning or after sunset. The best way to detect it is during its transit: when the planet crosses in front of the solar disk. The planet would appear as a dark spot moving across the Sun over a time interval of hours. By the mid-nineteenth century, many astronomers, both professional and amateur, were rigorously searching for Vulcan. The popular press was also driven to frenzy, carrying headlines a number of times on the successful finding of a new planet. All these claims later turned out to be false, mostly due to mistaking dark sunspots for shadows of Vulcan.

Another way to search for Vulcan was to look for uncharted stars near the Sun during solar eclipses. As Vulcan is expected to be near the Sun, it is difficult to spot except during the few minutes of darkness when the Moon covers the Sun. Telescopic observations were set up during the total solar eclipse of July 24, 1878 in search of Vulcan. Again, successful detections were claimed, but the supposed Vulcan later turned out to be stars.

In spite of the unsuccessful searches, scientists were convinced of Vulcan's reality because of their complete faith in Newtonian mechanics. Le Verrier continued to believe in the existence of Vulcan to the day he died on September 23, 1877. One way to account for the orbital anomaly of Mercury (which has since been revised to 43 arc seconds per century from 38 arc seconds per century) is that Newtonian physics is wrong. Some suggested that the inverse-square gravitational law was not exact, with the power of the distance dependency deviating slightly from the value of 2. But this is seen as philosophically unattractive to most scientists. The mystery was finally solved in 1916 when the German physicist Albert Einstein (1879–1955) developed a new theory of gravitation that was able to successfully predict the advance of the perihelion of Mercury (Chap. 13).

---

[5]The maximum elongation of Mercury is not a constant value because of its elliptical orbit and the inclination of its orbital plane relative to the plane of the ecliptic.

## 6.5 Search for Planet X

During the search for Vulcan, there was also an effort to search for a planet beyond the orbit of Neptune. It was believed that there exist irregularities in the orbits of Uranus and Neptune that cannot be fully explained by Newtonian mechanics. Such irregularities could be due to the gravitational pull of an unseen outer planet. This hypothetical planet has been called Hyperion, Hades, Planet X, and other names. In the early twentieth century, the Lowell Observatory in Arizona was used extensively in the search for Planet X. Success was proclaimed when American astronomer Clyde Tombaugh (1906–1997) discovered Pluto in 1930 and it was designated as the ninth planet in the Solar System. Later observations found that the mass of Pluto is too low to cause the orbit irregularities of Uranus and Neptune, and Pluto's orbit and physical properties are now known to be typical of the large number of small bodies found in the outer Solar System (Sect. 10.3). Pluto was removed from the list of planets by the International Astronomical Union in 2006. The discrepancy between theory and observations of the orbit of Uranus and Neptune disappeared when more accurate values for their masses were determined from space missions.

## 6.6 Lesson to Be Learned

The nineteenth century witnessed Newton's pinnacle of success. Due to the triumph of Newtonian mechanics in the prediction of Neptune, scientists in the nineteenth century were absolutely convinced of the existence of Vulcan as the explanation for the problem of the advance of Mercury's perihelion. All that was needed was better observations, and elaborate campaigns were mounted to search for Vulcan. These efforts persisted for over half a century. Although it is easy to follow a path of demonstrated success (such as the use of epicycles to explain the retrograde motion of planets and the use of Newtonian mechanics to explain anomalies in the orbits of Uranus and Mercury), sometimes a qualitative (or paradigm) change such as the Copernican or Einsteinian revolution is needed for science to move forward.

Although very much in the scientific mainstream of the nineteenth century, the search for Vulcan is now almost totally forgotten. It is never mentioned in textbooks and fifty years of scientific work has been erased from the collective memory of the scientific community. Science as taught today is a sanitized version of history, emphasizing the successes from hindsight but almost always ignoring the failures. We should remember that failures are very much an integral part of the scientific process. They should not be forgotten, but rather be used as lessons so that we do not repeat such errors again.

### Questions to Think About
1. What is the significance of Halley's discovery that comets orbit around the Sun?
2. Kepler believed that there should be only six planets because there are only five regular polyhedra. Why do we not believe in this kind of reasoning now?

3. What role does luck play in the advancement of science?
4. Bode's empirical law quite successfully explains the relative sizes of planetary orbits at least out of Uranus, but no physical theory was developed to explain it. Is it a case of pure coincidence?
5. What lessons should we draw from the loss of the planetary status of Ceres and Pluto?
6. The discovery of Neptune was considered as a confirmation of the success of the scientific method and the power of prediction. Were scientists wrong to extrapolate this success to the search for Vulcan? What lesson can we learn from the unsuccessful search for Vulcan?
7. In order to explain an observed phenomenon, a hypothetical planet Vulcan was proposed based on the faith that Newtonian physics was correct. Is there a hypothetical entity that we believe in today to explain an observational fact that could turn out to be fictitious?
8. Astronomers' faith in the existence of Vulcan was based on their faith in Newtonian mechanics, which turned out to be inadequate. How much faith we should put in today's physical theories?
9. Why are failures in science not taught in our science curriculum? What do students miss something by not learning about failures?

# Chapter 7
# Is the Sun a Star?

After Copernicus removed the idea that the celestial sphere rotates around the Earth once a day, Thomas Digges suggested that there is no need for stars to all lie on the surface of the celestial sphere and an individual star can be at any distance from Earth. The Italian monk Giordano Bruno (1548–1600) raised the possibility that stars could be distant suns. In his book *On the Infinite Universe and Worlds* published in 1584, Bruno suggested that since there are planets revolving around the Sun, there could be planets (including other Earths) revolving around other stars. For these and other beliefs, Bruno was condemned to death for heresy in 1600 (Sect. 18.3, Vol. 1).

Because of the dominant effects of the Sun on our lives, it has been considered the most important of all celestial objects and Sun worship was common among ancient cultures. To suggest that the Sun is just one of the thousands of stars is to remove its special status, with profound philosophical and religious implications. If stars are distant suns, stars must be very far from Earth as the Sun is much brighter than the stars. To test this hypothesis, we must answer the question: How far away are the stars? Because our perception of stars is based on their distribution on a two-dimensional celestial sphere with no sense of depth, new methods must be devised to determine the distance to stars, which is no easy task.

## 7.1 Brightness Drop-Off with Distance

We all know from everyday experience that as we move a candle or light bulb further away, its brightness as perceived by our eyes decreases. The exact form of the decrease of brightness with distance is not difficult to determine. Assuming that the light source gives out light evenly over all angles, the amount of light that one receives per angular area (the technical term for this is "luminous flux") is evenly spread out over the surface area of a sphere at that distance. Because the surface area

© Springer Nature Switzerland AG 2021
S. Kwok, *Our Place in the Universe - II*,
https://doi.org/10.1007/978-3-030-80260-8_7

of a sphere at distance $D$ is $4\pi D^2$, the luminous flux received by an observer from the light source decreases with the inverse-square of the distance.

The same principle can be applied to stars. Assuming that stars have constant intrinsic brightness (the technical term is "luminosity," defined as the amount of energy radiated by a star over a period of time), then their apparent brightness should vary when the Earth moves closer to or away from them. If the size of the Earth's orbit around the Sun is a significant fraction of the distances to the stars, then the brightness of stars would be expected to change as the Earth revolves around the Sun. The fact that this is not observed is a testament that stars are far away, many times more distant than the Sun. Exactly how much more distant is an important question and was subject to quests by astronomers from the seventeenth to nineteenth centuries.

## 7.2   Distance to Stars

If stars are distant suns, they appear less bright than the Sun because their emitted energy is spread over a larger area. If we restrict the flow of sunlight through a very small angular area, then the amount of sunlight that we receive will be reduced. In the seventeenth century, Dutch astronomer Christian Huygens (1629–1695) allowed sunlight to pass through pinholes of various sizes in a darkened room until the brightness of the Sun was reduced to a level similar to that of the star Sirius. In order to do this, he had to drill the smallest hole possible and then cover it with layers of opaque material. He estimated that he had to reduce sunlight by 800 million times to match the brightness of Sirius. Using the inverse square law, Huygens concluded that Sirius is $\sqrt{800,000,000} = 28,000$ times more distant than the Sun. We now know that Sirius is intrinsically 25 times brighter than the Sun and taking this into account would give a distance of 140,000 AU. The actual distance is 544,000 AU. Thus, this simple experiment was a reasonable way to estimate the distances to stars, although one must take into account the differences in their brightness to approximate their correct distances. In spite of its simplicity, the method nevertheless gave a good order-of-magnitude estimate of stellar distances.

Unlike stars which have constant brightness, the brightness of planets varies. The apparent magnitude of Mars varies from +1.86 to –2.94, Jupiter from –1.66 to –2.94, Saturn from +1.17 to –0.55. This variation of apparent brightness of planets can be understood by their changing distances from Earth as planets and Earth revolve around the Sun. During opposition, the superior planets Mars, Jupiter, and Saturn are at their brightest as they are closest to Earth.

In 1668, Scottish astronomer James Gregory (1638–1675) proposed a method to use planets to estimate distances to stars. Because planets shine from reflecting sunlight, the brightness of planets is proportional to the fraction of sunlight intercepted by the planet at its distance from the Sun. If a planet reflects all the sunlight that it receives, its apparent brightness as observed from Earth can be estimated based on the distances of the planet from the Sun and from the Earth.

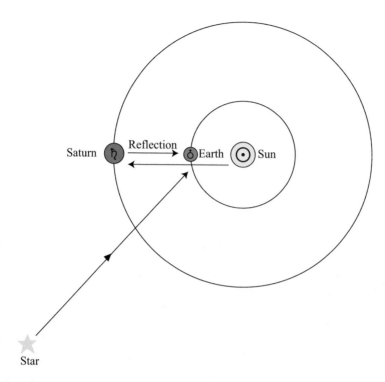

Star

**Fig. 7.1** Schematic diagram illustrating the photometric method for stellar distance determination. This method is based on the comparison between reflected sunlight from planets (which decreases with inverse fourth power of distance) to light emitted by stars (which decreases with inverse square of distance). Diagram is not to scale

Roughly speaking, light received by the planet from the Sun is reduced in proportional to the inverse square of its distance from the Sun, the light received by Earth from the planet is further reduced by another inverse squared distance. So for a planet in the outer part of the Solar System, the apparent brightness of a planet as viewed from Earth is approximately inversely proportional to the fourth power of its distance from the Sun.

Stars, however, shine on their own. If they have intrinsic luminosities similar to the Sun, their apparent brightness must decrease in proportional to inverse-squared distance. By considering the two different ways that brightness depends on the distance between a planet and a star, the distance to the star can be estimated (Fig. 7.1). For a star that has the same intrinsic brightness as the Sun but the same apparent brightness as Saturn, the star must be at a distance of ~400,000 AU, or 6 light-years away, based on the modern value of AU (Appendix D). For example, Saturn at its brightest has an apparent magnitude of −0.55, and the bright stars Sirius and Canopus have an apparent magnitude of −1.46 and −0.74 respectively, so the distances of these stars can be estimated using this method.

This method relies on comparison between the brightness of planets and stars and is now known as the photometric method. Applications of the photometric method were used by various astronomers to estimate the distances to bright stars. For example, Sirius was placed at 240,000 AU by Philippe Loys de Chésaux in 1744, at 500,000 AU by J.H. Lambert in 1760, and at 440,000 AU by John Mitchell in 1767. Although this method is subject to errors, such as the planet may not reflecting sunlight perfectly or stars not having the same luminosity as the Sun, the method nevertheless suggests that stars are very far away.

## 7.3  Are All Stars the Same?

After the direct measurement of stellar parallax (Sect. 2.4) in the nineteenth century, the distances to several nearby stars were determined. It was soon realized that stars are not exactly like the Sun. In 1841, Thomas Henderson found a parallax of ¼ arc second for Sirius, implying a distance of 825,000 AU (the modern value is 544,000 AU). This distance is much larger than the 28,000 AU estimated by Huygens (Sect. 7.2) based on the assumption that Sirius is similar to the Sun. The larger distance for Sirius means that Sirius must be intrinsically brighter than the Sun. This opens the possibility that not all stars are the same. Stars may be distant suns, but they are not the same as the Sun.

Modern knowledge puts Sirius as having a luminosity 25 times larger than that of the Sun and a distance of 8.6 light years. The second brightest star in the sky Canopus is 10,000 times brighter than the Sun and is located farther away from us at a distance of 310 light years. The bright summer star Vega is 40 times brighter than the Sun and has a distance of 25 light years (Appendix A). There are other stars that are intrinsically thousands of times brighter than the Sun. The apparent brightness of a star is a rough guide to its distance (fainter ones being more distant), but not exactly, as intrinsic brightness covers a wide range.

## 7.4  Demotion of the Sun

The Sun, which is the brightest celestial object, is basically not brighter than other stars once the Sun is placed at a distance typical of other stars. In fact, there are many stars that are intrinsically much brighter than the Sun. When compared to the average intrinsic brightness of other stars, the Sun is actually a very ordinary star, no different than or distinct from millions of other stars. The fact that the Sun has such dominant effects in our lives is just because of its proximity. Our existence on planet Earth relies on an ordinary star, and it is becoming increasingly difficult to argue that we are at a special place in the Universe.

**Questions to Think About**

1. What are the philosophical and religious implications of the hypothesis that the Sun is a star?
2. We may not be "at a special place in the Universe," but could we be "special in the Universe" because of the uniqueness of life on Earth?

# Chapter 8
# A New Way of Thinking

Early civilizations relied on religion and mythology to explain the world that they lived in. The idea that celestial motions and the structure of the Universe are governed by natural processes that can be understood by rational thinking was pioneered by the Greeks. The first major figure to explore this concept was Thales of Miletus (625–545 B.C.). He was followed by his two disciples Anaximander (610–546 B.C.) and Anaximenes (585–528 B.C.). From observations of the apparent daily horizon-to-horizon trajectories of the Sun and the stars, Anaximander hypothesized that the Sun and stars move in complete circles, and they continue to move beneath the Earth even when they can no longer be seen. He, therefore, deduced that the Earth must be a free-floating, isolated object in space (Sect. 4.3, Vol. 1). This process of observations, summary of patterns, formulation of a hypothesis, and prediction of a new conclusion, formed the basis of the scientific way of thinking.

The use of mathematics as part of the reasoning was developed by Pythagoras of Samos (570–495 B.C.). He discovered that the pitch of the sound is inversely proportional to the length of the string that produced it. Because the length of strings can be quantitatively measured, he argued that natural phenomena can also be quantitatively described by precise mathematical relationships. This led later astronomers such as Aristarchus, Hipparchus, and Ptolemy to employ the mathematical techniques of geometry and trigonometry to construct quantitative models of the Universe. The use of mathematics transformed astronomy from a descriptive to a quantitative and predictive science, and it influenced the development of science to this day.

The theoretical basis of logical arguments and method of inquiry was formalized by Socrates (470–399 B.C.). His student Plato (428–348 B.C.) believed that truth could be obtained by pure rational thought, the process we now call metaphysics. The philosopher who had the greatest impact was Aristotle (384–322 B.C.), who used metaphysical thinking to develop a comprehensive worldview that included structure of matter (the five elements), causes of motion (natural and violent), and structure of the Universe (superlunary and sublunary spheres) (Chap. 14, Vol. 1).

© Springer Nature Switzerland AG 2021
S. Kwok, *Our Place in the Universe - II*,
https://doi.org/10.1007/978-3-030-80260-8_8

Aristotle's worldview does not depend on empirical evidence but derives from common sense and his belief of "the way that it ought to be."

Greek philosophical thinking peaked between 500 and 300 B.C. This was a time when Greece was divided into individual city-states without a common ruler or a dominant religion. It is interesting to note that the greatest period of philosophical development in China occurred during the Period of Warring States (472–221 B.C.) when many different schools of thought flourished. As in the case of Greece, China was divided into seven states and did not have a common religion. These schools of thought that appeared during the Warring States Period include Confucianism, Mohism, Taoism, and Legalism. Although they all had their own worldviews, ways of individual behavior and modes of government, they have in common that they did not make use of mathematics and all descriptions were qualitative. After the unification of China in 221 B.C., China was ruled by a succession of dynasties, with emperors who declared they derived their mandates to govern from Heaven. In 134 B.C., Emperor Wu of the Han Dynasty (漢武帝, 156–87 B.C., reigned from 141 to 87 B.C.) decreed Confucianism to be the official philosophy and all other schools of thought were abandoned (獨尊儒術, 罷黜百家). The Confucian philosophy was adopted to create a strict social hierarchy of Heaven-Earth-Emperor-Father-Teacher and rules of personal conduct to impose conformity. The Confucian classics were the only texts that thought were worthy of study, and everything that needed to be learned was already in these texts.

This monopolization of thought meant that no new way of thinking was allowed or tolerated. As a result, China never developed a scientific way of thinking like the Greeks. Although Chinese astronomers accumulated a great wealth of observational data of celestial objects, they never developed any quantitative model of cosmology similar to that of Hipparchus or Ptolemy.

In Europe, the rise of organized religion replaced rational thinking as a path for the search for truth by revelation from a higher source. In 392 A.D., the Roman Emperor Theodosius I (347–395 A.D., reigned from 379 to 395 A.D.) declared Christianity to be the only legal religion in the Empire and closed all non-Christian temples. The burning of the Library of Alexandria and the closure of Athens Academy have been attributed to the rise of Christianity, although other factors may have also contributed to the demise of these institutions. The Bible became the only source of truth and the Greek way of thinking was labeled as "pagan" and incompatible with Christian doctrine. There was no need for an empirical study of Nature in the search for the truth as all the answers were already in the Bible. The Pope became the ultimate authority on the interpretation of the truth. This way of thinking dominated European society for over a thousand years.

# 8.1 The Scientific Method

The modern scientific method has its roots in rationalism where reason was proposed to be the main source of knowledge. Rationalism as part of the scientific process was championed by French philosopher René Descartes (1596–1650). This was followed by empiricism where empirical evidence is the ultimate source of truth, developed by English philosophers Francis Bacon (1561–1626) and John Locke (1632–1704). The scientific method was formalized in the twentieth century as a procedure that scientists follow in the search for the truth, or the closest approximation to the truth.

The scientific method consists of a number of steps: (i) it begins with observations, both qualitative and quantitative. (ii) If regularity or repeatable patterns are observed, then the observed empirical relationship is formulated. If quantitative data are available, these data could be fitted by a mathematical formula. (iii) A hypothesis is proposed to explain the observed patterns. If there is a quantitative mathematical empirical relationship, then one or more variables are proposed to reproduce the relationship. From the hypothesis and variables, predictions of future behavior of the pattern, or new phenomena could be made. These predictions can then be tested with new observations or experiments. If discrepancies are found, the original hypothesis needs to be revised, and the prediction and testing steps are repeated again from step (iii).

Central to the scientific method is the scientific theory. A theory is a generalized form of hypothesis in step (iii), which is created to explain a general class of phenomena. For example, the atomic theory explains the structure of matter, and the theory of evolution explains how organisms change over time. The word "theory" thus has a different meaning in science from the everyday usage of the word as meaning "a guess." A successful theory is the best that science can do as there is no absolute truth, just approximations of the truth.

# 8.2 Examples of Scientific Methods at Work

As an illustration of the scientific method at work, we can discuss the evolution of our cosmological model from the flat Earth-spherical sky model to the two-sphere universe model (Sect. 1.2). The Sun rises and sets every day and this pattern is repeated day after day and year after year. The flat Earth-spherical sky model assumes that the Sun moves in circular paths, across the spherical sky above the flat horizon during the day and in circular paths below the horizon during the night. This model is successful in explaining the phenomena of sunrise and sunset and the apparent shape of the trajectory of the Sun across the sky.

The correctness of this model can be tested by extending its predictions to observers at different locations. Quantitative data for the apparent trajectory of the Sun in the sky can be obtained by using a simple stick planted vertically in the ground and measuring the length and angle of the shadow of the Sun through the day and through the year. These data can be recorded and plotted by observers at different locations on Earth. If the Earth is flat, these trajectories would be the same, independent of the location of the observer on Earth. The length of the day, the time period between sunrise and sunset, would also be the same everywhere. These were not found to be the case.

The changes in the angle of the Sun's path relative to the horizon between observers at different locations suggest that the horizon is not the same for all observers. The Sun rises from the horizon at a steeper angle in Alexandria than in Rome (Fig. 5.2, Vol. 1). The fact that observers in different locations see different stars (Canopus can be seen in Alexandria but not in Rome) also does not support the flat-Earth hypothesis. One of the ways to have changing horizons for different observers is that the Earth is not a plane, but a sphere. This led to the development of the two-sphere universe model, a spherical Earth at the center of a larger spherical sky upon whose surface the Sun and the stars move (Fig. 1.2).

A mathematical model can be constructed to show how the horizon between observers at different latitudes will change. This model can also predict quantitively the inclination angle of the Sun's trajectory at any latitude. The time and orientation of sunrise and sunset can be calculated and compared to actual observations.

A further elaboration of the two-sphere universe model is made by introducing the concept that the path of the Sun through the stars (the ecliptic) is inclined relative to the equator of the spherical Earth. This successfully explains the change in the daily trajectory of the Sun throughout the year. The time of sunrise/sunset can be predicted on any day of the year.

Most importantly, the two-sphere universe model could predict the apparent paths of the Sun observed from any location on the spherical Earth. This included locations people had not yet traveled to, such as the southern hemisphere, and even extreme locations such as the north pole. This model therefore successfully explains almost all the observational facts known at the time and has the power to make predictions for the position of the Sun at different locations and at different times. The two-sphere universe model is therefore a more successful theory than the flat Earth-spherical sky model.

## 8.3    What Is a Good Scientific Theory?

The first requirement for a successful theory is self-consistency. It cannot contain contradicting components. This aspect of the theory can be analyzed with mathematics. The second criterion is its ability to explain all known phenomena within a class and predict not-yet-known phenomena. Newton's theory of planetary motion could explain the observed orbits of all known planets (six at the time), the orbits of comets, and predict the orbit of unknown planets (as in the example of the discovery of Neptune, Chap. 6). The third criterion is simplicity. If more than one theory can explain the same set of observations, then the simplest one is preferred. This criterion is known as Occam's razor. In summary, a good theory is one that can explain a wide variety of phenomena with a minimal number of hypotheses.

Can theories be overthrown? As we make more observations or perform new experiments, existing theories may reach their limits. For example, Newton's theory of gravity was very successful in explaining planetary motions. However, it failed to explain the advance of Mercury's perihelion (Sect. 6.4). This single failure requires it to be replaced by a new theory, in this case, Einstein's theory of gravitation. Scientific theories, by their nature, are not static but evolve over time to become more accurate in their descriptions of nature.

At a certain stage, theories that explain different classes of phenomena could be integrated to offer a unified explanation. Newton unified the laws of planetary motion as formulated by Kepler with the laws of motion of terrestrial bodies by Galileo into a single theory, therefore unifying the motions of both celestial and terrestrial bodies. The apparently separate phenomena of electricity and magnetism were integrated by James Clerk Maxwell (1831–1879) into a single theory. Einstein further unified Maxwell's theory of electromagnetism with his theory of motion to suggest that electricity and magnetism are just different manifestations of the same phenomenon observed by observers in different inertial frames of reference (Sect. 13.5).

One corollary of the above is that science cannot find the absolute truth, only better approximations of the truth. Einstein's theory of gravity can explain more phenomena than Newton's theory of gravity; therefore, Einstein's is a better approximation of the truth. It does not mean, however, that a better theory of gravity will not be developed in the future. There are also examples of two theories that are equally good in explaining a physical phenomenon. For example, the spectra of atoms can be explained by both matrix (Heisenberg) and differential equation (Schrödinger) formulations of quantum mechanics.

## 8.4 How Do We Know that a Theory Is in Trouble?

The first sign of a problem for a theory is that its predictions fail to fit the facts. One example is the once very popular phlogiston theory of combustion. In the eighteenth century, chemists believed that things burn because they contain a substance called phlogiston. When a combustible substance such as wood burns, it uses up part of its phlogiston content and the wood loses weight. However, the development of balances allowed accurate weight comparison before and after burning, and it was shown that some metals (e.g., iron) gain weight when burned. This contradicts the phlogiston theory and suggests that the combustion process must have obtained additional matter elsewhere, possibly from the air. This led to the discovery of oxygen in the air as an agent for combustion, resulting in the demise of the phlogiston theory.

Even a very successful theory may eventually run into its limits. As experimental or observational data improve, a theory may need to be fine-tuned or improved in order to fit the more accurate data available. Although Ptolemy's theory of epicycles could predict planetary positions hundreds of years into the future, more precise determination of planetary positions in later years made necessary the adjustment of parameters or even the introduction of further epicycles in order to fit the data. The increasingly complex nature of Ptolemy's model was a sign of its forthcoming doom.

Another feature of a theory's weakness is the appearance of coincidences. Scientists are extremely suspicious of coincidences as they suggest a hidden pattern that is yet to be discovered. In Ptolemy's model, the centers of the epicycles of the inferior planets are always in the direction of the Sun. For the superior planets, the line connecting the planet to the center of its epicycle and the Earth-Sun axis is always in parallel (Sect. 15.3, Vol. 1). These coincidences suggest a role of the Sun in the motion of planets that is not specified in the model. These coincidences are finally removed in a heliocentric model with elliptical orbits. So, in spite of the predictive power of Ptolemy's model, the theory has intrinsic deficiencies which need to be addressed.

## 8.5   What Is Not a Scientific Question?

Not all questions are answerable by the scientific method. The scientific method relies on observations. If an event cannot be observed, then it belongs to the arena of metaphysics, philosophy, or religion. For example, the question of what was there before the Big Bang is not a question that can be answered by empirical means and therefore is not a scientific question. There are questions in ethics such as "Should humans eat animals?" that may also lie outside the realm of science.

There is a common confusion in our education system as to what constitutes science. Students are often asked in a science class or exam questions such as "What is the name of this plant?" Because different languages have different names for the same plant, clearly there is no uniquely correct answer to this question, and the ability or inability to answer such questions does not reflect a student's ability to conduct science. Some concepts are a matter of definition. There is no correct answer to the question "Is Pluto a planet" because it depends on our definition of a planet, which is a matter of semantics.

Here are some examples of scientific questions: Why is the sky blue? Why is ocean water salty? How do we know the shape of the Earth? Where did humans come from? Although these questions may seem difficult, they are questions that can be investigated with the scientific method. We should note, however, that answers to these questions can change over time as our understanding deepens.

## 8.6 Application of the Scientific Method to Other Disciplines

Although the scientific method was first developed to tackle questions in natural sciences, it has also been applied to study patterns of human and social behavior. If business dealings and the exchange of goods are governed by certain collective behavior rules, then these rules can be discovered through the study of economics. If the development of human societies over time is guided by certain underlying principles, then hypotheses can be proposed for these driving forces of history. Humans live together in communities and the structures of these communities seem to be guided by certain rules, and the search for these rules represents the study of sociology. Because these subjects of study all employ the scientific method, they are also considered scientific disciplines.

The scientific method is not applicable to all arenas of human endeavor. Literature, art, and music are not subject to empirical testing. Their merit is judged subjectively by their perceived internal beauty and their ability to express human emotions and provoke human emotional responses. There is no "true" or "false" art, just "good" and "poor" art.

In contrast to popular belief, mathematics is not a scientific discipline as it is not subject to empirical testing. A branch of mathematics contains theorems deduced from a set of axioms, and these axioms are not testable. Euclidean geometry has an axiom that states "For any given point not on a given line, there is exactly one line through the point that does not meet the given line." A whole system of plane geometry can be developed from this axiom. However, if we change the axiom from "exactly one line" to "no line" or "infinite number of lines," then we have systems of non-Euclidean geometry. Any one of these systems is not truer than the others.

However, mathematics can be judged on its utility. Over history, branches of mathematics were developed to give a quantitative description of certain phenomena. For example, Einstein made use of non-Euclidean geometry to develop his general theory of relativity. Werner Heisenberg (1901–1976) used matrix theory to formulate quantum mechanics. The only criteria for good mathematics are inherent beauty and wide applicability.

## 8.7 Limitations of Science

At a basic level, our senses and our rationality may be the most powerful tools that we have for understanding nature, and the scientific method may be the best way for us to harness them to come to the clearest understanding of the world around us. Unfortunately, the nature of science is a source of great confusion among the general public. Although commonly perceived as such, science is not a collection of facts or classifications, but a methodology of thinking. What is generally accepted as "truth" may evolve over time, but the process is constant.

The scientific method is the most powerful tool that we have for understanding nature, but it is not omnipotent. Our current framework of science is based on a number of assumptions, including these: The world is rational, ordered, and understandable; there are patterns and laws behind every phenomenon governing our body, our mind, and our society; every biological process can ultimately be understood by chemistry, which in turn can be reduced to physics; the Universe is made of matter and governed by universal interactions (e.g., gravity). These assumptions have proven useful and have led to great progress in our understanding of Nature. Just as metaphysics has outlived its usefulness as a methodology for the understanding of the physical world, it is not impossible that we may reach a limit to what the scientific method can do in the future.

**Questions to Think About**
1. The geocentric model of Ptolemy quite accurately fits the motion of the planets. Before Copernicus, many astronomers believed that with more or better epicycles, the problem of planetary motions could be completely solved. Why did the geocentric model eventually lose out to the heliocentric model? In what ways is the heliocentric model a better scientific theory?
2. Give further examples of questions that cannot be answered by the scientific method.

# Chapter 9
# What Are Stars Made of?

In traditional Greek philosophy, matter on Earth was composed of four basic elements: earth, air, fire, and water. For objects in the heavens, Aristotle declared "that there is something beyond the bodies that are about us on this earth, different and separate from them." Heavenly objects were believed to be made of materials different from anything found on Earth. Aristotle named this ingredient of celestial objects "ether."[1]

Early astronomical studies focused on the beauty and harmony in the movements of the Sun, Moon, and planets against the background of fixed stars. For stars, the emphasis was on surveying the sky to produce more accurate catalogs of stellar positions and brightness. After Newton, celestial mechanics became the new fashionable subject. Applying new mathematical techniques at their disposal, astronomers concentrated on how planetary motions followed Newton's gravitational theory and laws of motion. They treated the planets as mathematical points moving along curved trajectories dictated by Newton's laws, and the physical nature of celestial objects was not on most astronomers' minds.

Stars are far away. Without the ability to go there to conduct physical experiments, how can we know the composition of the stars and planets? A pessimistic view of this question was expressed by the French philosopher, Auguste Comte (1798–1857). In his book *Cours de Philosophie Positive* in 1835, he wrote in relation to the stars: "We understand the possibility of determining their shapes, their distances, their sizes and their movements; whereas we would never know how to study by any means their chemical composition, or their mineralogical structure, and, even more so, the nature of any organized beings that might live on their surface."

Unbelievable as it may seem, just a few decades after this pessimistic statement, scientists started to identify chemical elements in the Sun and stars. This was achieved by progress in our understanding of the interaction between light and

---

[1] Sometimes spelled as "aether."

© Springer Nature Switzerland AG 2021
S. Kwok, *Our Place in the Universe - II*,
https://doi.org/10.1007/978-3-030-80260-8_9

matter—more specifically, how the color of light received from stars can tell us about the physical constituents of the Sun and the stars.

## 9.1  Color of Sunlight and Physical Objects

For thousands of years, visible light was the only source of information for the study of distant celestial objects. A physical understanding of the nature of light is therefore fundamental to the development of astronomy. In 1666, Newton let a narrow beam of light pass through a hole into his darkened room in Cambridge. He then passed the sunlight through a glass prism, and projected the light onto a wall. He noted that while the incoming sunlight appears white, it spreads out into the colors of the rainbow. He called this spreading of color a "spectrum."

Newton's experiment showed that sunlight is composed of a mix of colors, ranging from red to violet. We see rainbows in the sky because water vapor in the clouds, acting like a prism, spread out the white light of the Sun. Because sunlight is the source of illumination that allows us to see physical objects around us, our perception of objects is the result of interaction between sunlight and the objects. Because different objects interact with sunlight in different ways, their apparent colors can tell us about the nature of the objects.

Our eyes can perceive objects of different colors. Leaves are green, trees are brown, and the seas are blue. The realization that sunlight is a composite of different colors allows us to understand the origin of the color of objects. Our perception of the color of objects is the result of light's interaction with matter. When light strikes matter, it can either be absorbed or reflected. Different matter reacts differently when it encounters light. We see an object as having a red color because it reflects mostly red light but absorbs all other colors. A white object is made of materials that reflect all light and absorb little. Conversely, a black object is something that absorbs totally and reflects nothing. Metals are shiny because they reflect well and glass is transparent because it both absorbs and reflects very little.

An object can also self-radiate, that is, emit light on its own. A burning candle emits visible light and can light up a dark room at night. While the Moon and the planets shine from the reflection of sunlight, the Sun and the stars emit light on their own. The visible colors are not the only colors of light. There are colors beyond the reddest color that we can see (Sect. 14.4). If our eyes could see infrared light, we would see that everything in our surroundings is emitting infrared light. A dark room is actually bright as everything in it shines in the infrared. Glass may be transparent to visible light but it both absorbs and emits infrared light very well.

By knowing how different matter reacts differently to light through absorption, reflection,[2] and emission, we can use the colors of light that we receive to infer what

---

[2] The technical term for reflection is scattering. It refers to how a light ray changes direction upon hitting an object.

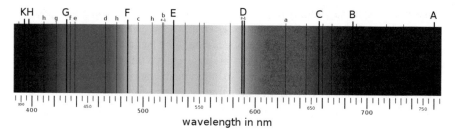

**Fig. 9.1**  Dark lines in the solar spectrum. Wikimedia commons

kind of materials are present. Because celestial objects are far away and we cannot perform a physical experiment to determine their composition, astronomers rely on the light that we receive from celestial objects to infer their composition. By observing the spectrum of asteroids and studying how rocks react to light, we can conjecture that asteroids have compositions similar to rocks on Earth. This was the beginning of the observational technique of spectroscopy: the detailed analysis of light and the interaction between light and matter.

## 9.2  Dark Lines in the Sun

In our everyday lives, we mostly see light of continuous colors. Until the recent development of artificial lighting, our primary source of light on Earth has been sunlight. The Sun emits continuously in colors of red to violet, giving a combined perception of white light, which can be spread out into rainbow colors by a prism. In 1802, the English chemist William Wollaston (1766–1828) repeated Newton's experiment, except that he replaced Newton's pinhole with a narrow slit. He found that in addition to continuous colors, there are also five dark lines imposed on the continuous colors of sunlight. He thought these dark lines represented boundaries between colors.

By employing narrow slits and a prism in front of a telescope, the Bavarian optician Joseph Fraunhofer (1787–1826) observed a much more extensive occurrence of dark lines in sunlight in 1814. The number of lines is so numerous they cannot be boundaries of colors. Fraunhofer suggested that these dark lines are inherent part of sunlight, and not the result of optical or atmospheric effects. He tabulated 574 dark lines and labeled the strongest lines with letters in the alphabet (A, B, C, etc.). Several Fraunhofer lines are shown in Fig. 9.1. Today, we refer to the distribution of colors, including both continuous colors and discrete lines, as the spectrum of light.

Fraunhofer also saw the same pattern of dark lines as the Sun in the spectra of the Moon and planets, therefore confirming that the light from the Moon and the planets are just reflected sunlight. When he turned his telescope, equipped with a spectroscope, to stars, he found that starlight also has dark lines, but the lines are not in the same place as in sunlight. The relative brightness of stellar dark lines also has different patterns from the Sun. Fraunhofer came to the conclusion that these dark lines are intrinsic to the Sun and the stars, although he had no explanation for their origin.[3] The variation of spectral patterns in stars suggests that each star has a unique spectral fingerprint that we could use to infer their chemical composition. This discovery is now recognized as the beginning of stellar spectroscopy.

## 9.3   Does the Sun Contain the Same Elements as Earth?

Chemists in the mid-eighteenth century noticed that hot flames contaminated with metals or salts produced distinct bright colors. This is the basic principle behind how colorful fireworks are produced. In 1859, the German physicists Gustav Kirchhoff (1824–1887) and Robert Wilhelm Bunsen (1811–1899) (creator of the Bunsen burner) discovered that when they heated different chemical elements, they could observe different characteristic patterns of bright colored lines through a spectroscope. By purifying the samples, they were able to identify the unique spectral patterns of elements such as sodium, lithium, and potassium.

Using these results from the laboratory, Kirchhoff and Bunsen detected lines of barium and strontium in distant flames of fires. This was the first demonstration that the spectrum of light can be used to remotely analyze the chemical composition of distant objects. This opens the possibility of determining the chemical composition of celestial objects, which are beyond the reach of experimental analysis. In 1861, Kirchhoff identified lines of sodium, calcium, magnesium, iron, chromium, nickel, barium, copper, and zinc in the spectrum of the Sun. These identifications represented the first evidence that the Sun is made of the same chemical elements as those of the Earth.

## 9.4   A New Element in the Sky

In 1868, the English astronomer Norman Lockyer (1836–1920) noticed a brilliant yellow-orange line in the solar atmosphere. The same line was also observed independently by French astronomer Jules Janssen (1824–1907) from India during the August 18, 1868, solar eclipse. Because this line was not known in the laboratory studies of chemical elements, Lockyer suggested that this line was due to a new element called "helium"—after helios, the Greek word for Sun. In 1895, Scottish chemist William Ramsay (1852–1916) examined gases released from the rare

---

[3] In the twenty-first century, astronomers are facing a similar challenge. Hundreds of dark lines in the interstellar medium remain unidentified. These lines are collectively called "diffuse interstellar bands" and are believed to be molecular in origin. But the carriers of these bands are unknown.

mineral cleveite and noticed the bright yellow line Lockyer observed. The missing element, helium, was found on Earth. That was the one and only time that an element was detected in the heavens before it was found on Earth.

The possibility that new chemical elements could be discovered in celestial objects generated a great deal of excitement. Following the success of the discovery of helium, a green line at the wavelength of 530.3 nm was found in the spectrum of the corona of the Sun during the total solar eclipse on August 7, 1869. Because this line did not correspond to any known atomic lines in the laboratory, it was named "coronium" based on its discovery in the solar corona. This new element turned out to be an illusion. Seventy years later, this green line was finally identified as originating from the common element iron, but in a very highly ionized form ($Fe^{13+}$, atom of iron with 13 of its electrons removed). Under the very hot conditions of the solar corona, the element iron can be stripped off many of its outer electrons, resulting in an ion of the element which is unfamiliar to us on Earth.

## 9.5   The Mystery of Nebulium

In the 1860s, English astronomer William Huggins (1824–1910) fitted a spectroscope to his telescope. He brought chemicals and a battery to the observatory to compare laboratory spectra with those from the stars. On August 29, 1864, he used his spectroscope to observe the planetary nebula NGC 6543 (Fig. 9.2). He found that the spectrum of the nebula has hardly any continuous light as is usually seen in stars. Instead, its spectrum is dominated by a few strong bright lines.

Huggins expressed his excitement over this unusual discovery in his memoir *The New Astronomy: a personal retrospect* in 1897:

> On the evening of August 29, 1864, I directed the telescope ... to a planetary nebula in Draco. The reader may be able to picture to himself ... the feeling of excited suspense, mingled with a degree of awe, with which, after a few moments of hesitation, I put my eye to the spectroscope. Was I not about to look into a secret place of creation?
> I looked into the spectroscope. No such spectrum as I expected! A single bright line only! ... The light of the nebula was monochromatic, and so, unlike any other light I had yet subjected to prismatic examination, could not be extended out to form a complete spectrum ... A little closer looking showed two other bright lines on the side towards the blue. The riddle of the nebulae was solved. The answer, which had come to us in the light itself, read: Not an aggregation of stars, but a luminous gas.

The spectrum of NGC 6543 is dominated by emission lines (Fig. 9.3). It is very different from the spectra of stars, which consist of dark absorption lines on top of continuous light. Huggins recognized that the emission lines must have originated from a collection of gaseous material. The line in the blue was later identified as an emission from the atom of hydrogen. The two green lines did not resemble any known atomic lines in the laboratory and remained a mystery for another 60 years. At a time that many nebulae were resolved into collection of stars (Chap. 12), Huggins found that there are nebulae which are completely gaseous in nature.

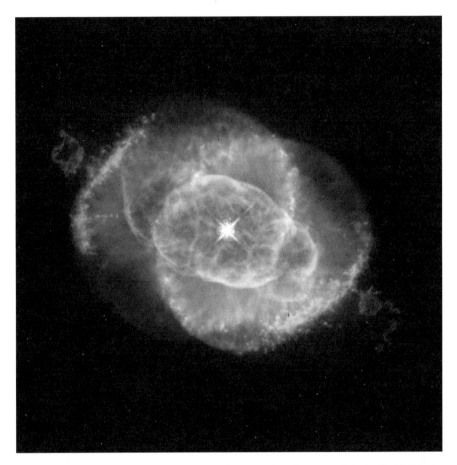

**Fig. 9.2** Discovery of emission line spectrum in stars. The planetary nebula NGC 6543 in the constellation of Draco was the first celestial object observed by Huggins to show an emission line spectrum. This is a color composite image created from multi-band imaging by *Hubble Space Telescope*

The spectroscopic observations of Huggins represented a major shift of the practice of astronomy from surveying the positions and brightness of celestial objects to the understanding of their nature. Success in the interpretation of nebular spectra required advances in the laboratory determination of the spectra of atoms and the theoretical understanding of the structure of atoms. This represents the beginning of the modern discipline of astrophysics.

By the early twentieth century, comparison between stellar spectrum and laboratory spectra of atoms has resulted in the identification of almost all stellar lines. However, some remained unidentified. The most significant were two very bright green lines in the spectra of planetary nebulae. Following the example of the discovery of helium, these two green lines were believed to be due to a new element "nebulium" unknown on Earth. It was only after the development of the quantum

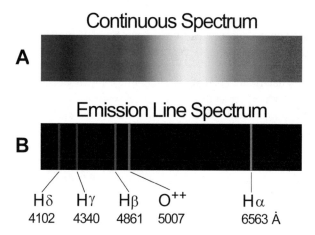

**Fig. 9.3** Comparison between the continuous spectrum of white sunlight and the emission line spectrum of a nebula. The strongest lines in the nebular spectrum are lines of hydrogen at the wavelengths of 656.3, 486.1, 434.0, and 410.2 nm, and two unidentified green lines at 495.9 and 500.7 nm. Figure adapted from Kwok, S. 2001, *Cosmic Butterflies*, Cambridge University Press

theory of atoms in the early twentieth century that American physicist Ira S. Bowen (1898–1973) was able to identify in 1928 that these two green lines are due to the common element oxygen, not some exotic new element. The reason that these lines of oxygen are not seen in the laboratory is that the condition of the terrestrial environment is very different from that in a gaseous nebula. Only in the very low-density (vacuum-like) conditions of a gaseous nebula can these oxygen lines be emitted by the atom.

The identification of mysterious "nebulium" lines as due to the common element oxygen has led to the realization that although we have the same materials on Earth and in space, they may manifest themselves differently due to different physical conditions of their environment.

## 9.6 The Beginning of Astrophysics

The spectra of gaseous nebula show us that their very different spectra are not due to difference in chemical composition, but rather that the elements are under different physical conditions. Although the spectral appearance of atoms is different in nebulae and in the laboratory, the laws governing them are the same. Stars are too far away for us to place a thermometer to measure their temperature, but temperatures of stars can be inferred through the measurement of ratios of atomic lines. In 1920, Indian physicist Meghnad Saha (1893–1956) showed that stellar spectra not only indicate composition of elements in a star, but can also be used to diagnose its physical condition. Intensities of these different lines depend on a star's temperature, pressure, and density. The spectral lines of atoms can therefore serve as probes into the physical conditions of stars. Through detailed analysis of stellar atomic lines, we learned that the outer atmospheres of stars can be of very different temperatures, ranging from cool stars of 3000 degrees Kelvin[4] to hundreds of thousands of degrees.

The ability to determine local physical conditions without carrying out an in-situ experiment was a major step in the science of astronomy. The great distances of stars are no longer an impediment. We can examine stars in the same way we examine the physical composition of a rock or the internal organs of a frog. The only constraint is our ability to see. As long as we can detect the light from a celestial object, we can use terrestrial instruments to analyze the light and infer the conditions of the object. The understanding of the interaction between light and matter therefore became the foundation of modern astrophysics.

## 9.7   Different Kinds of Stars

Through the different spectral patterns observed, extensive efforts were invested to classify stars into different "spectral types." Working with the extensive library of stellar spectra at Harvard College Observatory, American astronomer Annie Jump Cannon (1869–1941) and her colleagues produced catalogs of spectral properties of stars. This culminated in the publication of the Henry Draper Catalogue of 225,300 stars published between 1918 and 1924. The Harvard system of stellar spectral classification later allowed astronomers to relate the stellar spectral properties to their surface temperatures, masses, and luminosities. Stars that are more massive are more luminous and hotter. The star Vega has a mass about twice that of the Sun, a temperature almost twice that of the Sun, and is 40 times more luminous.

This classification scheme allowed later development of theories of stellar structure and stellar evolution. The physical properties of stars can be modeled by assuming that stars are spherical bags of hot gas held together by gravity. At each point inside a star, the attractive force of gravity is balanced by thermal pressure (due to gaseous atoms colliding with each other). A more massive star has a stronger gravitational pull and requires higher temperature to balance it. This balance between thermal pressure and gravity is what makes a star stable without collapsing into itself. Comparison of these stellar models with observations allow astronomers to use the spectral types of stars to infer surface temperatures of stars.

Stellar spectral analysis also suggests that stars do not stay the same forever. Although astronomers initially thought that stars evolve from one spectral type to another, it was later realized that stars are born differently. Stars that are hotter and more luminous than the Sun are born more massive. However, there are stars which are very red, large, and luminous ("red giants"), which we believe are stars at later

---

[4] See Sect. 14.5 for definition of the Kelvin temperature scale.

stages of evolution. In 5 billion years, our Sun will also become a red giant. Like humans, stars also have a life cycle of birth and death (Chap. 23).

## 9.8   Unification of Matter in Heaven and on Earth

By the early nineteenth century, scientists had come to accept the theory that all matter is made of atoms (Chap. 14). Atoms group together to form molecules, which are the basic units for chemical reactions such as combustion. There are a finite number of distinct atoms (called elements) and they represent the fundamental building blocks of all matter. Although almost all the elements were first identified on Earth, their presence was also detected in stars. In spite of the large distance of stars, the light that they emit contains signatures of chemical elements which can be detected through spectroscopy. The demonstration of this connection between heaven and Earth was the result of marriage between astronomy and the laboratory sciences of physics and chemistry. Without the laboratory work of Kirchhoff and Bunsen, the mysterious lines in the spectra of the Sun and stars would have remained unidentified. Chemical elements are universal: They not only compose everything on Earth, they also represent the constituents of stars.

The success of the discovery of a new element in the Sun, helium, prompted the search for other new elements in celestial objects not known on Earth. However, it was not to be. The lines attributed to coronium and nebulium turned out to be fictitious. These unidentified lines are due to the common elements iron and oxygen radiating under unusual conditions. Because helium is also found on Earth, there is no "heavenly" element that is unique to celestial objects. Contrary to the belief of Aristotle, celestial objects are not made of "ether" (Chap. 14, Vol. 1) but constitute the same ordinary matter as on Earth. There is nothing sacred about the Sun and the stars and they are just physical objects made of the same chemical elements that we have on Earth.

Whenever a line of research has proven to be successful, there is a temptation to follow the same path, as we saw in the examples of the discoveries of the planet Neptune and the element helium. After the success of Neptune, astronomers spent decades searching for Vulcan using the same techniques but ended in failure. After the discovery of helium, it was thought that further new elements could be discovered in the Sun and the stars, but so far none have been found. The new frontier turned out to be astrochemistry, where many molecules not known on Earth are found in interstellar clouds (Chap. 20). The Universe may be built from the same elements, but the different conditions in the Universe allow different constitutes to be synthesized from the same set of elements.

### Question to Think About
1. When we encounter an unexplained phenomenon, it is tempting to attribute it to an exotic origin. It took more than 50 years for the exotic entities of coronium and nebulium to be dismissed. Are there examples of scientific entities that are popular today but may turn out to be fictitious in the future?

# Chapter 10
# Origin of the Solar System

One of the basic premises of Aristotle's cosmological view is the distinction between heaven and Earth. Aristotle divided the cosmos into two regions. The first, the terrestrial world below the orbit of the Moon (sublunary), is unpredictable and constantly changing. The Earth has varying weather patterns and suffers from destructive hurricanes, volcanic eruptions, floods, and other unpredictable natural disasters. Above the orbit of the Moon, there is tranquility. The second, the super-lunary world consisting of the Sun, planets, and the stars, is unchanging and everlasting. Not only are the contents of these worlds different, they are also governed by different laws of motion. Motions on Earth are either "violent motions" caused by physical contact, or "natural motions" of falling objects. Terrestrial motions have a beginning and an end. In contrast, motions of celestial objects are continuous and eternal: they move uniformly in trajectories of perfect circles.

From the last chapter, we learned that the Earth and stars are made of similar substances. The same chemical elements that make up the Earth are also the basic constituents of the Sun and the stars. Objects on Earth also follow the same laws of motion that govern the movement of planets. There is no fundamental difference between heaven and Earth.

In the thirteenth century, St. Thomas Aquinas (1225–1274) integrated Aristotelian physics and Ptolemaic cosmology into Christianity. The question of origins was settled upon the biblical account of God creating the entire cosmos, making the Earth, the Sun, and the stars over a time sequence of seven days. After the unification of both the content and movement of heaven and Earth formulated by Newton, scientists began to raise questions about the origin of the Earth, the Sun, the Moon, and the planets. Were they made by natural processes from a pool of pre-existing ingredients? In the Book of Genesis, Earth was created on the third day of Creation, and the Sun, the Moon, and the stars on the fourth day. Is this the correct sequence of events? If the Solar System was not created by divine creation but by natural physical processes, what is the cause of formation?

© Springer Nature Switzerland AG 2021
S. Kwok, *Our Place in the Universe - II*,
https://doi.org/10.1007/978-3-030-80260-8_10

## 10.1  Formation of the Solar System

When we look at our present Solar System, it is obvious that it has some systematic patterns and its structure is not random. Here are some obvious questions that need to be answered. Why are there only eight planets? Why are the planets so widely separated from each other? Why are the orbits of the planets nearly circular, but not perfectly circular? Why do the planetary orbits all lie on the same (ecliptic) plane? Why do the planets all revolve around the Sun in the same direction?

Kepler tried to explain the first two questions with a metaphysical model of polygons and attributed the structure of the Solar System to divine design (Sect. 19.2, Vol. 1). Newton was also a religious man and saw the above systematic patterns as evidence of the wisdom of the Creator. Although Newton developed a physical mechanism that drives the planets to move like a clockwork, he did not feel the need to address the origin of the Solar System because it was created by God.

The first person to explore a physical theory for the origin of the Solar System was the French philosopher René Descartes. From the fact that planets revolve in near-circular orbits around the Sun, he proposed in 1632 that the Solar System descended from a rotating whirlpool of gases. A physical model of the formation of the Solar System was proposed by Immanuel Kant (1724–1804) in his 1755 book *Universal Natural History and Theory of the Heavens*. Kant suggested that the Solar System originated from a rotating and contracting proto-solar cloud. In 1798, Pierre Simon Laplace (1749–1827) suggested that as the proto-solar nebula cooled, the Sun condensed out of this gaseous cloud, and the Earth and the other planets formed from rings of gases in a flattened proto-solar nebula disk. As the solar nebula contracts, it also rotates faster. As the result of rotation, centrifugal forces change the shape of the nebula from a sphere to a disk. Because planets condensed out of the same rotating gas, they naturally would revolve around the Sun in the same direction. This idea is generally known as the nebular hypothesis.

In order for the nebular hypothesis to be a successful theory of the origin of the Solar System, it has to quantitatively explain the fact that the planets do not revolve around the Sun at the same rate, but instead they rotate differentially, and the angular speeds of the planets decrease with their distances from the Sun. From the observation of sunspots changing position with time, Galileo concluded that the Sun is rotating and argued that if the Sun rotates on its own axis, so can the Earth (Sect. 20.1, Vol. 1). Modern observations show that the Sun does not rotate as a solid body, since its rotational speed at the equator is faster than at the poles. Its rotation period at the equator is 25 days.

How did the original rotation of the nebula result in the different rotational speeds of the Sun and the planets? If the Solar System formed out of a proto-stellar nebula, then from Newtonian physics, the angular momentum[1] should stay the same after

---

[1] The angular momentum of an object with mass $m$ rotating at angular speed $\omega$ is $m\omega R^2$, where $R$ is the distance between the object and the center of rotation. For the planets, $R$ is the distance between the planets and the Sun.

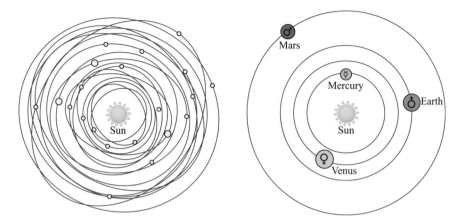

**Fig. 10.1** A schematic drawing illustrating how the terrestrial planets were formed. Through the process of collision and accretion, the four terrestrial planets were formed from a large number of smaller bodies

formation of the planets. From the masses and angular speeds of the planets, one can show that the planets account for 99% of the total angular momentum of the Solar System, most of which are taken up by Jupiter and Saturn. The Sun inherits less than 1% of the total.

These are very difficult questions to answer. The English physicist James Jeans (1877–1946) suggested in 1917 that the planets were formed as the result of a close approach of a star to the Sun. The tidal forces between the star and the Sun stripped out materials from the Sun to form the planets. This would solve the angular momentum problem, but such close encounters may be unlikely.

In the following one hundred years, various variations of the nebular condensation and external tidal perturbation theories were developed. As for the reason there are only eight planets, we now believe that initially there were thousands of small bodies (planetesimals) in the early Solar System. As they collided with each other, they gradually merged into fewer but larger planets (Fig. 10.1). The exact reason why the Solar System is segregated into two separate classes of planets: the terrestrial (rocky) and giant (gaseous) planets is not clear.

Theories of formation and evolution of the Solar System were difficult to test because we had only one example of formation of a planetary system—our own. The recent discovery of planetary systems around other stars now allows us to test theories of the formation and evolution of the Solar System.

## 10.2  Extrasolar Planetary Systems

Although it was commonly believed that other stars may also have planetary systems, planets revolving around other stars are difficult to detect because they are much fainter than their parent stars. Any direct imaging would be nearly impossible because the image will be dominated by light from the much brighter parent star. From Newtonian mechanics, we learn that the orbital motion of the planet Jupiter will cause the Sun to undergo revolution around the common center of mass of the two objects. By detecting the slight motion of a star induced by one of its planets, we can deduce the presence of a planet without actually observing it. If we measure the wavelength of a spectral line in a star with a high degree of accuracy, we can deduce the change of radial velocity of the star through the Doppler effect. If a periodic shift is observed, we can interpret this cyclic variation in velocity as due to the presence of an unseen orbiting planet (Fig. 10.2).

Modern telescopes equipped with high-precision spectrometers can now detect a change in radial velocities of stars as small as 1 m per second. The first planet detected outside of the Solar System was around the star 51 Pegasi, found by Swiss astronomers Michel Mayor (1942–) and Didier Queloz (1966–) in 1995 using this method. The physical properties of the planet and its orbit can be determined by measuring the period of the spectral shift.

Another commonly used method to search for extrasolar planets is planetary transit (Sect. 4.4). When a planet crosses the surface of its parent star, the light from the parent star will be partially blocked and becomes slightly dimmer for a finite period. By making high-precision photometric (brightness) measurements of stars,

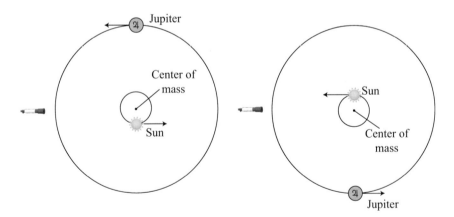

**Fig. 10.2**  Detection of extrasolar planets using the radial velocity method. When a massive planet (e.g., Jupiter) orbits around its parent star (in our case the Sun), it will cause the Sun to move in a circular orbit around the center of mass of the two objects. When observed by a distant observer, this movement of the Sun causes the spectral lines of the Sun to shift as the result of the Doppler Effect. Detection of the Sun's periodic shifts in radial velocity can be used as evidence of the existence of a planet even though the planet is never directly observed

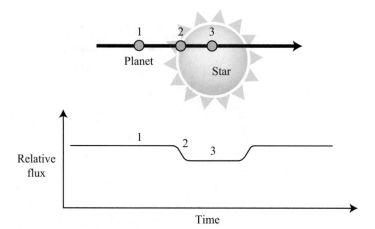

**Fig. 10.3** A schematic illustration of a planetary transit. When a planet crosses in front of a star, the brightness of the star will decrease. Using repeated photometric flux measurements over time, the presence of a planet can be revealed by examining the light curve of a star

periodic dimming will indicate the existence of a planet (Fig. 10.3). The physical properties of the planet can be estimated from the duration and degree of dimming.

Both the radial velocity method and the planetary transient method are examples of scientists inferring the existence of an object (extrasolar planet) without directly seeing it. Since 1995, thousands of extrasolar planets have been found, mainly through these two methods. In some cases, multiple planets around a single star can be found by matching the observed radial velocity curves with models. The examples of other planetary systems can provide useful tests of theories of planetary system formation. For example, it is now realized that planets may not stay in the same orbit after formation, and they do migrate from one orbit to another. Planetary systems are not static and can evolve over time.

## 10.3   The Outer Solar System

The general perception of the Solar System in the mid-twentieth century consists of the Sun, nine planets and their moons, asteroids (mostly located in the asteroid belt between Mars and Jupiter), and comets that intrude into the Solar System from the outside. The limit of the Solar System was considered to be at the orbit of Pluto. Using increasingly more sensitive detectors on large telescopes, astronomers found many rocky objects that are under the dynamical influence of the Sun but are located beyond the orbit of Pluto. Pluto is in fact only one of tens of thousands of objects in a thick ring around the Sun. This ring extends to 50 astronomical units and is named the Kuiper Belt in honor of Dutch astronomer Gerard Peter Kuiper (1905–1973), who independently predicted its existence with British army officer Kenneth Essex

**Fig. 10.4** A schematic drawing of the Kuiper Belt and the Oort Cloud. At the center of the Solar System is the Sun, with eight planets revolving around it on the ecliptic plane. Beyond the orbit of Neptune is the disk-like Kuiper Belt, which in turn is surrounded by a spherical Oort Cloud. The Oort Cloud serves as a reservoir of comets, which from time to time venture into the inner Solar System. The figure is not to scale and the comets are symbolic icons only

Oort Cloud                              Kuiper Belt

Edgeworth (1880–1972). The Kuiper Belt is probably the remnants leftover from the formation of the Solar System, where the inner part condensed into the major planets, leaving thousands of small, icy objects stranded in the outer parts.

The first Kuiper Belt object was discovered in 1992. Since then, nearly one thousand new members of the Solar System have been found, several of which are as big, or even bigger, than Pluto. These discoveries raised doubts about whether Pluto can genuinely be called a planet, as it is no different in character from the other Kuiper Belt objects. This led to the decision by the International Astronomical Union in 2006 to exclude Pluto from the list of planets. As the result of discovering a large population of new objects in the outer Solar System, it seems more logical to put Pluto into this group, although a large segment of the general population is not happy with this apparent "demotion" of Pluto.

We now believe there are over 100,000 objects with sizes larger than 100 km populating the Kuiper Belt. If we extrapolate to sizes as small as 10 m, the population of such small bodies in the Kuiper Belt is estimated to be in the quadrillions ($10^{15}$).

Is the Kuiper Belt the outer limit of the Solar System? No, beyond the Kuiper Belt is an immense sphere called the Oort Cloud (named after the Dutch astronomer Jan Hendrik Oort, 1900–1992) which extends from 3000 to 100,000 AU from the Sun (Fig. 10.4). The outer edge of the Oort Cloud extends to one-third of the way to our nearest stellar neighbor Proxima Centauri. This Oort Cloud is the home of the long-period (>200 years) comets, which when perturbed by the passage of a nearby star, can venture periodically into the inner Solar System. While on this journey, the proximity to the Sun causes their icy surface to evaporate, creating an extensive tail. For the other dormant comets that reside in the Oort Cloud, the Sun is too far away and its radiation too weak to cause any harm. Current estimates put the number of comets in the Oort Cloud in the trillions, or even tens of trillions.

At the time that astronomers are detecting galaxies millions of light years away, it turns out that we have not explored our own backyard thoroughly. The discovery of Kuiper Belt and Oort Cloud objects shows that we have still a lot to learn about the structure of our Solar System. Any theory of the formation of the Solar System will have to take into account the Kuiper Belt and the Oort Cloud, in addition to the planets in the inner regions.

## 10.4    The Question of Origins

Newton and his contemporary scientists in the seventeenth century were quite satisfied to accept the Universe as it is. There are rules that govern the movements of celestial objects, but they did not feel the need to ask questions about the existence of celestial objects. The significance of the nebular hypothesis of Kant and Laplace is not so much in the details, but the fact that they asked the question of origins. By the eighteenth century, scientists have gone beyond exploring the laws that govern the movement and physical constituents of celestial objects and began to wonder about their origins. Where did the Sun, the Earth, and the planets come from? When and how they were made? What were the primordial ingredients from which the Solar System emerged? Scientists were no longer satisfied by the assumption of everlasting celestial objects and began to ask about the cause of their existence.

Is there an end to our question of origins? Once we have answered the question of the origin of the Solar System as due to the condensation of a spiraling disk of gas, we must ask where this primordial solar nebula came from. Is there a limit to what physical processes can explain? Are we only pushing the question to a deeper level? Is the ultimate answer for the origin of everything beyond the realm of science?

**Questions to Think About**
1. Many people who grew up having been taught in school that there were nine planets are unhappy with the International Astronomical Union's removal of Pluto as a planet. Do you think that this decision of the Union is justified?
2. Why was the Aristotelian worldview so appealing to medieval thinkers? What was the consequence of this fascination?
3. Can you compare the influence of Aristotelian dominance on Middle Ages Europe to the influence of Confucian philosophical dominance on Chinese society from 134 B.C. (declaration of the Confucius doctrines as official ideology of the Han Dynasty) to 1911 A.D. (formation of the Chinese Republic)?
4. What are the philosophical implications of the discovery of extrasolar planets?

5. If extrasolar planets cannot be directly observed, how can astronomers be certain about their existence and physical properties? Is direct observation of an object a necessary condition for confirmation of its existence?
6. For thousands of years, we accepted plants and animals as they were, and the question of the origin of species did not arise until the nineteenth century. Why did ancient people feel the need to understand the movement of celestial bodies and the structure of the universe? Why did not they accept celestial bodies as they were? Why were they more curious about the Sun and the stars than about plants and animals?
7. Can you imagine a physical model that could lead to a chaotic heaven—for example, with sunrise and sunset occurring at seemingly random times? What would it be like to live in such a world?
8. The dominant questions in science today are the origin of life and the origin of the Universe. When these questions are answered, are there more fundamental questions that could be asked?

# Chapter 11
# The Plurality of Worlds

Although the center of the Universe shifted from the Earth to the Sun after the heliocentric theory of Copernicus, our perception of the Universe in the seventeenth century was still dominated by the Solar System. Cosmology at the time envisioned a spherical Universe with the Sun at the center, surrounded by the planets (one of which is the Earth) all revolving around the Sun on a two-dimensional ecliptic plane. This Solar System was in turn surrounded by a sphere of stars, each located at distinct distances from the Sun. The Universe was believed to be infinite and without an outer boundary.

This model puts the Sun in a special position in the Universe. With advances in observational astronomy, the possibility that the Sun is just one of the stars and there could be multiple solar systems was raised by various philosophers, including René Descartes. This view, together with the possibility of other inhabited planets, was popularized by the book *Entretiens sur la pluralité des mondes* (*Conversations on the Plurality of Worlds*) by Bernard le Bovier de Fontenelle (1657–1757) in 1686. This book was written in French (instead of Latin) for the layman and went through 35 editions in the author's lifetime. The book also generated controversy because it seeded doubts about God's role in this new cosmology.

While the arrangements of the celestial objects have changed in the new cosmology, the underlying principle that "all things were made for men" had not changed, as stated by William Derham (1657–1735) in his 1715 book *Astro-theology: a demonstration of the being and attributes of God, from a survey of the heavens*. Durham argued that the new cosmology is not in conflict with the Bible, as God must have a reason for creating the stars, and multiple worlds are more for the glory of God.

However, the plurality of worlds did raise the difficult question of the centrality and uniqueness of men. If there are multiple solar systems and other earths, why would God send his son to this world to die for our sins? In the words of Thomas Paine (1737–1809), if God looks after millions of worlds, why should He "quit the care of all the rest and come to die in our world, because they say one man and one woman had eaten an apple?" Is it possible that we are the only ones who have sinned

© Springer Nature Switzerland AG 2021
S. Kwok, *Our Place in the Universe - II*,
https://doi.org/10.1007/978-3-030-80260-8_11

and fallen, as proposed by Edward Young (1683–1765)? Another interpretation put forward by James Beattie (1735–1803) was that our sin and redemption were used as an example to inspire people in other worlds.

The existence of other worlds was also proposed to provide alternate destinations for sinners. Because the Earth is no longer the center of the Universe, there is no need to place Hell at the center of the Earth (Chap. 14, Vol. 1). Instead of being condemned to Hell in perpetuity, the fallen could inhabit different parts of the universe for definite periods, allowing for opportunities of repentance and purgatory. Progress in the science of astronomy has led to revisions in theology.

The debate over the plurality of worlds stretched into the nineteenth century. In 1853, William Whewell (1794–1866), Master of Trinity College, Cambridge and tutor of the Prince of Wales, wrote the book *On the Plurality of Worlds* in which he argued that the Earth alone has sentient beings, and the rest of the Universe is without life. While this may fit well with the Biblical account of Creation, it also raises questions about why God created planets, stars, and nebulae. Among the increasing number of celestial objects (already known to be in the millions by the nineteenth century), only the Sun and the Moon are of practical value to humans. What is the purpose of other objects in the Universe?

The question of humans' role in the Universe, while important, could not be settled by theology alone. The role of the human species must be viewed in the context of the precise place of the Sun in the Universe. Is the Sun at the center of the Universe, or it is just an ordinary star among millions? If so, what is the shape of the Universe and where is the Sun's location in it? These are the questions that we will address in this chapter.

## 11.1   What Is the Milky Way?

Since ancient times, people of all cultures have been aware that in addition to stars, there is a bright band of light that stretches across the nighttime sky (Fig. 11.1). The name of this band of light varies among civilizations. Our current name of the Milky Way probably originated from the Roman term "via lactea," which literally means "the road of milk." The Chinese named it the Silvery River, as it seems to dissect the sky and separate the constellations. The Inca civilization also considered the Milky Way to be a river flowing through the sky. In some North American Indian tribes, the Milky Way was considered to be a bridge between heaven and Earth. In Greek mythology, the Milky Way was heaven. Aristotle did not consider the Milky Way a celestial object and put it in the upper air as part of the sublunary world.

As explorers traveled south of the equator, they found that the Milky Way is more prominent in the southern skies. The observation of the Milky Way in both hemispheres allows us to have a more complete picture of the shape of the Milky Way and its relationship with other significant celestial planes and circles.

The Milky Way is a band (about 10–20° wide) of light in the sky and can be approximated by a circle that turns with the daily rotation of the stars. This circle

**Fig. 11.1**   The Milky Way. This picture was taken on June 16, 2020, near Williams, Indiana, USA. Credit: Zolt Levay Photography

(called the "milk circle" by the Greeks but now called the galactic equator), together with the ecliptic, the celestial equator, the meridian, the horizon, and the Tropics of Cancer and Capricorn, are circles in the celestial sphere that are represented in the armillary sphere (Chap. 7, Vol. 1). The galactic equator is inclined approximately 63° relative to the celestial equator and intersects the celestial equator at the right ascension of $6^h 50^m$ and $18^h 50^m$. The intersection points of a line perpendicular to this circle with the celestial sphere are the north and south galactic poles (Fig. 11.2). The north galactic pole is located in the constellation Coma Berenices at right ascension $12^h 51^m$, and declination +27°, and the south galactic pole in the constellation Sculptor at right ascension $0^h 51^m$ and declination −27°.[1]

As we can see from Fig. 11.2, the Milky Way extends from declination 60° north to declination 60° south. So when it is observed from mid-northern latitudes (e.g., Paris, New York, or Tokyo), part of the Milky Way (the northern-most parts) can always be seen (circumpolar), and part of the Milky Way (the southern-most parts) can never be seen (do not rise above the horizon). The plane of the Milky Way passes through the constellations of Cassiopeia in the north, through Cygnus, Sagittarius, and to as far south as Crux and Centaurus. The easiest way to see the Milky Way in the northern hemisphere summer is to look for the prominent summer triangle (consisting of the stars Altair, Deneb, and Vega), which is nearly overhead around midnight during summer when observed from mid-northern latitudes (Fig. 11.3). Deneb in the constellation of Cygnus is in the Milky Way and Vega (in Lyra) and Altair (in Aquila) are on either side of the Milky Way. In the northern hemisphere winter, look for the prominent constellation of Orion, which is west of the Milky

---

[1] The equatorial coordinates change slightly due to precession. The values used here are for the year 2000 coordinates.

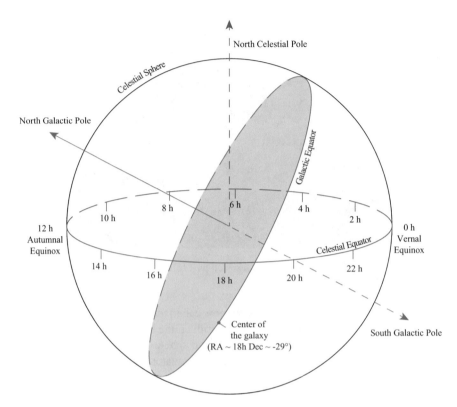

**Fig. 11.2** The galactic plane in the celestial sphere. The galactic plane (shaded area) is inclined ~63° relative to the celestial equator (red circle). The Earth is at the center with north pointing up. The axes perpendicular to the galactic plane are the north and south galactic poles. The galactic center is located at approximately right ascension 18$^h$ and declination −29°

Way. Both the Northern Cross (in the constellation of Cygnus) and the Southern Cross (in the constellation of Crux) are on the plane of the Milky Way (Fig. 11.4). The Milky Way is most prominent when viewed from the southern hemisphere, with its brightest parts near the constellation of Sagittarius which passes nearly overhead at 30° south latitude.

The Milky Way, being such a prominent feature in the sky, demands an explanation. Stars and planets are all point-like objects to the naked eye. Why is there a band of diffuse light in the sky? What is the relationship of the Milky Way to the ecliptic, where the planets lie? No model of the nature of the Milky Way was offered in the cosmological picture of Ptolemy, nor was it explained in the heliocentric model of Copernicus.

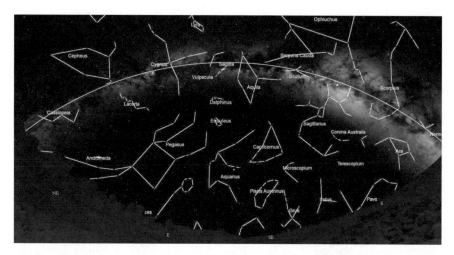

**Fig. 11.3** View of the Milky Way from Honolulu, Hawaii, USA. The Milky Way can be seen stretching across the summer sky (June 15, 2 am), passing through the constellations of Cassiopeia, Cygnus, Aquila, and Sagittarius. The line is the galactic equator. The directions on the horizon are marked. Star charts generated by Starry Night Pro. Version 7.6.6 © copyright Imaginova Corp. All rights reserved. www.StarryNight.com

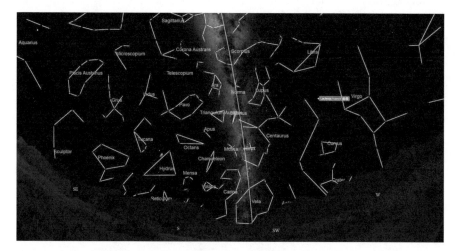

**Fig. 11.4** View of the Milky Way from Melbourne, Australia. The Milky Way is prominent in the South hemisphere winter sky (June 16, 12 am), passing through the constellations of Sagittarius, Scorpius, Centarus, Crux, and Carina. The line is the galactic equator. The directions on the horizon are marked. Star charts generated by Starry Night Pro. Version 7.6.6 © copyright Imaginova Corp. All rights reserved. www.StarryNight.com

## 11.2   Shape of the Milky Way

Our modern understanding of the nature of the Milky Way began with the invention of the telescope. In 1610, Galileo observed the Milky Way with a telescope and resolved it into a collection of faint stars. How does this collection of stars in a band in the sky fit into the scheme of the celestial sphere? If the Milky Way represents a concentration of stars, then the distribution of stars in the celestial sphere is not uniform. Just as the planets are confined to the plane of the ecliptic, stars also tend to cluster around the Milky Way.

The first speculation on the shape of the Milky Way was contained in the book *Original theory or new hypothesis of the universe* published by Thomas Wright (1711–1786) in 1750. He envisioned the Milky Way as a large collection of stars (including our Sun) and the appearance of the Milky Way is the result of us viewing it from the inside. The model he proposed was that the Milky Way represents stars distributed on the surface of a hollow sphere (Fig. 11.5). The Sun, being part of a thin layer at the surface, will see the rest of the stars on the surface in the form of a band.

This theory was tested by William Herschel in 1785 by counting the number of stars in different regions of the sky (Fig. 11.6). Assuming a uniform distribution of stars in space, a higher concentration of stars in a certain direction implies a larger size of the Milky Way along that direction. From this map, he deduced that the Milky Way is in the form of a flattened disk, with a diameter five times larger than its

**Fig. 11.5** Model of the Milky Way by Thomas Wright. The Milky Way is modeled as a layer of stars (including the Sun) on the surface of a hollow sphere. Picture adapted from Wright's book *An Original Theory or New Hypothesis of the Universe* (1750)

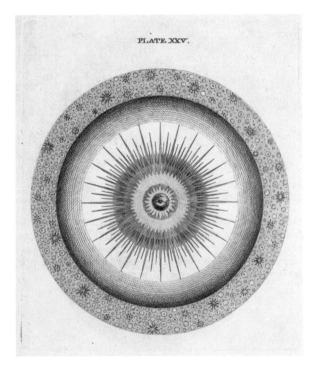

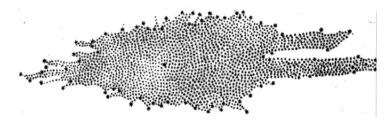

**Fig. 11.6** Distribution of stars as mapped by Herschel. Original illustration in *Philosophical Transactions*, 1785, Vol. 75

**Fig. 11.7** Kapteyn's model of the Milky Way has the Sun at its center

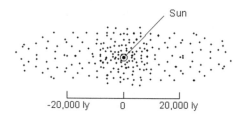

thickness, with the Sun located at the center. He estimated the size of the Milky Way in its longest dimension to be 10,000 light years, or 600 million astronomical units. This size is much larger than any previous estimate of the size of the Universe.

## 11.3   A Larger Universe

In 1906, Dutch astronomer Jacobus Kapteyn (1851–1922) studied distributions of stars in the sky using data collected from over 40 different observatories. By counting stars in different directions, he estimated distances statistically based on parallaxes and proper motions of nearby stars. Kapteyn found that stars thin out in all directions away from the Sun, as one would expect if the Sun is located near the center. If the Sun is near an edge of the Milky Way, more stars would be observed in the direction of the center and fewer toward the edge. In 1920, Kapteyn proposed that the Milky Way (at that time, believed to be the entire universe) has the shape of an ellipsoid with a diameter of ~60,000 light years. The Sun is located close to the center of the Milky Way (Fig. 11.7). Kapteyn's model of the Milky Way was published in *First Attempt at a Theory of the Arrangement and Motion of the Sidereal System* in 1922. He died shortly afterward in the same year.

## 11.4   The Distance Problem

After Thomas Digges suggested that stars do not all lie on the surface of the celestial sphere (Chap. 18, Vol. 1), astronomers were faced with the possibility that stars can be at different distances from the Sun. We live in the three-dimensional world. In order to place the location of stars in the three-dimensional universe, we need the coordinates of their apparent positions in the celestial sphere (right ascension and declination) as well as their distances. While their apparent positions can be determined relatively accurately (better than 1 arc second by the seventeenth century), the determination of distances was a difficult problem. Because stars are all point-like, a faraway bright star may look the same as a nearby faint star. For very nearby stars, their distances can be measured using the technique of parallax (Chap. 7), but this technique is limited to stars out to several hundred light years. For most stars, their parallaxes are too small to be measured.

Spectral classification of stars allowed us to put stars into different categories (Sect. 9.7). The spectral class of a star is related to its intrinsic brightness. If we know the distance (and therefore intrinsic brightness) of a nearby star of a certain spectral class, we can assume that a distant star of the same spectral class will have the same luminosity, and we can derive its distance from its apparent brightness according to the inverse-square law. A nearby star of a certain spectral class, therefore, serves as a "standard candle" with which distant stars can be compared and their distances derived.

For distant stars where spectroscopic observations are not feasible, we need other standard candles as guides to distances. A candidate for such standard candles was found in 1908 by American astronomer Henrietta Leavitt (1868–1921) when she was studying variable stars in the Small Magellanic Cloud. Variable stars are stars that change in brightness in a periodic manner. A class of variable stars, called Cepheid Variables, is easily recognizable by their pattern of light variation: they show a rapid rise to maximum brightness and a slow decline to minimum brightness. Leavitt found that the apparent brightness of Cepheid Variables in the Small Magellanic Cloud is correlated with their variation period. Assuming that these stars in the Small Magellanic Cloud are all at the same distance, this implies that the intrinsic brightness of stars is correlated with their periods. This correlation is known as the Cepheid period–luminosity relationship.

This relationship was calibrated by the Danish astronomer Ejnar Hertzsprung's (1873–1967) observation of nearby Cepheid Variables whose distances can be determined by the method of parallax. Because periods of variable stars can be measured easily by successive photometric observations, we can infer the intrinsic brightness of the object from its period. Cepheid Variables are also very bright (thousands of times brighter than the Sun) and therefore can be spotted at large distances. If a Cepheid Variable can be found in a stellar system (e.g., globular cluster), then one can infer the distance to the system by observing the period of the Cepheid Variable. Cepheid Variables, therefore, became a very useful tool for the determination of distances.

## 11.5    Removal of the Sun from the Center of the Milky Way

Globular clusters are highly spherical, compact systems of hundreds of thousands or even millions of stars (Fig. 11.8). These clusters are members of the Milky Way and their spatial distribution in the Milky Way can give hints about the location of its center. In 1909, Swedish astronomer Karl Bohlin (1860–1939) noticed that a large fraction of the globular clusters were concentrated in the direction of the constellation Sagittarius. In the direction opposite to Sagittarius, there are very few globular clusters. Assuming that these clusters are symmetrically distributed around the center of the Milky Way, Bohlin suggested that the center of the Milky Way is in the direction of Sagittarius. However, his idea was not widely accepted.

In order to test this idea, one needs to quantitatively measure the distances of globular clusters. Upon learning that there are Cepheid Variables in some globular clusters, American astronomer Harold Shapley (1885–1972) applied Leavitt's period-luminosity relationship to derive the distances of globular clusters. With these distances, he could plot the distribution of globular clusters in the Milky Way. From the three-dimensional distribution of globular clusters, he derived the center of their distribution, which presumably represents the center of the Milky Way (marked with a red cross in Fig. 11.9). It is clear from this plot that the Sun is far from the center of the Milky Way.

Shapley estimated the size of the Milky Way to be 300,000 light years, which is much larger than the size of 60,000 light years found by Kapteyn (Fig. 11.7). Most

**Fig. 11.8**  Globular cluster M13. Globular clusters are stellar systems containing hundreds of thousands of stars. The globular cluster M13 in the constellation of Hercules is located 23,000 light years away in the halo of the Milky Way galaxy. Image was taken with the CFH12K camera on the Canada-France-Hawaii Telescope. Image credit: CFHT/Coelum—J.-C. Cuillandre & G. Anselmi

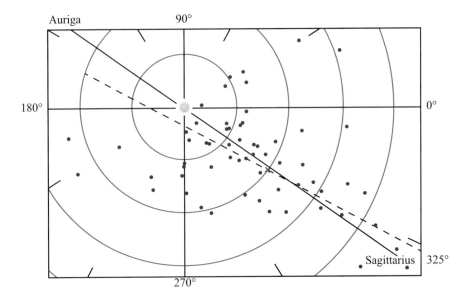

**Fig. 11.9** A map of the distribution of globular clusters projected on the plane of the Milky Way. The positions of globular clusters are plotted as blue dots. The yellow symbol in the middle is the position of the Sun. The derived center of the Milky Way is marked by the red cross. The direction from the Sun to the proposed Galactic Center is indicated by the solid line. The dashed line is the major axis of the distribution of globular clusters. The green circles are radial distances from the Sun at intervals of 10,000 parsecs (Parsec is a distance unit used by astronomers. One parsec is about 3.26 light years). The center of the Milky Way is about 20,000 parsecs from the Sun and is in the direction of the constellation Sagittarius. The numerical numbers are old galactic longitudes with markers shown every 30° of galactic longitude. Figure adapted from Shapley 1918, "Globular Clusters and the Structure of the Galactic System", *Publ. Astr. Soc. Pacific*, **30**, 42

significantly, he placed the Sun 60,000 light years from the center of the Milky Way (Fig. 11.9). With improved observations, we have now determined that the center of the Milky Way is in the constellation of Sagittarius at the right ascension of $17^h 46^m$ and a declination of $-29°$.

## 11.6   Philosophical Implications

The heliocentric theory of Copernicus demoted the Earth to one of the planets and placed the Sun, instead of the Earth, as the center of the Universe. Telescopic observations of the Milky Way showed that the Milky Way is made of stars and represents the entire known Universe. Progress in the understanding of the structure of the Milky Way increasingly raised doubts about the central location of the Sun in the Milky Way. The shift of the Sun from the center of the Milky Way represented another significant degradation in our place in the Universe. Not only the Earth has no privileged place in the Universe, neither does the Sun.

During the eighteenth century, we saw a gradual shift away from the Bible as the ultimate source of authority for truth. For questions relating to our place in the Universe, scientists increasingly relied on empirical means to seek the answer instead of relying on the Bible for guidance and answers. As for way of thinking, scholars were changing from Aristotle's "the way things should be" to "let us find out how things are." Nothing is sacred and anything is possible. By the early twentieth century, scientists were no longer certain that the Sun was at the center of the Universe, or that humans occupied a special place in the cosmos. The possibility of multiple suns, other earths, and extraterrestrial civilizations raised endless questions, with some answers that may not have been consistent with the commonly accepted religious doctrines at the time.

**Questions to Think About**
1. How can one reconcile the plurality of worlds with the teachings of the Bible?
2. Imagine you are a keen observer of the sky before the invention of telescopes. What would be your interpretation of the nature of the Milky Way?
3. The Milky Way is a prominent feature in the night sky. Why did Ptolemy and Copernicus not include it in their respective cosmological models?
4. Recognizing a pattern is often the first step to discovery. Discuss the steps that led Leavitt to discover the period–luminosity relationship of Cepheid Variables.

# Chapter 12
# The Nature of Nebulae

In the pre-telescope era, Ptolemy and Tycho Brahe listed approximately 1000 stars in their catalogs. The gradual improvements of telescopes have greatly increased the inventory of stars. In 1801, an extensive star catalog of 47,390 stars was compiled at the Paris Observatory. Between 1852 and 1859, the Bonn Observatory cataloged the positions and magnitudes of 320,000 stars. These catalogs were later supplemented by observations obtained in the southern hemisphere. The higher sensitivity that came with the invention of photographic plates in the nineteenth century also made the detection of stars much more efficient. In the 300-years period after Galileo's first telescope astronomical observations, the known content of the Universe expanded from thousands to millions of stars.

The increase in the detected number of stars necessitated a revision of the cosmological model of the Universe. At the beginning of the twentieth century, astronomers commonly believed that the Milky Way, containing millions of stars and nebulae, represents the entire Universe. The size of the Universe is defined by the size of the Milky Way, replacing the need for the Greek's crystalline celestial sphere of stars. As demonstrated by the sky survey of Herschel and star counts by Kapteyn, the Sun is located at the center of the Milky Way (and therefore the Universe).

An alternative model of the Universe was proposed by German philosopher Immanuel Kant (1724–1804), who in his 1755 book *Universal natural history and theory of the heavens* suggested that the Milky Way was only one of many systems of stars in the Universe. This model was later referred to as the island universe hypothesis.

Although all celestial objects except the Sun and the Moon appear to the naked eye as point-like objects, telescopic observations led to the discoveries of other extended sources. First the planets Venus, Mars, Jupiter, and Saturn were found to have finite angular sizes, then more and more distant objects were found to be extended in appearance. The nature of these nebulous objects and how they are different from stars introduced a new conceptual problem for astronomers. Is their nebular appearance the result of stars huddled together in groups (clusters), or are

© Springer Nature Switzerland AG 2021
S. Kwok, *Our Place in the Universe - II*,
https://doi.org/10.1007/978-3-030-80260-8_12

they collections of gaseous materials? Because the Solar System was believed to have formed from a gaseous nebula (Sect. 10.1), could these nebulous objects represent other solar systems in formation? This led to the debate between two schools of thought—the island universe hypothesis of Immanuel Kant and the nebular hypothesis of Pierre Simon de Laplace.


## 12.1   Discovery of Nonstellar Objects

The fact that not all stars are point-like can be traced back to ancient times. The Greeks knew that there were objects in the sky that are fuzzy in appearances. Ptolemy noted in his star catalog that five of his objects were "cloudy stars," because their appearances were different from typical stars. Halley believed that these objects could represent breaks in the celestial sphere and classified six such objects as nebulae (Latin for "fuzzy" or "cloud"). He thought of them as distinct objects and their light travelled a long way through space and was diffused through the intervening medium.

With better telescopes at their disposal, astronomers in the mid-eighteenth century were able to resolve some nebulae into clusters of stars, but others remained fuzzy clouds even when observed with powerful telescopes. In 1782, Charles Messier, a French comet hunter, published a list of 103 nebulae so that he and others could avoid mistaking them for comets. Number 31 in his catalog is the Andromeda Nebula (Fig. 12.1).

In 1783, William Herschel began a survey of the sky for new nebulae and star clusters. By 1802, he had cataloged 2500 nebulae. With his 48-inch telescope, he was able to resolve some cloudy nebulae into groups of stars. This led Herschel to believe for a while that all nebulae were distant systems of stars. He came to think of them as other Milky Ways as first proposed by Kant.

Many of Herschel's nebulous objects were later collected into the *New General Catalogue (NGC) of Nebulae and Clusters of Stars* published by Danish astronomer Johann Louis Emil Dreyer (1852–1926) in 1888. One of these objects, NGC 1514, was discovered by Herschel to have a bright central star with a faint luminous atmosphere. This convinced him that the object could not be an unresolved stellar cluster. Instead, Herschel suggested that NGC 1514 is a star surrounded by gaseous materials. He called this class of objects "planetary nebulae" as they resemble the greenish disk of planets such as Uranus or Neptune.

In fact, the Messier catalogue in 1784 already included four planetary nebulae: M27, M57, M76, and M97. Planetary nebulae are not star clusters, and they are also different from other diffuse nebulae such as the Orion Nebula in that they often have a well-defined symmetric appearance. Because they have a single star in the center, planetary nebulae are stellar objects with an extended circumstellar gaseous envelope. We will discuss the nature of planetary nebulae later in Chap. 23.

Although Herschel was an early believer in the island universe hypothesis, his discovery of planetary nebulae changed his mind. The island universe hypothesis

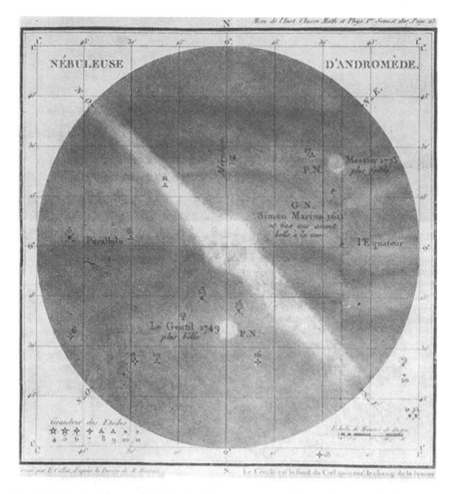

**Fig. 12.1** Messier's drawing of M31, the Andromeda Nebula. Also shown are M32 (below M31) and M110 (upper right). Published by Messier in 1807 in the *Recueil de l'Institute*, Vol. 8, p. 213

suffered a further setback with the discovery of spiral nebulae. In 1845, William Parsons, third Earl of Rosse, built a reflecting 72-inch diameter (1.8-m) telescope at Parsontown, Ireland. The telescope, Leviathan of Parsontown, remained the largest telescope in the world until 1917. With this large telescope, Lord Rosse discerned a spiral structure in the nebula M51 in the Messier catalog (Fig. 12.2). The detection of spiral structures gave much credence to the nebular hypothesis of Laplace as the spirals could represent a system of rotating gases.

The nature of nebulae can best be resolved with the determination of their distances. If they are stellar systems, then they must be very far away. If they are rotating gaseous systems, then they are much closer. In the case of planetary nebulae, they are certainly not very distant objects if their central stars are similar to normal stars.

**Fig. 12.2** Drawing of M51 as sketched by Lord Rosse as seen from his 72-inch telescope in 1849

## 12.2   The Mystery of Spiral Nebulae

Spiral nebulae such as M51 were thought to be rare. Near the end of the nineteenth century, astronomers realized that better images of celestial objects can be obtained if telescopes are placed on a high site, where the atmospheric interference is less than that near sea level. The invention of photography also revealed more details of nebulae than naked eye observations and hand drawings. The American astronomer James Keeler (1857–1900) used the 0.9-m reflecting telescope on Mt. Hamilton, California (1300 m above sea level), to discover many new spiral nebulae. These

**Fig. 12.3** Modern image of the spiral galaxy NGC 4535. Several spiral arms can clearly be seen around the nucleus of the galaxy. Image was taken with the MegaCam camera of the Canada-France-Hawaii Telescope. Image credit: CFHT/Coelum—J.-C. Cuillandre & G. Anselmi

nebulae have distinct arms spiraling from the center. Their appearance immediately suggests a rotating structure. A modern image of a typical spiral nebulae is shown in Fig. 12.3. Based on his count of three nebulae in each square degree of the sky, Keeler estimated that there could be 120,000 nebulae in the entire sky. He showed that spiral nebulae are not uncommon in the Universe.

## 12.3 Island Universes or Gaseous Nebulae?

Because of the success of the nebular hypothesis in explaining the origin of the Solar System (Sect. 10.1), it was quite natural to assume that the newly discovered spiral nebulae are external solar systems in the making. The alternative island universe hypothesis proposed by Immanuel Kant instead imagines that our Sun is one among many stars in the Milky Way and there are many other milky ways. Herschel initially considered the hundreds of nebulae he discovered in his sky survey to be distant systems of stars or island universes. However, he changed his mind to favor the nebular hypothesis after he discovered planetary nebulae.

After the first discovery by Lord Rosse of a spiral nebula in 1845, more spiraling clouds were discovered by astronomers using better telescopes. Spiral nebulae and elliptical nebulae are different from planetary nebulae or other irregular nebulae in two aspects. They look different and are found in different surroundings. Planetary nebulae and irregular nebulae are found near the plane of the Milky Way whereas spiral and elliptical nebulae are found outside of the plane. By the early twentieth century, the majority opinion was that spiral and elliptical nebulae are whirlpools of gas, representing other solar systems, or new star clusters in the making. They are nearby, relatively small, and are members of the Milky Way.

The clash of these two theories came to a head at the debate between Harold Shapley (1885–1972) and Heber Curtis (1872–1942) at the annual meeting of the National Academy of Science on April 26–28, 1920, in Washington, D.C. Shapley was a staff astronomer at Mt. Wilson Observatory, and later became the director of the Harvard Observatory for over 30 years. Curtis was the director of the Allegheny Observatory near Pittsburgh, a distinguished astronomer who devoted his life to photographic and spectroscopic study of nebulae. He later served as director of the University of Michigan Observatory.

The "debate" took the form of two lectures, both titled "The Scale of the Universe," given first by Shapley and then by Curtis. The lectures were followed by a discussion and this event is now referred to as the Great Debate. The debate centered around two issues: (i). How large is the Milky Way and (ii). What is the nature of nebulae? Shapley was there to promote the nebular hypothesis and to show that spiral nebulae are gas clouds within our Milky Way, whereas Curtis was there to defend the island universe hypothesis and to show that spiral nebulae are stellar systems beyond our own Milky Way. Based on his work with globular clusters, Shapley estimated the size of the Milky Way to be 300,000 light years (Sect. 11.5). A Milky Way of such an enormous size would have room to contain all the spiral nebulae. Curtis, however, argued for a much smaller Milky Way as described by the Kapteyn model and placed the spiral nebulae outside of the Milky Way.

Various observations were used to support the two sides of the argument. Cited in support of the nebular hypothesis was the observation in 1885 of a nova in the Andromeda Nebula (M31) that had temporarily outshone the nucleus of Andromeda Nebula itself. If M31 were a stellar system external to the Milky Way, the intrinsic luminosity of the nova would correspond to fifty million suns, which is far larger than the luminosity of a typical nova.

On the other side of the nova argument was that there were more novae in the Andromeda Nebula than in the Milky Way. If Andromeda were simply a nebula within our Milky Way, it would be difficult to explain why there were more novae in one small section of the Milky Way than in other sections. This supports the Andromeda Nebula as a separate stellar system with its own collection of novae. Novae were spotted mainly in the spiral arms of the Andromeda Nebula. Because novae are stars, this result is inconsistent with the idea that the spiral arms are gaseous disks. Curtis also observed that the novae in the Andromeda Nebula are on average about ten times fainter than typical novae in the Milky Way, which led him to estimate the distance to Andromeda to be 500,000 light years, assuming that these novae have the same intrinsic brightness as other novae in the Milky Way.

On spatial distribution, spiral nebulae are observed to crowd around the poles of the Milky Way and are not seen in the plane of the Milky Way, where most of the stars are concentrated. This argued for the island universe hypothesis.

Shapley's arguments against the idea that spiral nebulae are distant objects relied on the measurement of Dutch-American astronomer Adriaan van Maanen (1884–1946), who found that many spiral nebulae are rotating, with periods of tens of thousands of years. If the nebulae are rotating at that rate, they cannot be very far away otherwise their rotating speed would be close to the speed of light. Van Maanen concluded that spiral nebulae were several hundred light years in size, and no more than a few thousand light years in distance.

## 12.4   The Resolution

The main problem with the Milky Way models of Kapteyn and Shapley was their lack of consideration of the effects of obscuration of starlight by dust (small solid particles) in the plane of the Milky Way. When light encounters a small solid particle, it can be absorbed or scattered. The combined effect of absorption and scattering is called extinction. When there is dust between a star and an observer, the observer will see a fainter star. This effect was studied by Swiss astronomer Robert Trumpler (1886–1956). We must remember at that time most astronomers held the view that stars are the only objects in the Universe and the space between stars was completely empty. Many astronomers at the time found the idea that interstellar space is occupied by substances similar to dust or haze in the terrestrial environment to be distasteful.

Because of dust extinction, Kapteyn undercounted the number of stars in the direction of the center of the Milky Way. This suggests that the Sun is not located at the center of the Milky Way, as the true number of stars in the direction of the center of the Milky Way is much higher than observed. After properly taking into account the effects of dust extinction, the position of the Sun is pushed away from the center towards the edge of the Milky Way (Fig. 12.4).

Dust extinction also dims the light of Cepheid Variables in globular clusters, leading Shapley to think that they are farther away than they actually are. The same dust extinction also affects the observation of spiral nebulae in the plane of the Milky Way. Like stars in the Milky Way, dust also concentrates on the plane of the Milky Way. Because distant nebulae are obscured by dust, they are difficult to spot on the plane and therefore are seen mostly in directions away from the plane. When the effect of dust extinction is properly taken into consideration, modern analysis put the size of the Milky Way at about 100,000 light years, smaller than the 300,000 light

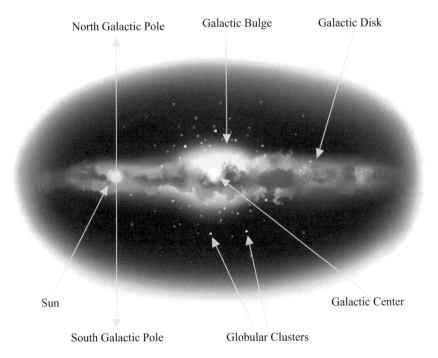

North Galactic Pole    Galactic Bulge    Galactic Disk

Sun                                            Galactic Center

South Galactic Pole    Globular Clusters

**Fig. 12.4** A schematic diagram of the structure of the Milky Way. The Milky Way has the shape of a flat disk with a spherical bulge in the middle. Globular clusters group around the center of the Milky Way and the Sun is located far from the center of the Milky Way

years estimated by Shapley, but still larger than the 60,000 light years estimated by Kapteyn. The distance of the Sun from the Galactic Center is now estimated to be 8200 parsec (26,700 light years), smaller than the 20,000 parsec estimated by Shapley (Fig. 11.9).

The bright nova that was observed in the Andromeda Nebula turned out not to be an ordinary nova, but a supernova, a different class of object much more luminous than novae. Instead of having the brightness of thousands of suns in the case of novae, supernovae at their peak can be as bright as billion times the brightness of the Sun. Although supernovae are rarer events than novae, the supernova in Andromeda is similar to the two supernovae within the Milky Way seen by Tycho in 1572 and Kepler in 1604 (Chap. 19, Vol. 1). Its high intrinsic brightness would require us to place the Andromeda Nebula at a much larger distance, outside of the Milky Way.

Shapley's argument against the island universe hypothesis very much depended on van Maanen's observation of the rotation of spiral nebulae. Van Maanen's observations were later shown to be spurious. Van Maanen was an observational astronomer of great reputation, and few doubted his results. As a matter of fact, his observations of the rotation of spiral nebulae were corroborated and confirmed by many other observers. His results were enthusiastically endorsed by prominent

**Fig. 12.5**  An example of a spiral galaxy viewed sideways. NGC 4565 is a spiral galaxy similar to our Milky Way galaxy. A spherical bulge can be seen at the center of the galaxy. The dark patches are the result of dust obscuration. If seen face on, this galaxy may look similar to NGC 4535 in Fig. 12.3. Image was taken with the CFH12K camera of the Canada-France-Hawaii Telescope. Image credit: CFHT/Coelum—J.-C. Cuillandre & G. Anselmi

astronomers, including James Jeans, who used van Maanen's results to derive masses of nebulae.

As spectroscopic observations of spiral nebulae improved, they were found to be dominated by light from giant stars. This proved that spiral nebulae are made up of stars, making them distant milky ways. Spiral and elliptical nebulae outside of our own Milky Way are now called galaxies. A modern image of a spiral galaxy viewed sideways is shown in Fig. 12.5.

Shapley and Curtis were both partly right and partly wrong. Shapley was correct in placing the Sun away from the center of the Milky Way, but he greatly overestimated the size of the Milky Way. Curtis was right in placing the spiral nebulae outside of the Milky Way. Shapley and Curtis were both partly wrong in their models of the Milky Way as they both failed to take into account the effect of dust extinction on starlight.

Less than five years after the "Great Debate," the nature of spiral nebulae was finally settled. In 1924, American astronomer Edwin Hubble (1889–1953) used the largest telescope in the world at that time (the 2.5 m Hooker Telescope in Mt. Wilson) to resolve the outer parts of spiral nebulae into collections of stars. He found a dozen Cepheid Variables in the Andromeda Nebula and even more in Triangulum (M33). Using Cepheid Variables as standard candles, he determined a distance of 1 million light years for the Andromeda Nebula, making it definitely a stellar system outside of our Milky Way.

Later in the twentieth century, astronomers adopted the term "galaxy" to refer to stellar systems that are located outside of the Milky Way. The term "nebulae" is reserved for gaseous objects within the Milky Way. The Milky Way is also a galaxy, within which we reside. The Andromeda and Triangulum galaxies are two of our neighbors.

## 12.5   Sun's Motion in the Milky Way

After the development of the correct model of the Milky Way (Fig. 12.4), astronomers began to wonder whether the Milky Way is static, or rotates like our Solar System? By studying the proper motion and radial velocity of stars, Swedish astronomer Bertil Lindblad (1895–1965) found that stars in the outer part of the Milky Way rotate slower than stars near the center. This behavior is qualitatively similar to Kepler's third law of planetary motion, where the outer planets in the Solar System rotate slower than the inner ones. The center of the rotation is situated in the constellation of Sagittarius, near the center of the distribution of globular clusters as suggested by Shapley. Around 1930, the Dutch astronomer Jan Oort (1900–1992) determined that the Sun rotates around the center of the Milky Way at a speed of ~200 km per second. With an estimated distance of 25,000 light years from the Sun to the center of the Milky Way, this rotation speed suggests that the Sun will revolve around the Milky Way once in $2\pi$ (25,000 light years)/(200 km/s) = 240 million years.

By the early twentieth century, we learned that not only is the Sun not at the center of the Milky Way, it is also not static. The Sun is revolving around the center of the Milky Way, just like planets revolve around the Sun in the Solar System. The only difference is the time scale. Instead of revolutionary periods measured in years, the Sun and the other stars are revolving around the Milky Way for hundreds of millions of years. Because the Sun is about 5 billion years old (Chap. 17), the Sun must have gone around the Milky Way about 20 times since it was formed.

With the knowledge of the rotation speed of stars around the Milky Way and Newton's law of gravity, Oort estimated the mass of the Milky Way to be around 100 billion times the mass of the Sun (Appendix V). Assuming an average star mass of one solar mass, there must be about 100 billion stars in the Milky Way. This is an astonishingly large number. We must remember that we can see only thousands of stars with our naked eyes. With the use of telescopes, the number of visible stars has greatly expanded. Hundreds of thousands of stars have been cataloged by the end of the nineteenth century, and the total number of stars in the Milky Way can be extrapolated to be in the millions. However, this number pales compared to the hundred billion that one estimates from the rotational speed of stars around the Milky Way. There is no way one could count this number, but the knowledge of stellar dynamics allows us to arrive at this result.

## 12.6 How Can Experts Be Wrong?

The twentieth-century debate on the nature of nebulae and the shape of the Milky Way demonstrates that science is an ongoing process. Even with the large amount of data collected with sophisticated instruments on large telescopes, experts can still draw different conclusions and propose drastically different models to explain the same data. Our current view of the size and structure of the Milky Way is the result of such debates.

Van Maanen was an experienced observer and was highly respected for his work, yet his results turned out to be fictitious. Science is done by humans and humans make mistakes. The process of science often involves a weighing of evidence, what to believe and what to dismiss. Theoretical ideas are also subjected to bias and preconceived ideas. Scientists, like most people, are also swayed by majority community opinion, as rocking the boat can be risky to one's career. The road to the truth can be tortuous, with many detours and dead ends, even for seasoned scientists who are careful and diligent.

## 12.7 Evolution of Our Understanding of the Universe

The first successful cosmological model was the two-sphere universe model, which can explain the apparent motions of the Sun and the stars. It later evolved to the sophisticated quantitative model of Ptolemy, which can accurately predict the future positions of all celestial bodies including the planets. In a major shift in paradigm, Copernicus changed the center of the Universe from the Earth to the Sun, and demoted the Earth to one of the planets. By studying the constitutions of the Sun and the stars and measuring the distances to the stars, we learned in the eighteenth century that the Sun is also not special; it is just one of millions of stars in the Universe. In the late nineteenth century, we concluded that the Sun is part of a large stellar system of the Milky Way galaxy, which consists of billions of stars and is thousands of light years in size. The privileged place of the Sun at the center of the Milky Way was removed in the early twentieth century and we discovered that there are other galaxies like our own beyond the Milky Way. In a later chapter (Chap. 16), we will discuss how we came to believe that the Sun is just one of a hundred billion stars in the Milky Way, which is again one of a hundred billion galaxies in the Universe.

**Questions to Think About**
1. Why is it difficult to explore a structure when you are yourself embedded within that structure?
2. Why was it so controversial to expand the overall scale of the universe?
3. Science is supposed to be an objective exercise. How did van Maanen and others see rotation in spiral nebulae when no such rotation exists? What role does subjectivity play in science?
4. The appearance of spiral nebulae suggests a rotating system. It is therefore natural that astronomers in the early twentieth century regarded them as new solar systems in formation. This interpretation turned out to be wrong. What lesson can we draw from this?
5. What are the topics of "great debates" in current science?

# Chapter 13
# Are All Motions Relative?

Celestial objects in our Universe are not static. Much of the early history of astronomy was devoted to the study of motion of the Sun, the Moon, the stars, and the planets. Extensive observations were made to record their change in positions with time. The regular patterns of motions of celestial objects provided humans with the first impetus to understand the working of Nature.

In Ptolemy's model, the Sun, the Moon, the planets, and the stars all move relative to a fixed Earth. After Copernicus, we have a cosmological model where the Earth and five planets revolve around a fixed Sun, which is at the center of a sphere of fixed stars. Later we learned that Sun and the stars are not really "fixed" and they all move relative to each other (Chap. 5). After that came the realization that the Milky Way is a stellar system and stars inside this system (including the Sun) are revolving around the center of the Milky Way (Sect. 12.5). There are also other stellar systems (island universes or galaxies) beyond the Milky Way and they are all moving relative to each other.

We now face a fundamental question: what are they moving relative to? Is there a "fixed frame of reference" with which we can measure the movement of the Sun, the stars, and galaxies? If so, how is this frame of reference defined? Before we can answer this question, we first have to recognize that there are different kinds of motion. Newton describes objects that are moving uniformly in a straight line as in an inertial frame of reference, and those which are accelerating or rotating as in a non-inertial frame of reference. The major distinction between the two is that we cannot tell we are moving when we are in an inertial frame of reference, but we can tell that we are in a non-inertial frame of reference due to the existence of inertial forces (Sect. 3.2).

© Springer Nature Switzerland AG 2021
S. Kwok, *Our Place in the Universe - II*,
https://doi.org/10.1007/978-3-030-80260-8_13

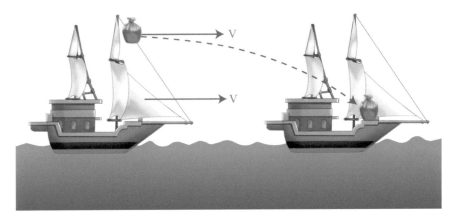

**Fig. 13.1** Principle of relativity. From the point of view of a person on a moving ship, an object dropped from the top of the mast will fall vertically to the bottom of the mast. A stationary observer on shore will observe that the object travels horizontally as well as vertically. The horizontal component of the object's motion ($V$) given by the ship will remain with the object even after it leaves the top of the mast (left). This observer will also observe that the object eventually falls at the bottom of the mast (right)

## 13.1   Principle of Relativity

Because a state of uniform motion is indistinguishable from a state of rest, the word "rest" loses its meaning. A man moving uniformly relative to another man can regard himself to be at rest and the other moving uniformly in the opposite direction. The laws of dynamics (motion of particles) studied empirically by Galileo were formulated mathematically by Newton. It was shown by Newton that if one replaces the particle velocity $v$ by $v + V$ where $V$ is the relative velocity between two observers, the mathematical form of the laws of dynamics stays the same. This invariance is called the Principle of Relativity.

From experience, it was discovered that not only are the dynamics of particles the same on a uniformly moving train as on the nonmoving train station, but all-natural phenomena behave exactly the same. For example, a liquid that boils at 150 degrees on a train station boils also at 150 degrees on a uniformly moving train. We can therefore generalize the principle of relativity to the following form: "one cannot tell by performing any experiment whether one is at rest or moving uniformly in a straight line." Alternatively, if two experiments are performed under identical conditions except that one is done in a laboratory at rest and the other in a uniformly moving laboratory, the two experiments should lead to exactly the same results. This can be illustrated by an example shown in Fig. 13.1. An object is dropped from the top of a mast of a moving ship. From the point of view of an observer on the ship, the object just drops straight down to the bottom of the mast. From the point of view of a stationary observer on shore, the object travels with the horizontal velocity of the ship and falls vertically according to the law of gravity. The trajectory of the object is a parabola, but it ends also at the bottom of the mast. Although the two observers have different interpretations of what happens, the results of the experiment are identical.

If we refer to the accumulated results of all experiments as the laws of nature, a third way of stating the principle of relativity is: "The laws of nature are the same in the laboratory at rest as they are in any uniformly moving laboratory."[1] The Principle of Relativity, which is so appealing philosophically, turns out to be true experimentally, at least between uniformly moving frames of reference. This set of uniform moving frames of reference is called the "inertial frame of reference." Under the principle of relativity, all inertial frames of reference are equivalent. We cannot perform an experiment to tell them apart and therefore we cannot prefer any particular one.

If the Principle of Relativity is true, one may ask whether there is a difference between the geocentric and heliocentric world views. The answer is that the Principle of Relativity does not apply in this case because Earth is rotating upon its own axis and is revolving around the Sun. Both types of motions involve acceleration and therefore both are non-inertial frames of reference.

## 13.2   The Need for a Fictitious Force

Earlier we defined an inertial frame of reference as the class of reference frames that are either at rest or moving uniformly in a straight line. Accordingly, we can define a non-inertial frame of reference as all accelerating and rotating frames. Although Newton's laws of motion are very successful, they create some unexpected strange results when applied to an observer in a non-inertial frame of reference. We will use an example of an artificial satellite to illustrate this anomaly.

From Kepler's third law, we learned that the periods of planets around the Sun are proportional to the 3/2 power of their distances from the Sun. Through the development of the laws of gravity and motion, Newton generalized planetary motions around the Sun to all motions around a central mass and predicted the possibility of artificial satellites around the Earth (Sect. 3.5). At an altitude of 400 km above the Earth (approximately the altitude of the International Space Station), an artificial satellite will have an orbital period of approximately 94 min (Appendix VI). If we put a satellite in a high enough orbit, it is possible that the satellite can have a period of 24 h, the same as the rotational period of the Earth. This special altitude, called geosynchronous orbit, is about 36,000 km. To an Earth-based observer, a satellite in geosynchronous orbit will appear to be stationary relative to the Earth. This object

---

[1] The Principle of Relativity was formally stated by Henri Poincaré in 1904 as "The Principle of Relativity, according to which the laws of physical phenomena must be the same for a stationary observer as for one carried along in a uniform motion of translation, so that we have no means, and can have none, of determining whether or not we are being carried along in such a motion."

hovers above the Earth with no apparent source of support. Because the satellite is attracted by the force of Earth's gravity, why does it not fall down?

Newton's solution to this awkward problem is that for an observer in a non-inertial frame of reference (in this case the rotating Earth), there is a fictitious outward force acting to prevent the satellite from falling. This fictitious force, called the centrifugal force, has no physical origin. Unlike the gravitational force caused by a body with mass, and the electric force caused by a particle with charge, there is no physical object that causes the centrifugal force. The centrifugal force arises when an object rotates relative to empty (or according to Newton, absolute) space.

Because the centrifugal force arises only in a non-inertial frame of reference, it is called the inertial force. On the surface of a rotating Earth, the magnitude of centrifugal force acting on a particle of mass $m$, is $m\omega^2 R_E cos\ \varphi$, where $\omega$ is the angular velocity of the rotating Earth, $R_E$ is the radius of the Earth, and $\varphi$ is the latitude. It is this fictitious centrifugal force that explains why the Earth has a larger radius at the equator than at its poles. Although the angular velocity of the Earth is the same at the equator and the poles, the centrifugal force is the largest at the equator and is zero at the poles. The difference in the centrifugal forces at different latitudes causes the Earth to bulge at the equator (or flatten at the poles). The fact that the Earth is not a perfect sphere is one of the experimental proofs of the rotation of the Earth (Sect. 2.1).

If the Earth is not rotating, a rocket launched from the North Pole along a longitude line will land exactly to the south. However, if the Earth is rotating, by the time the rocket lands, the ground would have turned to the east. To an observer at the North Pole, it would seem that the rocket has been deflected to the west (Fig. 2.2).

The theory that a projectile should always seem to be deflected to the right from the point of view of an observer in the Northern hemisphere was developed by Gaspard Gustave de Coriolis in 1835. To an observer on the rotating Earth, the projectile seems to be subjected to an invisible force of magnitude $2m\omega v$, where $v$ is the velocity of the projectile. This force, now called the Coriolis force, is again a fictitious force without a physical origin and is strictly the result of an observer being in a rotating frame of reference.

For Newton's laws of dynamics to work in a non-inertial frame of reference, we need to introduce artificial forces of nonphysical origin. Both the centrifugal and the Coriolis forces are needed to explain the apparent daily motion of the Sun. For an outside observer, the Earth is rotating on its own axis once every 24 h and revolves around the Sun once every 365 days. From the point of view of an observer on Earth, the Sun has a daily (diurnal) motion and a separate annual motion (Chap. 6, Volume 1). How can we explain the apparent daily motion of the Sun with Newtonian mechanics? Because we are in a non-inertial (rotating) frame of reference, the Sun is subject to two inertial forces: centrifugal and Coriolis forces. The centrifugal force is outward, and the Coriolis force is inward. However, the magnitude of the Coriolis force is twice that of the centrifugal force, therefore the Sun is subjected to a net inward acceleration of $\omega v$, where $\omega$ and $v$ are the angular and physical velocities of the Sun relative to the Earth. This inward force is what drives the Sun to move in an apparent circular orbit around the Earth. It is only through the introduction of these

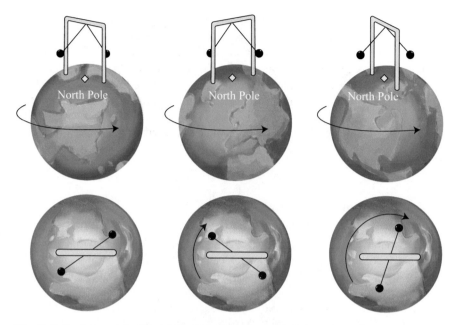

**Fig. 13.2** Pendulum at the North Pole. Top: From the point of view of an outside observer, the Earth is turning under the swinging pendulum. Bottom: From the point of view of an observer at the North Pole, the plane of the swing of a pendulum seems to rotate once a day

two inertial forces that we can explain the apparent motion of the Sun in the frame of reference of the Earth.

## 13.3 We Can Tell the Earth Is Rotating Without Looking Outside

Although it may seem that the two hypotheses of the turning of the celestial sphere of fixed stars and the rotating of the Earth are equivalent, in fact, they are not. The rotation of the Earth was illustrated beautifully by Bernard Léon Foucault (1819–1868) in 1851. Foucault hung a pendulum from the ceiling of the Pantheon in Paris and showed that the plane of the pendulum's oscillation rotated during the day. This can only happen if the Earth is rotating. The easiest way to visualize it is to perform the experiment at the North Pole (Fig. 13.2). From an observer outside the Earth, the plane of the pendulum remains in a fixed direction where the Earth below it rotates eastward with a period of a sidereal day. From the point of view of an observer at the North Pole, the Earth is not rotating but the plane of the pendulum rotates westward once a day. For locations at other latitudes $\varphi$, the period of rotation is 1 sidereal day/$sin\,\varphi$. At the latitude of Paris (48.9° N), the period is 31.8 h.

The Foucault's pendulum demonstrates that the geocentric and heliocentric models are not equivalent and the Earth is definitely rotating. But rotating relative to what? If the Earth is the only object in the Universe, we cannot imagine it to be rotating as the concept of self-rotation is meaningless. The fact that we can detect the rotation of the Earth without looking outside means that the Earth must be rotating relative to something. Newton's answer to this question was that the Earth is rotating against absolute space, a concept that is abstract and is not physical in the sense that it is not related to matter. It implies that absolute space can act on matter through the inertial force.

## 13.4   Origin of the Inertial Force

Newton found his need to introduce fictitious forces in a non-inertial frame of reference highly unsatisfactory philosophically and this disturbed him greatly. An isolated sphere rotating in absolute space will bulge at its equator, but a nonrotating sphere will not.

An alternative to the existence of absolute space is the physical origin of inertial forces. Irish Bishop George Berkeley (1685–1753) suggested that inertial forces are not the result of acceleration relative to absolute space but are caused by acceleration relative to the distant stars. This idea was promoted by Austrian philosopher Ernst Mach (1838–1916), who put it into more concrete terms. He considered that all motions, not just linear uniform motions, are relative. The laws of dynamics are the same in all frames of reference, not just inertial frames of reference. Inertial forces are manifestations of physical interactions between an object and distant stars. These statements are referred to as Mach's Principle.

Figure 13.3 illustrates the difference between the Ptolemaic and Copernican worldviews. In the Ptolemaic model (left of Fig. 13.3), the Earth is at rest and the sphere of fixed stars is rotating. In the Copernican model (right of Fig. 13.3), the sphere of fixed stars is stationary, and the Earth is rotating. In Newton's interpretation, the Earth is rotating relative to absolute space, and the Earth on the right will be subject to bulging of its equator while the Earth on the left will not. In Mach's point of view, the inertial force is caused by relative acceleration between the Earth and the fixed stars, and the bulging of the Earth will appear in both cases. In other words, there is no intrinsic difference between the Ptolemaic and Copernican models and the former is no less "true" than the latter.

In principle, Mach's idea can be tested experimentally. If we put two masses in relative rotation with each other as in Fig. 13.3, we can measure whether the centrifugal force appears or not. However, the effects are so small that it is not measurable. Mach's idea, therefore, remains untested.

Because Mach did not formulate a mathematical theory on the form of the interaction that causes the inertial forces, his statement is referred to as a principle, i.e., something that scientists may believe in for philosophical reasons but cannot prove. A "principle" in physics is analogous to a "conjecture" in mathematics, which

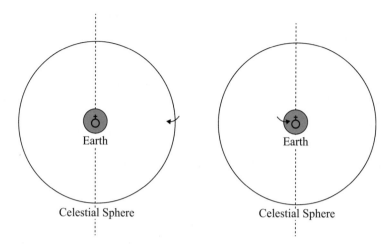

**Fig. 13.3** Comparison between the Ptolemaic (left) and Copernican (right) cosmologies. In the left panel, the Earth is at rest and the sphere of fixed stars is rotating, whereas in the right panel, the Earth is rotating, and the sphere of fixed stars is at rest. The dashed line is the N–S axis. Are these two views equivalent?

is a mathematical theorem that is yet to be proven. Like the principle of the universality of physics, the Principle of Relativity is a statement of faith.

## 13.5  Mathematical Formulation of the Principle of Relativity

Newton's laws of motion are the same in all inertial frames of reference. A dynamical experiment performed on a station at rest will have the same results when performed in a train moving uniformly on a straight line, and observers on the station or on the train will both agree that the trajectories of the projectiles are perfectly described by Newton's laws of motion.

In the late nineteenth century, the Scottish physicist James Clerk Maxwell (1831–1879) wrote down a set of equations that govern all electric and magnetic phenomena. Maxwell showed that electricity and magnetism are different manifestations of the same phenomenon, which may appear differently depending on the state of motion of the observer. A moving observer may interpret the result of the movement of a charged particle as due to an electric force, whereas a stationary observer will observe the same result but interpret it as due to magnetic phenomenon. The principle of relativity is experimentally confirmed: the result of an electromagnetic experiment does not depend on the movement state under which the experiment is performed.

However, if one tries to write the Maxwell's equations for a scientist in the train using the conventional way of adding velocities first used by Galileo (Sect. 13.1), it is found that their mathematical form is changed. For the Maxwell equations to look exactly the same for observers on the ground and on the train, the way to add velocities must be different. In 1904, the Dutch physicist Hendrik Lorentz (1853–1928) found a new way to add velocities that can make the Maxwell equations look the same for all inertial observers. This new way of adding velocities, now called Lorentz transformation, initially did not make any sense. It was Albert Einstein who gave a physical meaning to Lorentz transformation.

To do this, Einstein introduced two assumptions: first, the principle of relativity is valid; and second, the velocity of light is independent of the uniform motion of the source. In other words, the light beam emitted by the headlight of a fast-moving train will appear to have the same speed as light emitted on a stationary train station. The speed of light is independent of the state of movement of the source or the observer.

Einstein proceeded to show from these two hypotheses that all natural phenomena (including electricity and magnetism) are the same on the train and on the ground. However, these two hypotheses do have implications for Newtonian mechanics and our concepts of space and time, as first discussed by French physicist Henri Poincaré (1854–1912). For a fast-moving object with a velocity close to the speed of light, length is contracted and time is dilated. These results were published by Einstein in 1905 in a paper, "On the electrodynamics of moving bodies," now commonly referred to as the special theory of relativity.

## 13.6   Einstein's Theory of Gravity

After the publication of the special theory of relativity, Einstein proceeded to seek a mathematical formulation of Mach's Principle, in the hope of developing a theory that will show all motions are relative regardless of whether the objects are in uniform motion or in acceleration. In 1916, Einstein published a new theory to explain the effects of gravity as a curvature of space caused by the presence of mass, and particles moving through curved space following the shortest path (the geodesic). On a plane of flat space, the shortest distance between two points is a straight line. On the surface of a sphere, the shortest distance between two points is through the great circle (Fig. 13.4). The geodesic is the equivalent of a straight line on a plane and great circle on a sphere in an arbitrarily curved space. Einstein explained the orbits of planets without appealing to an action-at-a-distance force (Chap. 21, Vol. 1), and derived Kepler's laws of planetary motion following this geometric recipe. In effect, Einstein replaces Newton's gravitational force by a geometric theory of space-time distortion.

There are some differences between the Newtonian and Einsteinian descriptions. One of the predictions of the theory is the magnitude of the advance of the perihelion of Mercury (Fig. 6.4). The Einstein prediction of the value of 43 arc seconds per century nicely explained the anomaly that Newton's theory failed to explain (Sect. 6.4).

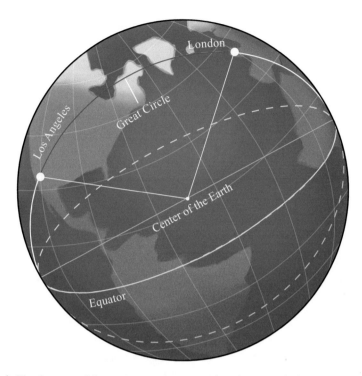

**Fig. 13.4** The shortest path between two points on a sphere is a great circle. A great circle is the geodesic for a sphere. When a plane flies from Los Angeles to London, it does not fly directly east, but instead flies north along the path of a great circle (red line) between these two points on a spherical Earth. The white line shows the rest of the great circle, which centers on the center of the Earth. The equator (in yellow) is also a great circle and it also has its center at the center of the Earth. The backsides of these two great circles are shown as dashed lines

Both Newton and Einstein's theories of gravity predict that the path of light will be bent when it passes near massive objects. This can be tested by observing the apparent positions of stars near the Sun during a total solar eclipse and comparing the positions of the same stars observed at night when they are far away from the Sun (Fig. 13.5). According to Einstein's theory, light from stars near the rim of the Sun will be deflected by 1.75 arc seconds, twice the value predicted by Newton's theory of gravity. In 1919, expeditions were made to Brazil and West Africa to perform experiments to test Einstein's prediction during the total solar eclipse on May 29. The proclaimed confirmation of Einstein's prediction by English astronomer Arthur Eddington (1882–1944) greatly enhanced the early acceptance of Einstein's theory of gravity. Eddington's experimental verification of the predictions of general relativity also made Einstein an international celebrity.[2] Einstein's predictions were further confirmed by later observations using the bending of radio waves

---

[2]Later analysis suggested that the 1919 eclipse results were less than conclusive due to the quality of data and uncertainties in the data reduction procedures. It has been suggested that Eddington's

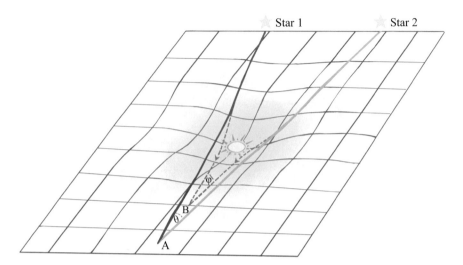

**Fig. 13.5** Bending of starlight through curvature of space. The mass of the Sun distorts the geometry of space near the Sun (shown as the grid) such that the angle between two stars will change from θ to φ

from background quasars with sky positions close to the Sun. Modern tests using the technique of long baseline radio interferometry have obtained results within 0.02% of the predicted values. Einstein's theory of gravitation has been shown to be more accurate than Newton's.

Although Einstein's theory of 1916 is commonly referred to as the general theory of relativity, in fact, it should be more accurately characterized as a new theory of gravity. Einstein was motivated to formulate a theory of Mach's Principle, but he was unable to achieve it. The general principle of relativity is yet to be successfully formulated mathematically.

## 13.7  No More Action at a Distance

One of the philosophically unsatisfying aspects of Newton's theory of gravity is that the force transmitted from the Sun to the planets is instantaneous. This "action-at-a-distance" force is also transmitted without a medium (Chap. 21, Vol. 1). In Einstein's theory of gravity, the Sun warps the space-time around it, and the planets travel in elliptical orbit according to the geodesic of this warped geometry (Fig. 13.5). The cause and effect from the Sun to the planets occurs at a finite speed, the speed of light. This finite speed of gravity has since been experimentally verified through the

---

haste in endorsing Einstein may have been motivated by theoretical bias and political considerations.

measurement of the bending of radio waves from a distant quasar by Jupiter as observed from Earth. No communication or interaction can occur faster than the speed of light.

One of the hypotheses of the special theory of relativity is that the speed of light is constant regardless of the speed of the light-emitting source. Although this hypothesis completely contradicts our common sense, we accept it because this hypothesis allows us to explain that the phenomena of electricity and magnetism are the same in all inertial frames of reference. In a way, we are replacing one seemingly unreasonable hypothesis of action-at-a-distance by another hypothesis of the constancy of the speed of light regardless of motion. This is the nature of science. The quality of a theory is not judged by how reasonable the hypotheses are, but by how well the consequences of the hypotheses can explain natural phenomena. This approach is fundamentally different from that of Aristotle, who begins from self-evident assumptions. Scientists are willing to accept the craziest assumptions, as long as they lead to predictions that fit observational results.

Although the Principle of Relatively is attractive philosophically, it is yet to be formulated mathematically for all frames of references. The ad hoc solution of introduction of inertial forces without a physical origin is philosophically unappealing, but we have yet to find an alternative.

**Questions to Think About**
1. If all motions are relative, is the heliocentric model more true than the geocentric model?
2. What are the nature and properties of absolute space in Newton's formulation of dynamics?
3. Einstein's theory of gravity is very much different, conceptually, and mathematically, from Newton's theory, suggesting that more than one formulation is possible to explain a physical phenomenon. Are other formulations of gravity possible?
4. It has been claimed that Eddington's observational confirmation of Einstein's prediction of the bending of light is not so much as a scientific success but a public-relations triumph (Earman, J., Glymour, C., 1980, *Hist. Stud. Phys. Sci.* 11, 49–85). To what extent do you think sociological factors contribute to the acceptance of scientific theories?

# Chapter 14
# The Nature of Light and Matter

The question of the nature of matter has a long history. Most ancient civilizations have pondered the question of the fundamental constituent of matter. Following the Greek philosophers Anaximander and Anaximenes, Aristotle proposed that terrestrial matter is made of four elements: earth, water, air and fire, and celestial bodies are made of ether. Elements were also linked to the concept of change. Heraclitus (535–575 B.C.) thought that it was natural for fire to go up and stone to fall down. The Chinese had a similar system of five fundamental elements (五行) that were left from the creation of the Universe: metal, wood, water, fire, and soil. According to the Taoist philosophy, these five elements not only makeup matter but also drive all natural processes. All terrestrial events, including functions in our body, are believed to be the result of transformations of these five elements. These five elements are also tied to celestial objects, with each corresponding to one of the five planets. The terrestrial–celestial integration forms the basis of Chinese astrology. The five elements, coupled with the concepts of yin (陰 related to the Moon, female) and yang (陽 related to the Sun, male), constitute the basis of the universal working of the Universe. Both the Greeks and the Chinese saw elements as constituents of matter as well as the cause of motion. The Chinese system is purely metaphysical and tries to be more encompassing than its Western counterpart.

After the scientific revolution, and the introduction of the concept of mass by Newton, matter became a physical concept and the search for the basic building blocks of matter became a problem that can be pursued by empirical methods. Mass is a property of matter, and the constituents of matter can be revealed by chemical and physical means. Matter can be transformed physically from one form to another (e.g., ice to water to steam), or changed into another type of matter through chemical processes (e.g., the burning of wood). The art of alchemy was a first step toward empirical studies of chemical transformation.

However, between the seventeenth and nineteenth centuries, there was still a lot of confusion regarding the nature of matter, light, heat, and air, and the interactions between these entities. For example, caloric was considered as a substance of heat, and light and caloric were listed among the fundamental elements by French chemist

© Springer Nature Switzerland AG 2021
S. Kwok, *Our Place in the Universe - II*,
https://doi.org/10.1007/978-3-030-80260-8_14

Antoine Laurent Lavoisier (1743–1794). It took many years before scientists realized that the Aristotelian concepts of earth, water, and air are compounds or mixtures of chemical elements, and fire is a chemical reaction (combustion) between matter and air. The flames in a fire are the result of light emitted by products of combustion (e.g., soot).

The discipline of astronomy is almost entirely built upon our observation of light from the celestial objects. An accurate understanding of the nature of light and how it interacts with matter is the foundation of modern astronomy. Our modern understanding of the physical structure of celestial objects is based on the technique of spectroscopy, the study of interaction between light and matter. The technique of spectroscopy reveals that light emitted by atoms comes in discrete colors. This discrete (or quantized) nature of matter allows a remote observer to unambiguously determine the nature of matter responsible for light received, opening up a new discipline of astrophysics (Sect. 9.3).

## 14.1   Fundamental Elements of Terrestrial Matter

Through the process of mining, early alchemists were aware of a number of distinct substances: gold, silver, mercury, copper, tin, lead, and iron, as well as sulfur, carbon, zinc, antimony, and arsenic as part of compounds. Some metals that have been used in antiquity were later separated and identified as distinct: bismuth (from lead and tin), cobalt (from bismuth), and nickel (from silver). These substances cannot be further broken down into something else, unlike rocks, wood, and water, which can be transformed to other substances through chemical reactions. For this reason, they are referred to as elements.

The concept of elements was formalized by Lavoisier. In his textbook *Elementary Treatise of Chemistry*, Lavoisier defined an element as a substance that cannot be broken down into a simpler substance by a chemical reaction. Although air was considered one of the four fundamental elements by Aristotle, scientists had long suspected that there is some physical substance in the air. While we cannot see or touch air, we know that there is something in the air as we cannot survive without it. By the eighteenth century, the search for the elusive substance in the air was a major pursuit of scientists. Recognizing that air is responsible for combustion, the Swedish chemist Carl Wilhelm Scheele (1742–1786) and English chemist Joseph Priestley (1733–1804) isolated the combustion-causing gas in 1771 and 1774, respectively, and this was named "oxygen" by Lavoisier. This replaced the hypothesis of phlogiston (a word from Greek, meaning "burnt"), which was previously believed to be the agent for combustion (Sect. 8.4). The other component in air that does not support combustion was later found to be nitrogen.

In 1671, Irish chemist Robert Boyle (1627–1691) noticed that a flammable gas is produced in a metal–acid reaction. In 1766, English physicist Henry Cavendish (1731–1810) isolated this gas and found it to produce water when burned. Lavoisier later gave this gas the name hydrogen, meaning "creator of water" in Greek. Because

| 1 | 2 | 3 | 4 | 5 | 6 | 7 | 8 | 9 | 10 | 11 | 12 | 13 | 14 | 15 | 16 | 17 | 18 |
|---|---|---|---|---|---|---|---|---|----|----|----|----|----|----|----|----|----|
| 1 H | | | | | | | | | | | | | | | | | 2 He |
| 3 Li | 4 Be | | | | | | | | | | | 5 B | 6 C | 7 N | 8 O | 9 F | 10 Ne |
| 11 Na | 12 Mg | | | | | | | | | | | 13 Al | 14 Si | 15 P | 16 S | 17 Cl | 18 Ar |
| 19 K | 20 Ca | 21 Sc | 22 Ti | 23 V | 24 Cr | 25 Mn | 26 Fe | 27 Co | 28 Ni | 29 Cu | 30 Zn | 31 Ga | 32 Ge | 33 As | 34 Se | 35 Br | 36 Kr |
| 37 Rb | 38 Sr | 39 Y | 40 Zr | 41 Nb | 42 Mo | 43 Tc | 44 Ru | 45 Rh | 46 Pd | 47 Ag | 48 Cd | 49 In | 50 Sn | 51 Sb | 52 Te | 53 I | 54 Xe |
| 55 Cs | 56 Ba | 57-71 | 72 Hf | 73 Ta | 74 W | 75 Re | 76 Os | 77 Ir | 78 Pt | 79 Au | 80 Hg | 81 Tl | 82 Pb | 83 Bi | 84 Po | 85 At | 86 Rn |
| 87 Fr | 88 Ra | 89-103 | 104 Rf | 105 Db | 106 Sg | 107 Bh | 108 Hs | 109 Mt | 110 Ds | 111 Rg | 112 Cn | | | | | | |

| 57 La | 58 Ce | 59 Pr | 60 Nd | 61 Pm | 62 Sm | 63 Eu | 64 Gd | 65 Tb | 66 Dy | 67 Ho | 68 Er | 69 Tm | 70 Yb | 71 Lu |
|---|---|---|---|---|---|---|---|---|---|---|---|---|---|---|
| 89 Ac | 90 Th | 91 Pa | 92 U | 93 Np | 94 Pu | 95 Am | 96 Cm | 97 Bk | 98 Cf | 99 Es | 100 Fm | 101 Md | 102 No | 103 Lr |

**Fig. 14.1** The Periodic Table. The first row of the periodic table has 2 columns, the second and third rows have 8 columns, the fourth and fifth rows have 18 columns, and the sixth and seventh rows have 32 columns. The blue numbers are the atomic number of the element

air and water can be broken down into other substances, they can no longer be considered fundamental elements. In their place, we have the new elements of hydrogen, nitrogen, and oxygen.

How many elements are there in nature? In addition to the 13 previously known elements (antimony, arsenic, bismuth, carbon, copper, gold, iron, lead, mercury, silver, sulfur, tin, and zinc), Lavoisier added 11 more: chlorine, cobalt, hydrogen, manganese, molybdenum, nickel, nitrogen, oxygen, phosphorus, platinum, and tungsten. But he also included light and caloric (a substance of heat) as fundamental elements. Both light and heat are forms of energy but are not considered to be "matter" as we define it today. Heat is now interpreted as the result of kinetic motion of gases. Caloric rays are a form of radiative energy and are part of electromagnetic radiation beyond visible light. The nature of light will be discussed in Sect. 14.3.

By the nineteenth century, the number of chemical elements known by chemists had grown to several dozen. All these elements have distinct weights and chemical properties. In 1865, English chemist John Newlands (1837–1898) noticed that when elements are arranged by increasing atomic weight, there is a repeated similarity in chemical properties of every eight elements, which he called the "law of octaves." In 1869 the Russian chemist Dmitri Mendeleev (1834–1907) arranged the 63 elements known at the time into a table of rows and columns. In the modern version of the periodic table (Fig. 14.1), elements that have similar chemical properties are put into columns, with most reactive metal-like ones on the left side of the table and nonmetals on the right.[1] Among elements in the same column, the heavier ones are placed on a row lower than those that are lighter. A new column of "noble gas" (mostly inert and not reactive) was added later after the discoveries of helium in 1868 (Sect. 9.4), argon in 1895, and neon, krypton, and xenon in 1898. In 1904, Mendeleev also added a row above the row of lightest elements (hydrogen and helium) consisting of coronium and ether, which he believed were even lighter elements. Both coronium and ether later turned out to be fictitious.

The creation of the periodic table was not trivial as the "periods" are not fixed. The first row of the table has two elements (hydrogen and helium), the next two rows have eight, but the fourth row (containing iron, nickel, copper) and fifth row (containing silver and tin) have more (each is now known to have 18 elements). The next row (containing gold and lead) has 32 members. Even the order of the elements was confusing because the only guide that the chemists had was the atomic weight and the concept of atomic number was not known at the time. Fitting the chemical elements into the table was like a jigsaw puzzle and was not easy.

A similar periodic table was also developed independently by German chemist Lothar Meyer (1830–1895). What distinguished Mendeleev from Meyer was that Mendeleev predicted new elements where he found gaps in his table. He suggested that there must be an element similar to aluminum but heavier, and another similar to silicon but heavier. These two predictions were confirmed with the respective discoveries of gallium in 1875 and germanium in 1886. In his lifetime, three of his predictions (scandium, gallium, and germanium) were confirmed, which greatly boosted the credibility of the periodic table. The filling of the gaps in the periodic table also gave chemists confidence that all the natural elements on Earth have been discovered.

The next step is to seek a fundamental understanding of the empirical form of the periodic table. The arrangement of the elements in the periodic table suggests that there must be an underlying physical basis. One possible hypothesis is that the elements are not indivisible but have internal components and the composition and arrangement of these components give rise to the periodic structure of the table.

## 14.2   Building Blocks of Matter

The elements in the periodic table are the basic building blocks of all matter on Earth. While each element is distinct from each other, what are their physical forms? The theory of atoms (from the Greek word "atomos," meaning indivisible) was developed by English schoolteacher John Dalton (1766–1844) in 1808. In his theory, all matter is made of large quantities of infinitesimally small atoms that are indivisible

---

[1]The original version by Mendeleev has the rows and columns reversed from the modern version. The current format of rows and columns is by American chemist Glenn Seaborg (1912–1999).

**Fig. 14.2** Examples of knots (top) and links (bottom) illustrating Lord Kelvin's knotted vortex theory of atoms. Figure adapted from an 1869 paper by Lord Kelvin

and indestructible. Each chemical element consists entirely of one kind of atom, which is identical in mass and properties and is distinct from other atoms by weight.

Central to the discipline of nineteenth-century chemistry is the belief that elements are fundamental and immutable. Chemical reactions do not destroy or create matter, but only represent a rearrangement of atoms. When two or more atoms get together, they form a molecule. This distinction between atoms and molecules was made by Italian chemist Amedeo Avogadro (1776–1856). Avogadro also proposed that air is made of a large number of molecules, which we now know to be mostly oxygen ($O_2$) and nitrogen ($N_2$) molecules. Rivers, lakes, and seas are made of water ($H_2O$) molecules. Our bodies are made up of many different kinds of molecules (proteins, fat, ..). The subject of chemistry is mainly about the study of molecules.

After the existence of atoms was accepted by the scientific community, the challenge was to explain why there are distinct elements. In 1867, Irish-Scottish physicist William Thomson (Lord Kelvin, 1824–1907) suggested that atoms were actually knots of swirling vortices in the luminiferous ether and proposed that different kinds of linked vortex knots would correspond to different chemical elements (Fig. 14.2). It was well known that distinct but infinite numbers of knots can be created from strings, and each of these distinct knots could represent a single element. The knots also have the property that they cannot be transformed into each other. Because ether was believed to be prevalent throughout the Universe as the carrier of light waves, Kelvin's theory fitted nicely with the scientific thinking at the time. The fact that it was a beautiful mathematical theory also added to its appeal. The knot theory eventually went out of favor as the result of futile searches for luminiferous ether (Sect. 14.7) and the discovery of internal structures of atoms.

The understanding of the nature of atoms gradually emerged with the realization that atoms are not truly indivisible. Although atoms are neutral in electric charge, they can be separated into two parts, a positively charged nucleus surrounded by negatively charged electrons. In 1911, New Zealand-British physicist Ernest

Rutherford (1871–1937) found that most of the mass of an atom resides within a very small nucleus, and English physicist Henry Moseley (1887–1915) determined that the positive charge of the nucleus (which we now call the atomic number) uniquely defines the element. It is the atomic number, not atomic weight as it was first proposed by Mendeleev, that determines the order of the elements in the periodic table. Because the atomic numbers increase with an exact interval of one, Moseley could easily identify gaps in the periodic table and correctly predicted the existence of elements with atomic numbers 43, 61, 72, and 75, which were later identified as new elements technetium, promethium, hafnium, and rhenium, respectively.

The mass of the atomic nucleus is contributed by two different subatomic particles: the proton, which is positively charged, and the neutron (which has no electric charge). The atomic number of an element corresponds to the number of protons in the atomic nucleus. The positive charge of the protons inside the nucleus is balanced by exactly the same number of negatively charged electrons outside of the nucleus.

A first step in understanding the structure of atoms was the solar system model of Ernest Rutherford, who suggested in 1911 that electrons revolve around the atomic nucleus like planets revolve around the Sun. In 1885, a Swiss schoolteacher, Jacob Balmer, derived a mathematical formula to fit the wavelengths of the lines seen in the spectrum of hydrogen. Balmer's formula was generalized by Swedish physicist Johannes Rydberg (1854–1919) in 1888 for all lines of hydrogen:

$$\frac{1}{\lambda} = R_H \left( \frac{1}{n^2} - \frac{1}{m^2} \right)$$

where $\lambda$ is the wavelength (a quantitative measure of color, see Sect. 14.4) of the spectral line, $R_H$ is the Rydberg constant (with a value of $1.09677583 \times 10^7 \text{ m}^{-1}$), and $m$ and $n$ are integers with $m > n$. For the Balmer series (spectral lines of hydrogen in the visible range), $n = 2$. For the first two lines of the Balmer series ($m = 3$ and 4), the wavelengths are 656 and 486 nm, respectively. The mysterious series of lines of the hydrogen atom are captured in a single simple formula.

The Rydberg formula also predicts other series of spectral lines of hydrogen. Beginning in 1906, a series of lines corresponding to $n = 1$ was discovered by American physicist Theodore Lyman (1874–1954) in the ultraviolet parts of the electromagnetic spectrum. This series of lines is now called the Lyman series of hydrogen. The first line of the series ($n = 1$, $m = 2$) occurs at the wavelength of 122 nm in the ultraviolet. The series of lines corresponding to $n = 3$ was discovered by German physicist Friedrich Paschen (1865–1947) in 1908. The first line of the Paschen series, corresponding to $n = 3$, $m = 4$, is at a wavelength of 1875 nm in the infrared. The predictions of the Rydberg formula are therefore confirmed.

In order to explain why atoms would emit light of fixed wavelengths and not continuous colors, English physicist John William Nicholson (1881–1955) suggested in 1912 that discrete lines of light are emitted when electrons jump from one fixed (quantized) orbit to another. This led to the theory of the hydrogen atom

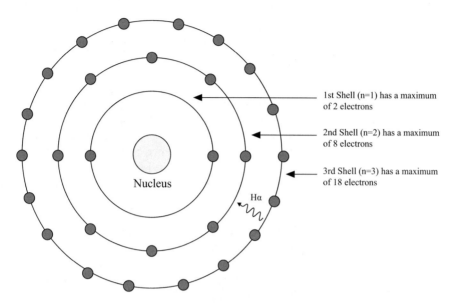

1st Shell (n=1) has a maximum of 2 electrons

2nd Shell (n=2) has a maximum of 8 electrons

3rd Shell (n=3) has a maximum of 18 electrons

Nucleus

Hα

**Fig. 14.3** A simplified schematic drawing of the Bohr atom. Electrons are in orbits around the atomic nucleus like planets in the Solar System. The innermost orbit can have a maximum of 2 electrons, the next 8, and the 3rd orbit 18. When electrons jump from one orbit to another, light of a specific wavelength is emitted

developed by Danish physicist Niels Bohr (1885–1962), who put together a detailed model of the orbits of electrons in a hydrogen atom and accurately reproduced the Rydberg formula for the spectrum of hydrogen. For example, the Balmer line Hα at the wavelength of 656 nm is the result of an electron in the hydrogen atom jumping from the third shell (principal quantum number $n = 3$) to the second shell ($n = 2$). The Balmer line Hβ at 486 nm is an electronic transition from $n = 4$ to $n = 2$ (Fig. 14.3).

According to Bohr's theory, each orbit is only allowed a maximum of $2n^2$ number of electrons. For $n = 1$, this maximum number is 2, for $n = 2$, this number is 8, and for $n = 3$, this number is 18. The physical basis of the periodic table was finally explained by the quantum model of the atom, whose rules of electron configurations explain the periods of 2, 8, 18, and 32 in the periodic table. The quantum theory of atoms, therefore, provided the complete physical basis for the discipline of chemistry.

## 14.3 The Confusion between Light and Heat

In the eighteenth century, there was a lot of confusion about the nature of light and heat. Both light and caloric were included by Lavoisier in his list of fundamental elements. As our eyes can respond to light and our ears can hear sound, our body can also sense heat. Our sense of sound is the result of vibration of air, without which no sound can be heard. Our everyday sense of hot and cold are also the result of air touching our body. However, we can also feel heat, even in the absence of air. We feel the heat of the Sun even though the Sun is separated from the Earth by huge distances over a near vacuum. On high mountaintops where the atmosphere is half of what it is at sea level, the warmth of the Sun is the same, if not more. Scientists in the eighteenth and nineteenth centuries recognized this other form of heat, and it is referred to as "heat rays" to distinguish it from "light rays" that can also be transmitted over a vacuum. To emphasize the difference between these two kinds of radiation, scientists at that time believed that while the Moon is also bright, it does not carry heat and therefore moonlight was considered "cold light" while sunlight is "warm."

We now understand that our sense of warmth from sunlight is the result of light interacting with the surface of our body. It is the same process by which a rock can be heated up under the Sun as sunlight is absorbed into the rock, and the light energy is converted to the kinetic motion of atoms inside the rock. Upon the receipt of sunlight, a rock and our body will be heated up, and as a result each will also shine on its own, although the light emitted is mainly in the infrared range (Sect. 14.6).

In 1800, William Herschel split sunlight into a color spectrum by passing the light through a prism and placed three thermometers under different colors of the split sunlight to measure their temperature. He noted that the highest temperature was recorded under red light. However, when he placed a thermometer beyond red light, he found that the temperature of that thermometer was higher than the temperatures of thermometers not placed under sunlight. He, therefore, detected energy in light beyond the visible. At that time, it was a common belief that light and heat are two different fundamental elements and Herschel thought that his experiment had confirmed the separation of these two entities. Herschel called sunlight beyond the red color "caloric rays."

After hearing about Herschel's discovery of "caloric rays," German chemist Johann Wihelm Ritter (1776–1810) looked for rays at the other end of the solar spectrum. In 1801, Ritter discovered radiation beyond the violet that causes chemical changes. This was given the name "chemical ray." The belief that the Sun emits "chemical," visible, and "heat/caloric" rays persisted throughout the nineteenth century.

These interpretations of sunlight arose because of the different effects of light on us. Light is reflected off a surface and therefore we notice the effect of "illumination" of light with our eyes. We feel the "heat" through our sense of touch, and we can detect chemical changes induced by light. By noting the different effects of light, scientists erroneously concluded that they have different origins. In fact, matter responds to light through both reflection and absorption, and responds differently to different colors of light as the result of the quantum structure of matter.

This confusion was not sorted out until many experiments later demonstrated that the solar "heat ray" has the same polarization, interference, and absorption bands as visible sunlight. As a result, there is no fundamental difference in nature between chemical rays, visible light, and caloric rays. They are all part of the electromagnetic radiation (Sect. 14.4) emitted by the Sun, and the caloric and chemical rays are now called infrared and ultraviolet radiation. The concept of "heat ray" therefore dropped out of scientific usage into the domain of science fiction, as used by H.G. Wells in *The War of the Worlds*.

The clarification of the concepts of light and heat did not occur until the late nineteenth century. Heat is a consequence of interaction between light and matter. When sunlight strikes our skin, we have a sensation of warmth. This sense of heat is not the result of a separate caloric ray from the Sun. However, a full understanding of the relationship between light and heat only became possible after the formulation of quantum physics, and a new understanding of the quantum interactions between light and matter.

## 14.4  Expansion of the Concept of Color

The theory of James Clerk Maxwell that united the phenomena of electricity and magnetism (Sect. 13.5) also predicted the existence of an electromagnetic wave that travels at the speed of light. Electromagnetic waves in the radio wavelengths were experimentally verified by German physicist Heinrich Hertz (1857–1894) in 1888. The coincidence of the speeds of electromagnetic waves and light led Maxwell to conclude that light is in fact an electromagnetic wave. We now understand that light is a manifestation of electricity and magnetism. The unification of the concepts of electricity, magnetism, and optics (study of light) is now considered to be the second most important step in unification in physics, after the unification of celestial and terrestrial motions by Newton.

The visible light spectrum is made up of different colors ranging from red to violet. We have now expanded the definition of colors beyond the visible range. X-ray, ultraviolet, visible, infrared, microwave, or radio are different colors of light (or more precisely electromagnetic radiation). The only distinction among them is their wavelengths (the distance between the repeating peaks of the waves), with radio waves having wavelengths of the order of meters and visible light with wavelengths of fractions of a micrometer. Within the range of visible light, different colors are characterized again by different wavelengths. The rainbow colors from red to violet correspond to a decreasing wavelength scale, with the red color having a wavelength of about 0.7 μm and the blue color about 0.4 μm. Just outside the visible range, on the long wavelengths side is the infrared, and on the short wavelength side is ultraviolet. X-rays have even shorter wavelengths than ultraviolet, with wavelengths on the order of nanometers (Fig. 14.4).

Alternatively, colors can be specified by frequencies measured in units of cycles per second (Hz). Frequency and wavelengths are two ways to describe the same thing, but in a reciprocal way. The multiplication product of frequency ($\nu$) and wavelengths ($\lambda$) is a constant (the speed of light) so one can easily derive one from the other. Mathematically, the relationship between the two is

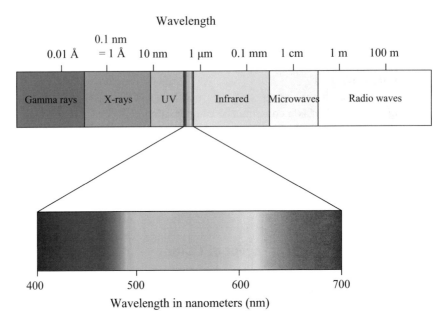

**Fig. 14.4** The electromagnetic spectrum. The visible region (bottom panel) is only a very small part of the electromagnetic spectrum (top panel)

$$\nu \cdot \lambda = c,$$

where $c$ is the speed of light ($3 \cdot 10^8$ m per second). In other words, electromagnetic waves with a high frequency have short wavelengths.

Frequency as a unit of color is more commonly used in radio waves, e.g., we refer to FM radios having frequency coverage from 88 to 108 MHz (1 MHz = $10^6$ Hz). Usual cellular phone communication frequencies are 900 MHz and 1.8 GHz (1 GHz = $10^9$ Hz).[2] Expressed in wavelengths, FM radio waves have wavelengths of the order of 3 m, whereas radio waves from cellular phones have wavelengths of 33 cm and 17 cm. As a matter of convention, scientists rarely use frequency as a measure of color in the visible.

## 14.5   Heat and Temperature

The concept of heat has its origin in our body's senses of hot and cold. In order to quantify these sensations, the concept of temperature was introduced. For example, the Fahrenheit temperature scale was designed to cover the range from the lowest ($0 \, ^\circ$ F) to the highest ($100 \, ^\circ$ F) temperatures that are likely to be encountered in

---

[2] Newer cell phone networks (e.g., 5G) use higher frequencies.

Europe. Because matter (e.g., water) can change from solid to liquid to vapor states upon heating, the Celsius scale was introduced to mark the freezing (0 °C) and boiling (100 °C) points of water.

By the eighteenth century, physicists have observed that there exist certain relationships between pressure, volume, and temperature of gas. When a volume of gas is heated, it expands (known as Charles' law). When a volume of gas is compressed, its pressure increases (known as Boyles's law). Although these empirical laws were well established, there were no explanations for them. In 1811, Italian chemist Amedeo Avogadro (1776–1856) proposed that a volume of gas is proportional to the number of molecules inside. Avogadro did not find much acceptance of his idea during his lifetime; it was only after his death in 1856 that his molecular hypothesis gained recognition in scientific circles.

If a fixed volume of gas contains the same number of molecules, can the concepts of pressure and temperature be related to the physical properties of molecules? In the late nineteenth century, Austrian physicist Ludwig Boltzmann (1844–1906) proposed that air is composed of fast-moving molecules, which are moving randomly at different speeds. The fraction of molecules moving at certain speeds obeys a formula, which has only one parameter—the temperature. Because this concept of temperature is due to moving molecules, it is called the kinetic temperature and Boltzmann's theory is known as the kinetic theory of gas. The kinetic theory of gas was not widely accepted until the early twentieth century.

If the speed of molecules is related to temperature, then there should be a temperature at which all molecules stop moving. It would be logical to assign a temperature of zero to a molecular speed of zero. This point of "absolute zero" becomes the starting point of a new temperature scale called Kelvin. A degree Kelvin is the same in magnitude as a degree of Celsius, just the starting point is different. The freezing point of water (0° Celsius) becomes 273° Kelvin.

## 14.6 Everything Shines

One of the most significant discoveries in the nineteenth century was that gravity, motion, light, and heat all carry different forms of energy and these energies are interchangeable, subject to the condition that the total amount of energy is conserved. Water flowing from a high place to a low place can drive a water wheel downstream. This is an example of conversion of gravitational (potential) energy into mechanical (kinetic) energy. In a steam engine, water is boiled into steam, and the heat in the steam can push a piston and turn a flywheel. This is an example of the chemical energy of fuel (wood or coal) being converted to heat and then to mechanical energy.

Light also carries energy. When light is absorbed by matter, its energy must go somewhere. Most of the time, the light energy is converted to heat. As a result of sunlight shining on a pond, the water molecules become agitated and move faster. Similarly, light shining into a room through a window causes the air molecules in the room to move faster in random directions. When these fast-moving air molecules hit our skin, we get a sensation of heat. A sensation of warmth corresponds to high temperature, which in turn corresponds to a collection of molecules moving fast.

Unlike air, which consists of randomly, fast-moving molecules, solids consist of atoms tightly bound with each other. Some simple solids are made of atoms arranged in a highly organized regular form (a crystalline structure). For example, salt is structured in a simple cubic form with the sodium and chlorine atoms located at alternate corners of the cube. Other solids, e.g., rock, may have more complicated or even random (amorphous) structures. No matter what the structure is, when light enters a solid, it causes the atoms to vibrate and warm up.

When light strikes an object, it can be reflected or absorbed. The color of the object is determined by the chemical structure of the object and by how much light it reflects or absorbs. A white object reflects most of the incoming light and a black object absorbs most of the incoming light (Sect. 9.1). A perfect absorber is called a "blackbody" in physics. When we see a red object, it means that the object reflects most of the red light but absorbs the remaining colors (green, blue, etc.) of sunlight.

The energy of light absorbed by an object is converted to heat, causing the temperature of the object to rise. The object will in turn radiate to release the amount of energy it absorbs. In 1900, German physicist Max Planck (1858–1947) derived a formula governing the amount of radiation at any wavelength emitted by a blackbody of a certain temperature based on the quantum nature of light. Planck hypothesized that the energy $(E)$ of light is given by the formula $E = h\nu$, where $\nu$ is the frequency of light and $h$ is a constant now referred to as Planck's constant. Planck's formula shows that all objects with a finite temperature will emit light in all colors (from X-ray to radio) and the amount of light emitted in each color (wavelength) is dependent on the temperature of the body. The hotter the object, the more its radiation is shifted to the blue, and a cooler object will radiate more in the red. The amount of light emitted has a peak for each temperature, and at both the highest and lowest wavelength ends the amount will diminish to zero (Fig. 14.5).

The relationship between the peak of the radiation and the temperature is very simple:

$$\lambda_{max} \, (cm) \cdot T \, (K) = 0.3$$

This is known as Wien's law. For example, the Sun has a temperature of about 6000 degrees Kelvin and will have its radiation peak at 0.00005 cm, or 0.5 μm, which is in the visible region. Our bodies are much cooler, about 37 degrees Celsius (or about 310 degrees Kelvin), and we shine mostly in the infrared with a peak at about 0.001 cm, or 10 μm.

The fact that our eyes can see visible light is not an accident. Through the process of evolution, our ancestor species developed eyesight in the visible spectrum to take advantage of the colors where the Sun shines most strongly. In other words, our visual ability is a consequence of the temperature of the Sun.

Strictly speaking, the temperature given in Planck's radiation law is called the radiation temperature. In our everyday environment where there is a very high

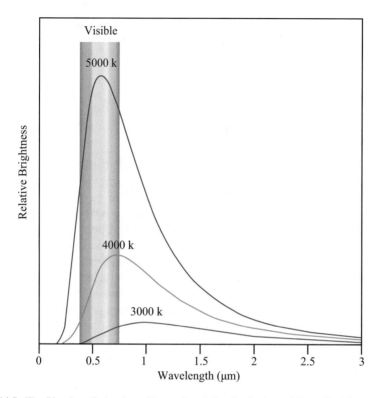

**Fig. 14.5** The Planck radiation law. The peaks of the distributions shift to the blue at higher temperatures. The area under each curve corresponds to the total amount of flux emitted. The visible region is indicated by the rainbow colors

density of particles and frequent collisions between molecules in the air and matter, the radiation temperature that determines radiation by matter is the same as the kinetic temperature of the air. So we often do not make a distinction between the two. This condition is called thermodynamic equilibrium. However, such equilibrium conditions may not apply to the low density of space.

Not only living things like animals and trees shine, but also rocks and stones, tables, and chairs. This may seem surprising. Our visual perception is based mostly on reflected light. When we move a chair to a windowless dark room, we can no longer see the chair. This does not mean that the chair does not give off light; it is just giving off light in a color that we cannot see!

Everyday objects are luminous in two different ways. They are bright in the visible because of reflected sunlight and at the same time they are also self-luminous as they are radiators of infrared light. While the Sun shines in all colors, the Moon is bright in the visible spectrum to varying degrees at different phases because its visual brightness is due to reflected sunlight. Only the parts of the Moon facing the Sun will be visually bright, but the whole Moon is always bright in the infrared, even during the New Moon.

Wien's law gives us the ability to easily estimate the temperature of a star by just measuring its color. By measuring the brightness of a star at several different wavelengths and estimating the wavelength at which the star's brightness peaks, one can use Wien's law to obtain its temperature.

## 14.7   The Search for Ether

The physical nature of light has occupied the minds of physicists since the time of Newton. From the observation that light travels in straight lines and can reflect off surfaces, Newton envisioned light as an ensemble of particles. However, the observed phenomena of diffraction and interference of light are better explained by light behaving like a wave, an idea championed by French physicist Augustin Fresnel (1788–1927). There are many examples of waves in our everyday life, including a wave in a pond or a sound we hear. Both types of waves require a medium to propagate, water for the former and air for the latter. The medium for the transmission of light waves was hypothesized to be luminiferous ether. Because there is light coming from stars, luminiferous ether must permeate throughout the Universe in order for starlight to reach us. The search for ether[3] therefore became a high priority in nineteenth-century physics.

Although the ether hypothesis requires a large amount of ether in the Universe, ether is difficult to detect because it does not interact with matter. However, because the Earth revolves around the Sun and therefore travels through ether in opposite directions over the seasons, it should be possible to detect the presence of ether through a change in the speed of light at different times of the year. As the Earth moves through the stationary ether, it should feel the effect of an ether wind, the same way that we feel the wind on our faces when we are running on a calm day. Many experiments were performed in the nineteenth century but none were successful in detecting the expected ether wind. The most well-known attempt was made by American physicists Albert Michelson (1852–1931) and Edward Morley (1838–1923) in 1887, who tried to detect the difference in speed of light traveling in perpendicular directions. No difference was detected, and the effect of the ether wind was not found.

However, the physics community was not convinced of the absence of ether and the search for ether wind continued with larger and more sophisticated apparatus. A claim of positive detection was made in 1925 by Dayton Miller (1866–1941), who won the Newcomb Cleveland Prize of the American Association for the Advancement of Science for this work. Later experiments disputed the result of Miller and confirmed the null result of Michelson and Morley, which is consistent with Einstein's hypothesis that the speed of light is independent of the movement of its

---

[3] Here "ether" is a transmission medium of light, which is borrowed from but not related to one of the five elements of Aristotle.

emitter (Sect. 13.5). The physics community eventually came to accept that it may not be necessary for light to be transmitted by a medium, and the need for the concept of luminiferous ether finally disappeared from physics research after the mid-twentieth century.


## 14.8   Quantum Theory of Light and Matter

As for the nature of light, the debate continued. The wave theory suffered a setback with the discovery of the photoelectric effect by German physicist Heinrich Hertz (1857–1894) in 1887. When light strikes a material (e.g., most metals), electrons are ejected from the material, but only if the frequency of light is higher than a certain value. Below this value, no electron is ejected even if the intensity of light is very high. This is known as the photoelectric effect. For most materials, the photoelectric effect requires visible or ultraviolet light to eject an electron. The photoelectric effect was explained by Einstein who proposed that light behaves like a packet of discrete energy (called a photon) in which energy is proportional to its frequency ($E = h\nu$). When the photon energy exceeds the binding energy of the material, an electron is ejected with its kinetic energy being equal to the difference between the photon energy and the binding energy of the material. For photons with energy lower than the binding energy, an increase in light intensity only increases the number of photons impacting the material but does not overcome the critical energy required.

By the early twentieth century, it became clear that the nature of light cannot be described by the concepts of wave or particles alone. Light behaves like waves sometimes (e.g., in interference and diffraction), but like particles at other times (e.g., the photoelectric effect). This was finally reconciled by the quantum mechanical concept of wave-particle duality. The French physicist Louis de Broglie (1892–1987) found that not only light, but electrons also behave partly like a wave and partly like particles. Although electrons were first perceived to be particle-like entities, they also show wave-like interference patterns. Nature at the microscopic level cannot be adequately described by analogies of our everyday experience as in the cases of waves or particles, and all matter show wave-particle duality.

The wave-particle duality represents an example of a scientific model contrary to our everyday experience. Starting with Aristotle, our early scientific models to explain the pattern of behavior of Nature all draw upon examples in our everyday experience. Now we have learned to accept a theory, although seems strange at first sight, because its prediction can explain observational facts. Often this requires the utilization of ever more sophisticated forms of mathematics. Einstein's theory of gravity as discussed in Sect. 13.6 is an example.

## 14.9  Science and Utility

One of the greatest successes of science has been its ability to explain our everyday experiences. Through our senses, we can see colors of light and feel heat. These perceptions can be explained by theories about the nature of light and how light interacts with matter. However, when we examine the nature of light in detail, we find that we cannot explain all its behavior with a simple theory. Light sometimes behaves like waves, and sometimes like particles. Such duality of property is alien to our common sense.

Both waves and particles travel with finite speeds. So, it may not be so difficult to accept that light also travels at a finite (albeit high) speed. However, it is contrary to common sense that its speed is constant irrespective to the speed of its emitter. The light sent out by the headlight of a fast-moving train has the same speed as the light emitted by a stationary lantern on the platform. The hypothesis of the constancy of the speed of light was needed by Einstein to preserve the principle of relativity and to allow the phenomena of electricity and magnetism to work the same way in all inertial frames of reference (Sect. 13.5).

The quantum theory of matter is considered one of the greatest achievements of modern science. In this model, a rock, a tree, or human bodies are all made of atoms, which are essentially hollow. The atomic nucleus where most of the atom's mass lies occupies a very small fraction of the volume of the atom. The hard surface of a rock that we touch is mostly empty inside. It seems contradictory to common sense that one can sit on a chair without collapsing even though the space inside the chair is mostly vacuum.

Many modern students make fun of Ptolemy's epicycles as being absurd. Are modern theories any less absurd? Why do we accept these strange hypotheses of wave-particle duality and constancy of speed of light? The simple answer is that because they work; the same reason that we accept Newton's theory of gravitation as a force without an agent acting at a distance (Sect. 21.3, vol. 1). Newton's theory can accurately describe the motion of planets as well as the trajectory of projectiles on Earth. We accept the hypothesis of action-at-a-distance because it explains our observations well and is widely applicable to many phenomena. Einstein's theory of general relativity can explain planetary motion more accurately than Newton's but is based on entirely different premises. Einstein's theory of gravity interprets planetary motion as objects following the shortest paths in a curved space-time wrapped by the presence of mass. This hypothesis is no less strange than Newton's, but we now prefer Einstein's theory of gravitation over Newton's because it works better.

Utility is therefore one of the criteria we use to judge a scientific theory. The hypothesis itself can be as strange as it can be as long as its predictions agree with observational data. Not only does a theory have to work, it should also be able to be falsified. We must be able to test a theory to see if any of its predictions are in contradiction to observations. No theory is sacred.

**Questions to Think About**

1. The first element of the periodic table and the most common element in the Universe–hydrogen, was not isolated until the eighteenth century. Why was its discovery so late?

2. Discuss the philosophical significance of the discovery of the chemical composition of air.

3. Is it a coincidence that the wavelengths of the spectra lines of hydrogen can be represented by such a simple empirical formula? Would the development of the quantum theory of atoms be much delayed if the formula were much more complicated?

4. For thousands of years, human perception of light was limited to the visual range. It was not until 200 years ago that we became aware that visible light is only a very small part of the entire spectrum of light. What are the implications of this expansion of vision?

5. Given the natural light background in our everyday lives, what would be our view of the world if we possess (i) infrared vision; or (ii) X-ray vision?

6. Why is unification so important to physicists?

7. How can something be simultaneously a particle and a wave?

8. How do you define something as real? Why do we consider electrons to be more real than ether?

9. Ether was thought to be real by most physicists in the nineteenth century, but this concept is now discarded. Are there any modern concepts that may turn out to be fictitious in the future?

10. In the late twentieth century, string theory (a mathematical variation of the knot theory of atoms) became popular as a model for elementary particles. Can it succeed where the knot theory failed? Is there a unique way to represent the structure of matter?

# Chapter 15
# The Human–Star Connection

The belief that there is a human–star connection can be traced back to antiquity. Among all the celestial objects, the Sun's effect on human lives is the most evident. Light from the Sun allows humans to see our surroundings and to function during the day, and the Sun provides the warmth that keeps us alive. The annual movement of the Sun is the cause of the seasons, whose temperature variations dictate the living patterns of the human population. The position of the Sun in the zodiac constellations gives a reliable signal to farmers for when to plow, seed, harvest, and store their grains. It is not an accident that many ancient cultures worship the Sun god, because the Sun is the most important object in their lives.

There are also clear signs that the Moon is connected to our lives. The Moon not only provides illuminations at night when the Sun is absent, but it also causes the periodic rise and fall of the tides. The correlation of the coming and going of tides with Moon phases was well known to fishermen since ancient times.

Because of the Sun's influence on human activities, it was widely believed that the planets and the stars also had effects on life on Earth. The practice of astrology was common in both eastern and western civilizations. The high degree of interest in astronomical observations partly originated from a desire to decipher celestial messages and foretell fortunes by the stars. In many cultures, transient events such as comets and eclipses are believed to be harbingers of forthcoming calamities. The conjunctions of planets (where the five planets are seen within small angular separations on the ecliptic) are also believed to forecast oncoming doom.

The need for astrological forecasts requires accurate predictions of future positions of the Sun, the Moon, and the planets. After the translation of *Almagest* into Latin, planetary tables were prepared in 1270 under the patronage of Alfonso X, King of Castile, based on Ptolemy's model. The Alfonsine Tables were later improved by Regiomontanus (1436–1476) in 1474, whose tables were used by Columbus in 1504 to predict a lunar eclipse, with the result of intimidating the American natives into giving food to his men. After Copernicus, Erasmus Reinhold (1511–1553) published the Prutenic Tables in 1551 (sponsored by Duke Albert of Brandenburg of Prussia) using Copernicus' model. It was the Prutenic Tables, not

© Springer Nature Switzerland AG 2021
S. Kwok, *Our Place in the Universe - II*,
https://doi.org/10.1007/978-3-030-80260-8_15

his heliocentric theory, that cemented Copernicus' reputation as an astronomer in the sixteenth century. However, these tables are not more accurate than those based on the model of Ptolemy. After developing his own laws of planetary motion, Kepler made use of elliptical orbits and law of areas, rather than epicycles, eccentrics, and equants, to make planetary predictions. His Rudolphine Tables, named after his patron Emperor Rudolph II, published in 1627, were more accurate than previous ones.

Astrology was deeply embedded in society well into the seventeenth century in Europe and the nineteenth century in China. Events in the heavens are directly related to earthly matters and astrological forecasting was taken extremely seriously. Both the European and Chinese rulers derived their legitimacy through their respective connections to God and heaven, as European kings were crowned by the Church and Chinese emperors declared themselves "sons of heaven." This mandate from heaven kept their subjects submissive and under control. After astronomical progress changed people's perception of heaven and earth and their place in it, the feudal hierarchy was finally dismantled.

We now view astrological influence on us as a myth and superstition. Except for the Sun and the Moon, planets and stars are too far away for their radiation and gravity to have any practical effect on Earth. Even when all five planets line up in the sky, their combined gravitational effects on Earth are minimal. By the eighteenth century, astrology was discredited by the scientific community. It, therefore, came as a surprise when we later found out that there are direct physical links between stars and us.

## 15.1   What Powers the Sun?

During the time of Aristotle, stars were thought to be divine objects that shine naturally. With the philosophical shift to a mechanical view of the Universe in the nineteenth century, scientists felt a need to seek an explanation of how stars shine using known laws of physics. Through the analysis of spectra of the Sun, astronomers learned that the Sun and the stars all have chemical compositions similar to elements on Earth. Unlike the planets which shine from reflected sunlight, stars shine with their own power and such powers are immense. From the amount of radiation that we receive from the Sun on Earth and the current value of the astronomical unit (AU) (Chap. 4), astronomers determined the energy output of the Sun to be $3.8 \cdot 10^{26}$ Watt. For comparison, the typical output from a power station is less than $10^9$ Watts. Some stars are even more luminous, generating power that is thousands of times that of the Sun.

How do the Sun and stars generate such a large amount of energy? From our experience on Earth, we learned that energy can be generated in a variety of ways. Our ancestors took advantage of gravity causing water to flow from high to low ground to generate energy with water wheels. Windmills are built to harness the power of wind. Animals and humans can move around and function with the consumption of food. Chemicals in food are converted into other chemicals through reactions in our bodies, releasing energy in the process. By burning wood, we are releasing the chemical energy content in wood. The burning of coal, oil, and natural gas are all examples of chemical power. The development of the steam engine was a major driver of the industrial revolution. Chemical fuels such as coal were used to boil water into steam. The thermal energy contained in hot steam was used to drive mechanical devices into motion. Internal combustion engines rely on the combustion of gasoline (a fossil fuel) to generate power. Transportation, heating, and other functions in our modern society very much depend on the utilization of chemical power.

The ultimate source of gravitational and chemical powers on Earth is the Sun. Sunlight evaporates water from the oceans that condenses into water droplets in clouds. The water contents in clouds are deposited in the forms of rain or snow in high mountains which later flows down to lower ground as rivers. The difference in gravitational energy between water in high and low altitudes can be turned into power by a water wheel or a hydropower station. Wood is produced from trees that convert solar energy to chemical energy through photosynthesis. Coal, oil, and natural gas are remnants of once-living organisms that still hold chemical energy within.

After the Second World War, we mastered another form of energy generation: nuclear energy. By splitting a heavy atom into smaller components (in a process called fission, the basis of an atomic bomb), energy can be released. Nuclear power plants have been constructed making use of heavy elements such as uranium and plutonium as fuels. These chemical elements have been buried inside the Earth since the formation of the Earth and can be mined. Nuclear fuels, therefore, do not rely on the current output of the Sun.

## 15.2 Source of the Sun's Energy

Not only does the Sun generate a lot of energy, it has been generating energy for a long time. A great reservoir of energy must reside inside the Sun, gradually releasing energy throughout the lifetime of the Sun. What is the form of this energy and how is it being converted to light?

Let us first assume that the Sun generates its energy through chemical burning. Because the amount of chemical energy released by molecules is approximately known, it can be estimated that if the Sun's energy were to come from burning something, it would have run out of fuel in a few thousand years. Another possible source of energy is gravitational energy: The Sun exists as a stable body through a delicate balance between gravitational attraction and internal thermal pressure. As a ball of self-gravitating gas, the laws of physics dictate that as the Sun contracts, it releases energy. Theoretically, the Sun could shine by continuously contracting, releasing its gravitational energy in the process. It is estimated that this process could allow the Sun to shine for 20–40 million years. From geological studies, we have learned that the Earth is 4.6 billion years old (Chap. 17). Because the Sun must be older than the Earth, it must have a source of energy that can last billions of years. This time interval is larger than the length of time that can be supplied by the release of gravitational energy, so another source of energy must be sought.

One of the fundamental beliefs of nineteenth-century chemistry is the immutability of elements. However, by the early twentieth century, physicists observed that certain chemical elements are intrinsically unstable, and they naturally decay into another element over time (Sect. 17.2). This suggests that chemical elements are not eternal but could be transformed from one to another. Atoms were also found not to be indivisible particles but to consist of a small nucleus surrounded by multiple electrons (Sect. 14.2). The atomic nuclei themselves were made of two separate particles: protons and neutrons. The number of protons (called the atomic number) defines the element and the number of neutrons defines the isotope of a particular element. An element can have a different number of neutrons and therefore a different atomic weight (the total number of protons and neutrons). For example, the most common form of the element oxygen ($^{16}O$) has 8 protons and 8 neutrons and an atomic weight of 16. One isotope of oxygen has 8 protons and 10 neutrons and an atomic weight of 18 ($^{18}O$).

If some of the neutrons can be changed into protons, that will allow the transformation of one element to another. This process is very similar to chemical reactions of molecules. Molecules are made of atoms (e.g., a molecule of water has two hydrogen atoms and one oxygen atom) and through the exchange of atoms, chemical reactions can occur, leading to the synthesis of a different molecule. Like chemical compounds changing from one to another through chemical reactions (which represent the re-arrangement of atoms), the nuclei of atoms can transform from one to another through nuclear reactions that release energy. This can be done either by splitting a heavy element into smaller ones (in a process called "fission") or combining several simple atomic nuclei into a larger one (in a process called "fusion").

Let us consider the example of the fusion of hydrogen into helium. By forcing four hydrogen nuclei ($^{1}H$) together, they can be transformed into one helium nucleus (with two protons and two neutrons). However, the mass of the final product ($^{4}He$), has a slightly smaller mass than the total of four hydrogens. The difference in mass ($\Delta m$) is released in the form of energy ($=\Delta mc^2$, where $c$ is the speed of light). If fusion can be utilized in stars, then the amount of energy possibly generated will be much greater than by chemical or gravitational means. Because hydrogen is the most common element in the Universe and the Sun is primarily made of hydrogen, fusion of hydrogen atoms in the Sun can generate enough power to keep it shining for billions of years.

However, nuclear fusion reactions require very high temperatures and densities. Calculations of stellar structures suggest that such conditions are achievable at the center of the Sun. In 1939, the German-American physicist Hans Bethe (1906–2005) showed that the Sun should be able to maintain its luminosity for 10 billion years by fusing hydrogen into helium in its core. The process is like the ignition of a hydrogen bomb, except that it is not a one-time event but is continuously and steadily operating in the interior of the Sun. The energy released by nuclear fusion reactions in the Sun is in the form of gamma rays, a very high energy form of light (Sect. 14.4). As the gamma rays generated in the core travel through layers of the Sun's envelope, they collide with atoms and electrons in the envelope and are gradually degraded into lower energy light. By the time the light emerges from the surface of the Sun, it is in the form of visible light. The problem of the source of the Sun's power was finally solved.

## 15.3   Direct Confirmation of Nuclear Fusion in the Sun

Nuclear fusion operates deep inside the Sun but all we can observe about the Sun is light coming from its surface. The surface of the Sun is opaque to visible light and we cannot see directly into the interior of the Sun. How can we be sure that it is the nuclear fusion that powers the Sun? Fortunately, the nuclear reactions transforming hydrogen into helium also produce a byproduct that can travel through the core to the surface of the Sun. The observation of this byproduct will allow us to directly test the hypothesis of nuclear reactions being responsible for the Sun's luminosity.

In addition to gamma rays that are released from the nuclear fusion of hydrogen into helium, the reactions also release a large quantity of particles called neutrinos. Neutrinos can penetrate anything practically without hindrance. They can emerge from the interior of the Sun, pass its surface and reach the Earth in eight minutes. Beginning with the pioneering work of American physicist Raymond Davis (1914–2006) in the 1960s, physicists have been building ever more sensitive detectors of neutrinos. At first, the observed flux of neutrinos seemed far below that expected by theory, so either the theory or the observations were wrong. The theoretical modelers of the Sun tried to adjust the model parameters to make the results fit, but the gap between the observed and theoretical values was too great. Something more fundamental was amiss. It was only in the 1970s that it was realized that neutrinos are not massless as it was assumed in the past, but in fact have a small finite mass. This opens the possibility that solar neutrinos can change form (or in physics jargon "favors") during their passage from the Sun to Earth, resulting in a lower detection rate when observed from Earth.

With the constructions of facilities such as the Super-Kamioka Neutrino Detection Experiment located under Mount Ikeno in Japan, and the Sudbury Neutrino Observatory 2100 m underground in Vale's Creighton Mine in Sudbury, Canada, physicists finally detected an observed flux of neutrinos that is consistent with the stellar structure and nuclear reactions models for the Sun. The reconciliation between theoretical predictions and observed neutrino fluxes required a new theory of the physics of neutrinos. The resolution of this "solar neutrino paradox" is an example of the scientific method at work: theories being modified to account for observations.

Previously, advances in physics have driven the understanding of celestial objects, but in the case of the solar neutrinos, it was astronomical observations of the Sun that motivated a new theory of fundamental physics of elementary particles to be developed. Instead of using a particle accelerator, the Sun has served as a laboratory for the understanding of fundamental physics.

## 15.4   The Solar Terrestrial Connection

The Sun not only has a radiative output, it also streams out particles in the form of a solar wind. When the charged particles of the solar wind interact with the Earth's magnetic field, it creates a magnetosphere surrounding the Earth. The phenomenon of Aurora Borealis (northern lights in the north hemisphere, or Aurora Australis, southern lights in the southern hemisphere) is caused by solar wind particles colliding with oxygen and nitrogen in the upper atmosphere. The oxygen atoms radiate a green light and the nitrogen atoms produce a purple light.

On occasion, the Sun can send out massive, fast-moving particles in the form of a solar storm. When these particles hit the Earth's magnetosphere, it can generate large electric currents that can affect radio communication and power transmission.

Although the effects of the solar wind on human activities are not as obvious and direct as sunlight, they nevertheless show that our parent star can impact the Earth in more than one way. By sending spacecraft into Earth's orbit, we can intercept the solar wind and study its content. The direct examination of the solar wind particles provides another means of observation beyond astronomical imaging and spectroscopy, confirming that the Sun is a physical object made of atoms similar to those on Earth.

## 15.5   Origin of Chemical Elements

After the identification of chemical elements as the fundamental building blocks of matter (Chap. 14), it was natural to ask the question, "where did they come from?" One obvious answer is that the chemical elements were "just there." They were either put on Earth by God in the process of Creation or were simply present at the beginning of the Earth and the Universe. The English chemist William Prout (1785–1850) was the first person to ask the question "what is the origin of chemical elements?" He speculated that the chemical elements were created from the simplest element hydrogen. Although Prout raised this fundamentally important philosophical question, he had no suggestion for how this may occur.

Could all the chemical elements that we have on Earth today be produced by nuclear reactions over the history of the Universe? The idea that nuclear processes may be responsible for the synthesis of elements was considered by Ukrainian physicist George Gamow (1904–1968) in the 1940s. Because the Universe was hot and dense during the Big Bang, nuclear reactions could have taken place then.

The second element of the periodic table helium could be built from converting four hydrogen atoms into helium atoms through the fusion process.

While this process can work for a few light elements, it was soon determined that it is unlikely that all chemical elements except a few light elements could have been made by the Big Bang. The possibility that chemical elements can be made in stars started with the detection of the element technetium (Tc) in the atmospheres of stars. In 1952, American astronomer Paul Merrill (1887–1961) obtained a spectrum of a red giant star and found the spectral signature of technetium. Because spectral lines of atoms are commonly detected in stars, this should not be a surprise—except that technetium is not a stable element. Over a period of around one million years, technetium will naturally decay into ruthenium or molybdenum. Because red giant stars are much older than the lifetime of technetium, its detection in red giants means that it could not have been inherited from birth but has to be manufactured locally within the lifetime of stars. This demonstrates that stars are factories of chemical elements.

The theory of nucleosynthesis of chemical elements in stars was first proposed by British astronomer Fred Hoyle (1915–2001) in 1946 and independently formulated by Hoyle and Canadian physicist A.G.W. Cameron (1925–2005) in the 1950s. In a paper titled "Synthesis of Elements in Stars" that Burbidge, Burbidge, Fowler, and Hoyle published in *Review of Modern Physics* in 1957, they outlined the nuclear reactions responsible for the synthesis of heavy elements in stars. Similar calculations were published by Cameron in typewritten form as a report from the Chalk River Laboratory in Canada. Using nuclear processes that can occur in the interior of red giant stars, both Hoyle and Cameron were able to explain the relative amounts of elements as observed in the spectra of stars. Through slow neutron capture and beta decay nuclear reactions, many heavy elements such as yttrium, zirconium, barium, lanthanum, cerium, praseodymium, neodymium, samarium, europium, etc. are also synthesized late in the life of a star. Some heavy elements (thorium, uranium, and plutonium) are synthesized during supernova explosions.

## 15.6   Chemical Composition of the Human Body

The basic units in human bodies are cells, which are in turn made of molecules. Our basic body functions of respiration, digestion, and muscular and sensory functions, are all driven by chemical reactions. Our body is therefore a giant chemical factory. The cells in our body are made of simple inorganic molecules such as water, and complex organic molecules such as protein, DNA, sugar, and fat. Each of these molecules may contain thousands of atoms, with the most common being hydrogen (61%), oxygen (26%), carbon (10%), nitrogen (2%), calcium (0.2%), phosphorus (0.1%), and sulfur (0.1%). We absorb these atoms through eating of food, drinking of liquid, and breathing of air. These molecular intakes are broken down by our body and the products are used to fuel our activities and to build or replace our body parts. In addition to these common elements, our bodies also need a variety of other elements to function. The hemoglobin in our blood contains iron, our thyroid gland needs iodine, zinc is needed in the production of insulin, calcium helps build our bones, sodium plays a role in sending nerve impulses, chromium helps cells to draw energy from blood sugar, copper help make red blood cells, molybdenum helps break down toxins, and manganese is needed in enzymes. We cannot live without many elements that are synthesized in stars.

Because every atom in our body was once inside a star, we are all made up of stellar material. The stellar manufactured atoms are ejected into interstellar space to form interstellar clouds. Interstellar clouds condense to form stars and solar systems (Chap. 23). Tiny pieces of solid materials aggregate to become planets. Living organisms evolved from materials present on Earth or imported from asteroids and comets (Chap. 19). Chemical elements generated from inside the core of a star found their way to the Earth and eventually to our bodies.

The understanding of the origin of the elements was one of the greatest scientific achievements in the twentieth century. Although ancient people had imagined a star–human connection in the form of astrology, the theory of stellar nucleosynthesis establishes a direct physical connection between stars and humans by tracing the atoms on Earth and in our bodies to the interior of stars. Its social implications are comparable to Newton's unification of terrestrial and celestial motions, and Darwin's theory of how biological species evolved. This link has put the existence of celestial, terrestrial, and living objects on a physical basis and the evolution of stars, Earth, and humans as coupled physical processes.

## 15.7   Universality of Physics and Chemistry

The immutability of elements was one of the pillars of nineteenth-century chemistry. It was the fundamental principle behind the explanation of chemical reactions, leading to a very successful description of the composition of air, water, minerals, and other parts of the terrestrial world. However, this principle was discarded when nuclear reactions were discovered. One element can decay into another element, two or more elements can combine to form another element, and an element can capture a neutron and then transform into a heavier element. The effect of this on our view of the world was comparable to the destruction of the concept of the immutability of the heavens by Tycho's discovery of supernovae (Chap. 19, Vol. 1). The view that part of our world is constant and everlasting is gradually being replaced by the concept of change.

It was the demise of this principle of immutability of elements that allowed us to understand the nature of the power of the Sun, a question that had puzzled scientists for over a hundred years. The chemical elements were not all present at the beginning of the Universe but were synthesized mostly in the interior of stars. The abundance of heavy elements was built up over time through many generations of stars. The success of the theory of stellar nucleosynthesis is the result of an interplay between nuclear physics and astronomy.

The successes in understanding energy generation in the Sun and the synthesis of elements greatly enhanced our belief in the universal applicability of physics. Not only are the laws driving the planets and terrestrial projectiles the same, the rules governing the structure of matter are also the same across the Universe. The extrapolation to the beginning of the Universe for the synthesis of light elements such as helium and lithium is based on our belief that the laws of nuclear physics work across billions of years in time.

## Questions to Think About

1. Ancient scholars did not feel the need to ask what makes the Sun shine. What were the events that motivated modern scientists to ask this question?
2. Is it necessary for chemical elements to have an origin? Why cannot the elements be just there at the beginning of the Universe?
3. The theory of the origin of elements shows how heavy elements can be built up from simple ones. But there must be a first ingredient from which everything else came. Where did this first ingredient come from? Is it possible scientifically to answer the question of the origin of everything?
4. How can we be sure that the atoms on earth are the same as the atoms in space? Could the spectral signatures be a coincidence?
5. If atoms in the sky were to obey different laws as atoms on earth, how would it affect our understanding of the universe around us?
6. Scientists used to believe in the immutability of elements, probably as strongly as we believe in the conservation of energy today. Is it possible that the principle of conservation of energy could be overturned one day?

# Chapter 16
# Is the Universe Finite?

In Ptolemaic cosmology, the Universe has a finite size defined by the celestial sphere upon which the fixed stars lie. Within this celestial sphere are the five planets and the Sun revolving around the Earth. Assuming that the interior volume of the celestial sphere is filled by orbits and epicycles of the planets, Ptolemy derived a size of the Universe that is 20,000 times the radius of the Earth (Chap. 15, volume 1). After the development of the heliocentric theory, the hypothesis of the daily rotation of the celestial sphere was replaced by the hypothesis of the rotation of the Earth. This removes the need for the stars to lie at the same distance and rotate together, which in turn opens the possibility that the stars may have different distances from Earth and that the Universe could be infinite in size (Chap. 18, Vol. 1). While Copernicus replaced the Earth with the Sun as the center of the Universe, this cannot be true in an infinite Universe. An infinite Universe has no center, and therefore whether the Universe is finite or infinite has implications for our place in the Universe.

In 1826, German astronomer Heinrich Wilhelm Olbers (1758–1840) asked the question, "why is the night sky dark?" He reasoned that if the Universe is infinite in extent and is uniformly filled with stars, then the night sky should be bright. This is based on the simple argument that although stars' brightness drops off with the inverse-squared of distance, the number of stars at each distance increases with the distance squared. So, a shell of stars at each distance will contribute equal amounts of brightness to the night sky. In an infinite Universe, the total brightness should be infinite. This is known as Olbers' paradox or the dark night sky paradox.

In 1895, German astronomer Hugo von Seeliger (1849–1924) showed that Newtonian gravitational theory is not compatible with an infinite Universe with a uniform distribution of matter. He concluded that either the Universe is finite or Newtonian gravitational theory does not apply on the large scale. This was known as Seeliger's paradox. At that time, there was not enough observational knowledge about the large-scale structure of the Universe for us to answer the question of whether the Universe is finite or not. The solutions to Olbers' and Seeliger's paradoxes have to wait until observations of distant galaxies and their distribution and movement can tell us more about the state of the Universe.

© Springer Nature Switzerland AG 2021
S. Kwok, *Our Place in the Universe - II*,
https://doi.org/10.1007/978-3-030-80260-8_16

## 16.1   Looking Back into the Past

In our everyday experience, the reception of distant light seems instantaneous. With the observations of stellar parallax, we realized that stars are far away. Even the closest stars are light-years away (Appendix I). This implies that the light that we see from a star now was emitted years ago. Using Cepheid Variables and other standard candles, we learned that galaxies are even farther away. Our galactic neighbor, the Andromeda Galaxy, is located at a distance of 2.5 million light-years from us, and other galaxies are even more distant. When we look at a galaxy, we are also looking back in time. If we can build a very powerful telescope that can peek into distances of billions of light-years, are we also looking back in time at what the Universe was like billions of years ago? Is there a limit to how far back we can look?

The issue of space and time, or distance and age, is of fundamental importance. If the Universe is unchanging and infinite in extent, we will find the same Universe no matter how far back we look. But if we look at distant objects and find a Universe different from what it is now, we may conclude that the Universe is evolving. The ability to look at distant objects (and therefore events in the past) will give us clues about the structure and history of the Universe.

Our problem is that the visible sky is two dimensional and we need a sense of depth to know how far away a celestial object is. Unlike stars, galaxies are too far away for parallax measurements, and other means of distance determination have to be developed. For very nearby galaxies where individual stars can be resolved and spectroscopic observations can be obtained, the spectral type (and therefore the intrinsic luminosity) of the star can be determined and the distance derived. For galaxies that contain Cepheid Variables, the distance of the parent galaxy can be determined from the Cepheid period–luminosity relationship (Sect. 12.4). Several classes of astronomical objects such as novae, planetary nebulae, brightest red giant stars, and supernovae, have all been used as standard candles for galactic distance determination.

With the development of stellar spectroscopy, the Doppler effect becomes a very powerful tool to study the motion of stars (Sect. 2.5). In addition to determining the speed of stars in the plane of the sky through the measurement of proper motions, astronomers can also measure the movement of stars along our line of sight to the star. The speed of stars, either approaching or receding, is called radial velocity. Proper motions and radial velocities together provide us with the knowledge of stellar motions in three-dimensional space. By the end of the nineteenth century, not only the distributions of stars can be mapped, but the motions of stars can also be measured.

## 16.2   Expansion of the Universe

The technique of using Doppler shifts to measure radial velocities can also be applied to nebulae if the wavelength of a spectral line can be measured. In 1912, Vesto Slipher (1875–1969) of Lowell Observatory in Arizona obtained a spectrum of the Andromeda Nebula (M31) and found it to be moving toward the Earth at a very high speed of 300 km per second, much larger than the typical radial velocities of tens of km per second observed in stars. Further observations of 24 spiral nebulae by Slipher found that many are moving at high speeds, some with velocities higher than 1000 km per second. With a larger sample, Slipher detected a pattern showing that spiral nebulae generally have their spectral lines shifted to red. This implies that spiral nebulae are mostly moving away from us, and the Andromeda Nebula was one of the few exceptions.

The high-speed movement of spiral nebulae is very different from the motions of stars in the Milky Way. This distinction was used by Curtis to argue that spiral nebulae are not members of the Milky Way, giving support to the island universe hypothesis (Sect. 12.3).

By 1924, Hubble had discovered Cepheid Variables in the Andromeda Nebula and determined its distance to be one million light-years. The Andromeda Nebula therefore cannot be a member of the Milky Way but is a separate stellar system outside of the Milky Way. We now call stellar systems outside of the Milky Way galaxies. Hubble proceeded to use this technique to determine the distances of a few nearby galaxies. When Cepheid Variables are too faint to be used in distant galaxies, he used the brightest stars in the galaxy as standard candles. By assuming that the brightest stars in galaxies all have the same intrinsic brightness, he could determine the distances of these remote galaxies.

At the same time, Hubble also directed his assistant Milton Humason (1891–1972) at Mt. Wilson to measure the spectral line shift (generally to the red) of galaxies found by Slipher. The 2.5-m telescope at Mt. Wilson was much more sensitive than the 0.6-m telescope at Lowell Observatory available to Slipher, and Humason was able to measure the Doppler shifts of spectral lines of much fainter galaxies. He also found that most of these Doppler shifts are to the red, and this is now commonly referred to as "redshifts" of galaxies. Most significantly, Humason discovered galaxies with redshifts much larger than those found by Slipher.

By 1929, Hubble had determined the distances of 24 galaxies out to a distance of six million light-years. When combined with the radial velocities of these galaxies measured by Humason, Hubble found a linear relationship: the farther away a galaxy, the faster it is moving away from the Earth (Fig. 16.1). This plot makes use of data that Hubble and Humason obtained, as well as those measured by Slipher. This linear relationship is often called Hubble's law, and the slope of this linear relationship is called the Hubble constant.

Hubble presented this distance–velocity relationship purely as an empirical law without speculating on its meaning. The interpretation was provided by Georges Lemaître (1894–1966), a Belgian astronomer and Jesuit priest. Two years before

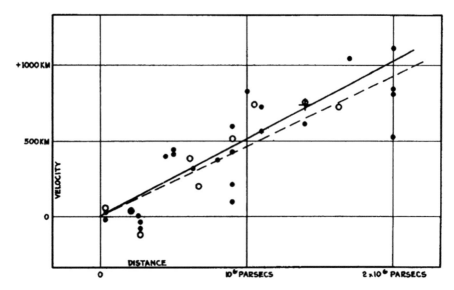

**Fig. 16.1** The velocity–distance relationship as originally published by Hubble. The slope of the line in this graph (now called the Hubble constant) represents the rate of expansion of the Universe. The original value of the slope as determined by Hubble was off by almost a factor of 10 because of incorrect distance calibration of Cepheid Variables. The black dots are the 24 nebulae with velocities measured. The open circles represent data when nebulae are put into groups. The two lines represent two different ways of fitting by Hubble. The original figure appeared in Hubble, E. 1929, *Proceedings of the National Academy of Sciences*, **15**, 168–173

Hubble published his graph, Lemaître had derived a solution to Einstein's theory of general relativity which showed that the Universe could be expanding. Previous solutions to Einstein's equations were all for a static Universe, as there was no evidence to the contrary. Lemaître interpreted the general recession of galaxies as the result of the expansion of the Universe.

At the XXXth General Assembly of the International Astronomical Union in Vienna in 2018, the Assembly voted to rename Hubble's law as the Hubble–Lemaître law, in order to properly recognize the role of Lemaître in the discovery of the expansion of the Universe.

The Lemaître interpretation suggests that the redshifts observed in galaxies are not the result of the Doppler effect of moving objects. The galaxies themselves are not moving; it is the space in which they are embedded that is stretching out. It is this stretching of space that causes the wavelength of light emitted by distant galaxies to be longer. The common analogy is that galaxies are like painted dots on a balloon. As we inflate the balloon, all the dots seem to be moving away from each other. The fact that all the galaxies seem to recede from us does not mean that we are at the center of the Universe. As one of the dots on the balloon, we have the perception that every other dot is moving away from us, but in fact, we are all moving apart from each other. Another analogy is raisins in a cake. As the cake is being baked in an oven, the raisins inside the rising cake are separated more and more from each other.

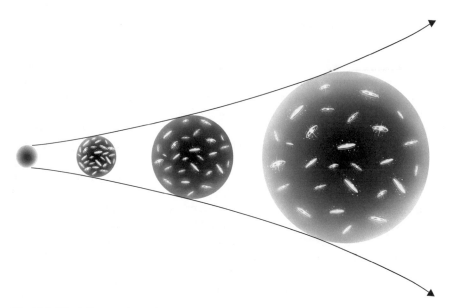

**Fig. 16.2** The balloon analogy of an expanding universe. The dots on the two-dimensional surface of an expanding balloon represent galaxies in the three-dimensional Universe expanding in the fourth dimension. From the point of view of an observer on one of the galaxies, all other galaxies are moving away

We live in a three-dimensional world and space is expanding in the fourth dimension. This is equivalent to dots on a two-dimensional surface of a balloon that is expanding in the third dimension (Fig. 16.2). Not only we are not at the center of the Universe, there is no center of the Universe in this model, just as there is no central dot among the many dots on the surface of a balloon. Our three-dimensional Universe is without a boundary, just as the two-dimensional surface of the Earth has no boundary. No matter how far we travel on this Earth, we will never encounter an "end."

The interpretation by Lemaître of the Humason–Hubble distance–velocity relationship as the result of the expansion of the Universe was one of the most successful applications of Einstein's theory of general relativity. Einstein's theory not only explains all the terrestrial gravitational phenomena previously explained by Newton, it also gives a better account of the planetary motions, in particular the advance of the perihelion of Mercury, and is now demonstrated to be applicable to the Universe as a whole. This wide applicability is why Einstein's theory of general relativity is considered one of the most successful physical theories of the twentieth century.

## 16.3   Large-Scale Structure of the Universe

After Copernicus, we realized that our Earth is one of the planets in our Solar System. By the early twentieth century, we also realized that our Sun is one of the stars in our Milky Way and our Milky Way is one galaxy among other galaxies. Are there even bigger hierarchical structures in the Universe? Having determined distances to the Andromeda Galaxy and some of the nearby galaxies, Hubble used the term "Local Group" in 1936 to describe some of the nearest galactic neighbors of the Milky Way. Using the 18-inch wide-field telescope at Mt. Wilson, Fritz Zwicky (1898–1974) and Walter Baade (1893–1960) performed a survey of the sky to map out the distribution of galaxies. From their sample of hundreds of thousands of galaxies, they found that galaxies tend to cluster together. The entire northern sky was surveyed at Mt. Palomar using the 48-inch wide-field telescope, resulting in a collection of photographic plates known as the Palomar Observatory Sky Survey. From the examination of these survey plates, American astronomer George Abell (1927–1983) published a catalog of 2712 clusters of galaxies. Some of the larger clusters can have over a thousand members and extend over 300 million light-years in size (Fig. 16.3).

The most well-known example is the Virgo Cluster located at ~50 million light-years away in the constellation of Virgo. It contains ~2000 galaxies. As early as the nineteenth century, William and John Herschel were already aware that there is a higher concentration of nebulae in the constellation Virgo near the north galactic pole. The Virgo Cluster extends over 15 degrees in the sky, or about 15 million light-years across. The total mass of the Virgo Cluster is estimated to be $10^{15}$ times the mass of the Sun.

Are there structures larger than galactic clusters? In 1953, French astronomer Gerard de Vaucouleurs (1918–1995) suggested that our Local Group, the Virgo Cluster, and some other galactic clusters could be members of an even larger entity called supercluster. The sizes of superclusters are hard to quantify but they are generally ten times larger than individual clusters. By mapping the distributions of clusters and superclusters, astronomers hope to find the large-scale structure of the Universe. This can be achieved by performing deep surveys of the sky by finding faint galaxies and measuring their red shifts. This will push our observations farther in distance and help us obtain a true map of the Universe in three dimensions. For example, the Two-degree-Field Galaxy Redshift Survey conducted at the 3.9-m Anglo-Australian Telescope between 1997 and 2002 mapped over 200,000 galaxies in two slices of the sky to a distance limit of about 2.5 billion light-years. These survey maps show that galaxies are not uniformly distributed in space but are concentrated in clusters linked by "filaments" and with empty spaces ("voids") in between. The physical origin of such large-scale structures is a subject of intense current studies.

Galaxy clusters and superclusters are still relatively nearby structures. How about the very large structure of the Universe? In Lemaître's interpretation of the velocity–distance relationship, galaxies are distributed in three dimensions, but the Universe

**Fig. 16.3** Abell 2151 cluster of galaxies. Located 470 million light-years away, the Abell 2151 cluster of galaxies contains elliptical, spiral, and irregular galaxies. Image taken with the MegaCam camera on the Canada-France-Hawaii Telescope. Image credit CFHT/Coelum—J.-.C. Cuillandre & G. Anselmi

is expanding in the fourth dimension. However, the Universe can be either "closed," "flat," or "open" (Fig. 16.4). A "closed" Universe can be visualized in the balloon analogy (Fig. 16.2), where the Universe is a finite fourth-dimensional sphere. Galaxies are distributed on their three-dimensional surface just like the dots on the two-dimensional balloon surface. Alternatively, the Universe can be a three-dimensional plane ("flat"), or a three-dimensional hyperboloid (with a negative curvature, "open"), expanding in the fourth dimension. If the Universe is a four-dimensional sphere, then it is finite but has no boundaries. On the finite surface of the Earth, one could travel in the same direction and eventually come back to the place one started. However, this cannot happen in our Universe as it is constantly expanding, making the journey longer and longer with time.

**Fig. 16.4** A schematic diagram illustrating the possible geometries of the Universe. These three surfaces are two-dimensional analogs of our three-dimensional universe. Our Universe could be closed (with a positive curvature like a sphere, left panel), flat (like a plane, middle panel), or open (with a negative curvature like a hyperboloid, right panel)

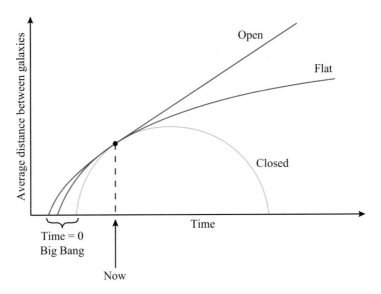

**Fig. 16.5** A schematic diagram illustrating how we can determine the geometry of the Universe. By measuring the rate of expansion in the past (at time left of "now"), one can distinguish between the different geometries of the Universe

If the Universe is "flat" or "open," it is infinite in extent and without boundary. Figure 16.5 gives a qualitative description of how the time evolution of the Universe depends on the geometry of the Universe. If there is enough matter in the Universe, the self-attraction of all the matter will slow down and eventually stop the expansion of the Universe, leading to it collapsing back onto itself. This corresponds to a closed Universe. If there is too little matter in the Universe, the Universe will keep on expanding and this corresponds to an open Universe. If the amount of matter is just "right" to balance self-attraction and expansion, then the expansion of the Universe will slow down but will not stop. This corresponds to a flat Universe.

The geometry of the Universe can be determined by measuring changes in the rate of expansion with time. If we look at distant galaxies, we are also looking at galaxies during the earlier age of the Universe. The rate of expansion at an earlier time is dependent on the geometry of the Universe (Fig. 16.5). Relativistic cosmological models suggest that if the redshift-distance relationship (Fig. 16.1) deviates from a linear form, exactly how much it deviates will tell us whether the Universe is open or closed. It, therefore, provides a great deal of incentive to extrapolate this plot to large distances.

Our current attempts to map the distribution of clusters of galaxies and voids are analogous to the mapping of the continents and oceans on the surface of the Earth during the age of exploration. Our effort to determine distances to galaxies is analogous to the determination of longitudes in the eighteenth century. We rely on objects such as Cepheid Variables or supernovae as "standard candles." Assuming that we know the intrinsic brightness of the "standard candles," we can derive the distances to their host galaxies by measuring their apparent brightness. However, no "standard candles" are perfect and we cannot ever be certain that these "standard candles" behaved the same way in the past as they do now. If their properties have evolved, then they can no longer be considered "standard candles."

The application of Einstein's general theory of relativity allows us to interpret galaxies' velocity–distance relationship as the result of a Universe expanding in the fourth dimension. Is this fourth dimension real? We live in a three-dimensional world and have no sense beyond three-dimensional space. The introduction of the fourth dimension allows us to elegantly explain the observed motion of galaxies in a mathematical model. The concept of a fourth dimension is therefore useful, just as the introduction of the concept of electrons is useful in understanding atomic spectra.

## 16.4   The Beginning of Time

Because of the finite speed of light and the fact that galaxies are far away, light from galaxies takes a long time to reach us. The image we see of a galaxy today with a telescope is actually the image of it millions of years ago. Because the Universe is expanding, galaxies must have been closer to each other earlier in time. It is natural to ask whether there was a beginning of the universe when everything was very close together. If what we see the Universe doing today reflects what it has been doing for billions of years, we could imagine a time when all galaxies were very close together. The possibility that there was a beginning, a moment of creation of the Universe, is a natural extrapolation of our observations today. From the slope of the distance–velocity relationship (Fig. 16.1), one could derive a current rate of expansion of the Universe. Assuming that this rate has been constant, one could derive an age of the Universe.

The rate of expansion of the Universe derived from the original graph of Hubble's distance–velocity relationship gave a very young age of the Universe. In fact, the derived age of the Universe was younger than the age of some stars and also the age of the Earth as it was known in the early twentieth century (Chap. 17). Later revisions to Hubble's distance determination of galaxies have decreased the rate of expansion and lengthened the age of the Universe. The currently accepted estimate of the age of the Universe is about 14 billion years, certainly older than the age of the Solar System (4.6 billion years) and older than the ages of known old stars.

A finite age of the Universe can easily explain Olbers' paradox. Because we can only look back to a finite time and distance, the observable Universe is finite independent of whether the Universe is finite or infinite in space. The number of stars (or galaxies) that we can see is finite and therefore the night sky is dark.

The possibility that there was a beginning in time was unappealing to some. The philosophy of dialectical materialism of German philosophers Karl Marx (1818–1883) and Friedrich Engels (1820–1895), which was the official philosophy of the Soviet Union, rejects this view of cosmology on philosophical grounds. An ever-lasting Universe without beginning or end was viewed as preferable. Such a version of cosmology was developed by British astronomers Fred Hoyle (1915–2001), Hermann Bondi (1919–2005), and Thomas Gold (1920–2004) in the form of a steady-state theory. In this theory, matter is continuously created to compensate for matter dilution as the result of expansion. Hoyle even coined the term "Big Bang" in 1948 to mock the idea of the creation of the Universe. However, the term "Big Bang" is now commonly used to refer to the cosmology of an expanding Universe with a beginning.

By extrapolating the laws of physics to a state of very high density near the beginning of the Universe, Gamow concluded in 1946 that the early Universe should have been dominated by light. From this, he predicted that remnants of such light should still be with us today in the form of isotropic radiation permeating all space. In the early days after the Big Bang, the Universe consisted of protons and electrons, and the scattering of light by electrons prevented light from escaping. The Universe was entirely opaque until approximately 400,000 years after the Big Bang when it cooled enough that protons and electrons could combine to form neutral hydrogen atoms. The light that emerged from the surface of the last scattering represents the last relic radiation of the Big Bang.

As we discussed in Chap. 9, all matter absorbs and radiates light. The light emitted by a perfect absorber (called a blackbody, Sect. 14.6) can be characterized by a single parameter, called the radiation temperature. The first prediction of the radiation temperature of the remnant of this radiation from the Big Bang was estimated by American physicists Ralph Alpher (1921–2007) and Robert Herman (1914–1997) to be about 5 degrees Kelvin (K). The accidental detection of isotropic radiation coming from all directions in the microwave wavelength region in 1964 by American physicists Arno Penzias (1933–) and Robert Wilson (1936–) of Bell Laboratories confirmed this prediction, and the Big Bang hypothesis has become the widely accepted model of cosmology. This radiation, now called cosmic background radiation, has been repeatedly measured to a high degree of precision and has a modern value of 2.73 K.

Because the early Universe must have been very hot and dense, nuclear reactions could be expected to take place. The synthesis of elements in the first few minutes of the Big Bang was formulated by Gamow and his student Ralph Alpher in a 1948 paper titled "The Origin of Chemical Elements." The light elements of hydrogen, helium, and lithium are believed to have formed in the Big Bang, before there was any star or galaxy. The first elements produced by Big Bang nucleosynthesis provide the basic ingredients for stellar nucleosynthesis in stars (Chap. 15).

## 16.5 An Evolving Universe

Many ancient civilizations viewed the nature of the Universe as either static or cyclical. Having witnessed the cyclical behavior of celestial objects, it was natural for ancient astronomers to develop a concept of time that is cyclical. On a daily scale, there are repeated cycles of the rising and setting of the Sun. On a monthly scale, there are repeated cycles of the moon phases. Over the period of a year, there are the cycles of the seasons. Planets all have their own synodic periods, and the 584-day period of Venus was important to the Mayans. From the existence of these cycles, our ancestors speculated that the cosmos may also have a temporal cycle on a much longer time scale.

However, on the human time scale, the cosmos seems to be static. Aristotle believed in the immutability of the heavens, where the Earth, the Sun, the planets, and the stars have always been there and always will be. The cosmology of Aristotle is a static one, where the Universe has always existed and will continue to exist to eternity. Hundreds of years before Galileo, Chinese astronomers were already aware of the coming and goings of sunspots. They were also aware of the appearance of new (or guest) stars before Tycho Brahe. In the view of Chinese astrology, the heavens are constantly changing, and these changes were interpreted as signs or messages from heaven to humans. However, we now believe that these changes are random (novae and supernovae) or cyclic (sunspots) events and not results of evolution.

The change from a static to an evolving model of the Universe allows us to contemplate questions of cosmic history. In the twentieth century, astronomers have determined the structure of the Milky Way and discovered galaxies external to the Milky Way. Although these new developments have greatly expanded the content of the Universe, the static nature of the Universe has not changed. It was the discovery of the Humason–Hubble distance–velocity relationship and Lemaître's interpretation of this observational relationship as evidence for the expansion of the Universe that led to an evolving picture of the Universe. Expansion of the Universe implies that the Universe may have had a beginning when it was dense and hot. As a result of expansion, the Universe gradually became more diluted in density and cooled in temperature. Astronomers today believe that stars and galaxies were formed as the result of cooling and the uneven distribution of matter in the early Universe. This clustering of matter and subsequent gravitational attraction created matter density high enough for nuclear reactions to occur and shining stars to form. Some astronomers have suggested that this occurred as early as one billion years after the Big Bang. If matter distribution was completely uniform at the beginning, stars and galaxies may never have formed. The existence of stars, the Earth, and humans are dependent on the exact early conditions of the Universe.

**Questions to Think About**

1. Compare the similarities and differences in philosophical implications of the Copernicus revolution and the discovery of the expanding universe.

2. If there is no boundary in space, why is there a beginning in time? If time and space are equivalent, should time also have no boundary?

3. How is the concept of time defined? What is the meaning of the beginning of time?

4. The distance–velocity relationship was found by Hubble (1929) not long after Einstein's formulation of the theory of general relativity in 1916, allowing Lemaître to interpret the relationship as evidence for the expansion of the Universe. If Einstein's theory had not been developed, how would astronomers explain this empirical relationship?

5. The separations between stars in the Galaxy are very large. Galaxies are also far apart. Why is the Universe so empty and why does matter occupy only a very small fraction of the volume of the Universe?

6. Our attempts to understand the large-scale structure of the Universe inevitably rely on assumptions that we must accept on faith. How far can we extrapolate our local knowledge of physics to distant space and a long time in the past?

7. What is the philosophical significance of the shift in cosmological view from Newton's static Universe to an evolving one?

8. Was the formation of the first stars a coincidence? What if the Universe did not form stars, consequently there are no humans to observe the Universe?

9. Today, Hubble is well known for the discovery of the expansion of the Universe, but in fact, he only discovered the distance–velocity relationship. Hubble continued to resist Georges Lemaître's expansion interpretation for many years, well into the 1950s. How does the scientific community assign credit and what role do public relations play in the scientific process?

10. How do we know that something is real? The introduction of a fourth spatial dimension allows us to give an elegant mathematical interpretation to the velocity–distance relationship, but is the fourth dimension "real" or just a convenience?

# Chapter 17
# Early History of the Earth

In the previous chapters, we have outlined the journey of our realization of the expanding spatial size of the Universe. In this chapter, we will discuss how our concept of temporal history has also been expanding. Human history is based on generational oral and written accounting of events of the past. The oldest reliable written history is probably by the Chinese, who can trace written records back to the Shang Dynasty (~1500 B.C.). Earlier human history is mostly based on myths and legends. Nevertheless, we can safely assume that human civilization has been around for thousands of years.

After Europe adopted Christianity as its dominant religion, the Biblical account of Creation became the authoritative source of the history of the cosmos. According to the Book of Genesis, humans, animals, the Earth, the Sun, and the rest of the celestial objects were created within a period of six days. Various biblical scholars from St. Basil the Great (fourth century), St. Augustine (fifth century), Venerable Bede (eighth century), Kepler (sixteenth century), and Newton (seventeenth century) gave estimates of the time of Creation to be around 4000 B.C. In 1650, the Archbishop James Ussher of Ireland (1581–1656), calculating from the biblical genealogy in the Old Testament, gave the precise time of the Creation as nightfall preceding Sunday, October 23, 4004 B.C., putting the Earth at less than 6000 years old at that time.

The biblical view was gradually being challenged during the Age of Enlightenment. Scientists began to seek independent estimates of the age of the Earth. In 1778, the French naturalist Georges-Louis Leclerc, Comte de Buffon (1707–1788), assumed that the present Earth cooled from molten rock and used the laboratory-measured cooling rate of heated rocks to estimate an Earth age of 75,000 years, which is considerably older than the Biblical age of the Earth.

By the nineteenth century, geologists realized that the Earth has not always been the same and they suspected that the mountains and seas have been subjected to slow processes of erosion and deposition over time scales much longer than thousands of years. The Scottish geologist Charles Lyell (1797–1875) promoted the idea that the observed geological changes are the result of natural causes. Through rainfall,

© Springer Nature Switzerland AG 2021
S. Kwok, *Our Place in the Universe - II*,
https://doi.org/10.1007/978-3-030-80260-8_17

freshwater leaches different chemical elements from the rocks and carries them to the oceans. Because oceans have no outlet, dissolved salt accumulated over time and the oceans became salty. If there is no removal mechanism, the age of the Earth can be estimated. In 1899, Irish geologist John Joly (1857–1933) compared the degree of salinity and accumulation rates and gave an estimate of 80–90 million years for the age of the Earth.

Because the Earth condensed out of the solar nebula, the Solar System must be at least as old as the Earth. The new geological age of the Earth came into conflict with the estimation of the age of the Sun. In 1856, the German physicist Hermann von Helmholtz (1821–1894) suggested that the Sun shines as the result of the release of energy through gradual contraction of the Sun. By assuming that the Sun is a ball of gas held together through a balance of gravitational self-attraction and thermal pressure of the gas, Helmholtz concluded that the Sun had a lifetime of about 22 million years. Using similar arguments with estimates of the amount of heat possessed by the Sun, William Thomson (who later became Lord Kelvin, 1857–1933) came to an estimate of 20–400 million years in 1864. He favored the lower end of the values and in 1897 revised the estimate down to 20–40 million years. Because of Lord Kelvin's academic stature and forceful personality, the belief that the Earth is less than 100 million years old remained the dominant view of the scientific community through the late nineteenth century.

Although Lord Kelvin's arguments were rigorous and elegant, they were unfortunately incorrect. Advances in nuclear physics identified nuclear fusion as the source of the power of the Sun (Chap. 15), and the geological determination of the age of the Earth could proceed without the worry of the Earth being older than the Sun.

## 17.1   Methods of Age Determination

The fact that tree trunks have rings was known in ancient times. By the fifteenth century, it was recognized that tree rings are due to the annual growth of a tree and each ring corresponds to one year in a tree's life. By cutting down a tree and counting the number of rings, one can determine how old the tree is. The oldest tree that this method has been used on is a bristlecone pine from California, which was about 5000 years old. This method can even be applied to trees that are dead, e.g., wood from historic buildings. The spacing between rings is not uniform as the temperature and humidity are not the same every year. For example, a severe winter will leave a distinctive mark on the tree ring. By placing two specimens side by side and matching the ring patterns, one can extend the coverage of the period (Fig. 17.1). Using multiple specimens of deadwood, one can extend the age determination up to about 14,000 years. We note that this age is already older than the biblical age of 6000 years. Because the tree-ring method is simple and reliable, it is used to calibrate the other age determination methods.

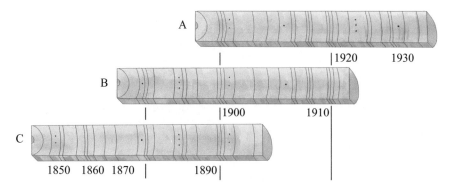

**Fig. 17.1** Method of tree-ring dating. By matching the ring patterns of trees of different eras, the range of tree-ring dating can be extended. Image based on illustrations produced at the Laboratory of Tree-Ring Research at the University of Arizona

In addition to dating, tree rings can also be used to trace past climates. For example, past cold weather and droughts can be identified to the year that they occurred through tree rings. Using tree rings and other methods, precise climate records can be determined for hundreds to thousands of years in the past (Chap. 22).

## 17.2  Radioactive Dating

Contrary to the belief that chemical elements are immutable, scientists at the beginning of the twentieth century learned that certain elements are intrinsically unstable. They tend to decay over a period of time into another stable element. The phenomenon of radioactivity was first noted by French inventor Abel Niépce de Saint-Victor (1805–1870) in 1857 and confirmed by French physicist Henri Becquerel (1852–1908) in 1896. Three examples of radioactive decay are illustrated in Fig. 17.2. In the example of alpha decay, the emitted particle (alpha particle) is identified as the helium nucleus. In beta and positron decays, the emitted particle is the beta (electron) and beta plus (positron) particles. All natural radioactive decays are accompanied by the release of energy, often in the form of gamma rays.

Although the decay of individual atoms occurs randomly, a sample of atoms will have a constant decay rate. The decays occur at an exponentially decreasing rate and the time scale of decay is measured by the half-life of the decay, which is the time needed for half of the parent nuclei to decay into daughter nuclei. The half-life for each unstable element is different (Table 17.1).

The absolute age of a material since its formation can be obtained by measuring the ratio of parent to daughter nuclei. By measuring the present ratio of the isotopes and setting the beginning of the clock by assuming an initial ratio, the value of the ratio of the two gives the time elapsed. One of the most useful isotopic ratios is carbon 14 ($^{14}C$) to carbon 12 ($^{12}C$) ratio. Carbon-14 is produced in the atmosphere

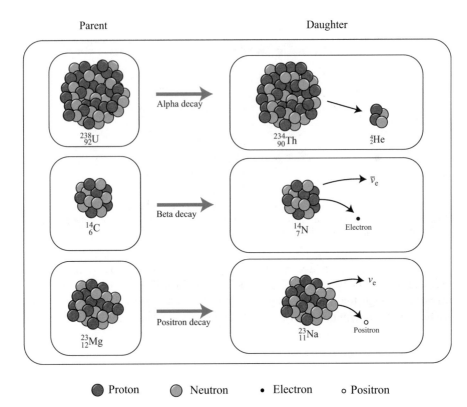

**Fig. 17.2** Schematic picture of radioactive decay. Three different types of decay are illustrated here. Top: alpha decay where the atomic number and atomic weight both decrease by 2. Middle: beta decay: the atomic number increases by 1 and the atomic weight remains the same. Bottom: positron decay: the atomic number decreases by 1 and the atomic weight remains the same. The number of protons and neutrons in the graphic of alpha decay is for illustration only and does not correspond to the actual numbers

**Table 17.1** Examples of radioactive clocks

| Parent | Daughter | Half-life (years) |
|---|---|---|
| Samarium-147 | Neodymium-143 | 106,000,000,000 |
| Rubidium-87 | Strontium-87 | 48,800,000,000 |
| Rhenium-187 | Osmium-187 | 41,600,000,000 |
| Thorium-232 | Lead-208 | 14,000,000,000 |
| Uranium-238 | Lead-206 | 4,500,000,000 |
| Potassium-40 | Argon-40 | 1,250,000,000 |
| Uranium-235 | Lead-207 | 704,000,000 |
| Samarium-147 | Neodymium-143 | 108,000,000 |
| Iodine-129 | Xenon-129 | 17,000,000 |
| Aluminium-26 | Magnesium-26 | 740,000 |
| Carbon-14 | Nitrogen-14 | 5730 |

http://pubs.usgs.gov/gip/geotime/radiometric.html
http://en.wikipedia.org/wiki/Radiometric_dating

by cosmic rays and its abundance is roughly constant over time. When a living organism is alive, its ratio of $^{14}C$ to $^{12}C$ is the same as that of the environment, through the constant exchange of organic and inorganic carbon molecules with the environment. After the organism died, it stops absorbing $^{14}C$ and the remaining $^{14}C$ decays to nitrogen-14 ($^{14}N$) with a half-life of 5730 years. By measuring the present $^{14}C$ to $^{12}C$ ratio, one can determine the age of carbonaceous materials up to 50,000 years ago. This method is known as carbon dating and is very useful in dating objects over the period of human history.

For longer periods, different pairs of isotopes need to be used. Through the recognition that lead is the final decay product of uranium, with a half-life of ~4.5 billion years (Table 17.1), the U.S. chemist Bertram Boltwood (1870–1927) in 1907 used the measured ratio of lead and uranium in rocks to derive an age of 400–2000 million years for the Earth. Using a similar method, the British geologist Arthur Holmes (1890–1965) increased the age of the Earth to 1.6–3 billion years in 1927. This was further increased to 3.4 billion years by Ernest Rutherford (1871–1937) in 1929. Using lead isotopes, Holmes arrived at a value of 3.35 billion years in 1947. Refinement of the radioactive dating method led to the identification of the oldest rocks (which are in Greenland) as 3.8 billion years old. By extrapolating backward in time to an age where there should have been no lead produced by radioactive decay, one could estimate the age of the Earth to be 4.6 billion years. In 1956, American geochemist Clair Patterson (1922–1995) measured the abundance of various lead isotopes in the Canyon Diablo meteorite and derived a more precise estimate of 4.55 billion years for the age of the Solar System.

We should note, however, that the technique of carbon dating assumes that the amount of $^{14}C$ in the environment is constant, but this is not necessarily true. In recent times, both the burning of fossil fuel and tests of nuclear bombs have changed the amount of $^{14}C$ in the air. This change in initial conditions makes it necessary to calibrate carbon dating using other techniques. For example, the ages of corals and marine sediments can be measured by both $^{14}C$ and thorium/uranium dating ($^{230}Th/^{234}U$) methods.

## 17.3   A Physical Connection between Heaven and Earth

Aristotle divided the Universe into superlunary (celestial) and sublunary (terrestrial) worlds and maintained a strict separation between the two (Chap. 14, Vol. 1). The celestial world is pure, holy, and unchanging, whereas the terrestrial world suffers from calamities such as floods, volcanos, hurricanes, and earthquakes. Transient objects in the sky such as comets and meteors were considered to be atmospheric phenomena and are different from the external celestial objects of the Sun, Moon, planets and stars.

The Earth was considered as an isolated body with no physical contacts with the other celestial bodies. Even after Copernicus and Newton, scientists maintained the view that the only direct contact between celestial objects and the Earth is light. We receive direct radiation from the Sun and can see reflected sunlight from the Moon, planets, asteroids, and comets. Stars are distant suns, whose radiation we can also observe and measure. Light is the only direct link between heaven and earth.

The first hint that a physical link may exist between heaven and Earth was the study of meteors and meteorites. Meteorites have been recognized as rocks fallen from the sky by many ancient civilizations. Hundreds of reported cases of meteorite falls were recorded in the imperial and provincial records of China, Japan, and Korea. The Romans considered meteorites gifts from gods and housed them in temples. In the Bible, meteorites are mentioned in Acts 19:35: "Fellow Ephesians, doesn't all the world know that the city of Ephesus is the guardian of the temple of the great Artemis and of her image, which fell from heaven?"

On a typical moonless night, several meteors (streaks of light across the sky) can be seen in an hour. Because the occurrence of individual meteors does not happen on a fixed schedule, meteors were generally believed to be an atmospheric phenomenon. Chinese astrologers, however, believed that meteors are fallen stars and carry astrological messages.

Several times a year, meteor showers can be seen radiating from certain stellar constellations. The best-known examples are the Perseid and Leonid meteor showers, which occur in August and November, respectively. Meteor showers can be extremely spectacular, with hundreds to thousands of meteors per hour streaming seemingly from a point in the sky (Fig. 17.3). The strength of the showers shows an apparent periodicity. Approximately every 33 years, the Leonid shower is observed to be particularly strong (called a "storm"). In 1863, American astronomer Hubert A. Newton (1830–1896) showed from historical records that the period of the Leonid meteor showers follows the sidereal year,[1] not the tropical year, therefore demonstrating that the meteor showers are caused by a celestial event.

It was a common belief in the eighteenth century that meteorites were created in the atmosphere by lightning, or by the accretion of volcanic dust. In 1794, the German scientist Ernst Friedrich Chladni (1756–1827) suggested the meteorites may be related to meteors with an origin not in the clouds but outside of the Earth. One of the clear signs that meteors and meteorites are related happened on May 14, 1861. A brilliant fireball (a meteor that is brighter than any of the planets) accompanied by loud explosions was seen and heard all over southern France. Shortly after, villagers of Orgueil collected over 20 pieces of black rock. It was clear that the celestial phenomenon of meteors has left behind remnants on Earth, and we can hold a piece of a celestial object in our hands. It was not until the nineteenth century, when chemical analysis of meteorites showed that their contents are different from terrestrial rocks, that the extraterrestrial origin of meteorites was accepted.

The link between meteor showers and comets was established in the mid-nineteenth century when the Perseid and Leonid meteor showers were traced

---

[1] Because of the precession of the equinox, the length of a sidereal year is not the same as the tropical year (see Sect. 11.4, Vol. 1).

**Fig. 17.3** Engraving showing the 1833 Leonid meteor shower. The meteors seem to radiate from a single point in the constellation of Leo

to the comets 109P/Swift–Tuttle and 55P/Tempel–Tuttle, respectively. We now realize that meteors are the result of remnants of comets and asteroids crossing the path of the Earth and entering the atmosphere. When the Earth's orbit intersects with the path of the comet, the Earth will pick up debris from the comet (Fig. 17.4). Friction generated between these falling solids and the Earth's atmosphere burns up

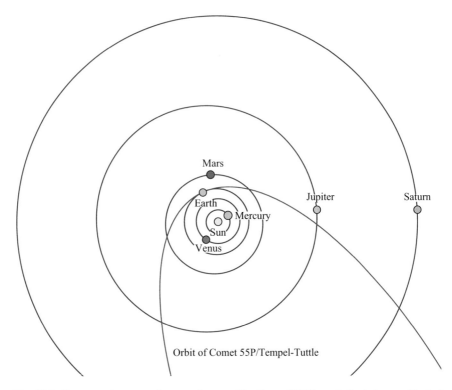

**Fig. 17.4** Comets as the cause of meteor showers. The Comet 55P/Tempel–Tuttle has an elliptical orbit (shown in red) with a period of 33 years. When the Earth crosses the orbit of a comet, we will witness a meteor shower. We will experience a meteor storm when a periodic comet returns near the Sun

a fraction of the particle and lights up the sky. The remaining pieces that survive the atmospheric entry are found on the Earth's surface as meteorites.

Because asteroids and comets are Solar System objects that orbit around the Sun, meteorites are the only celestial objects that we can pick up, touch, and examine without going into space. By performing physical and chemical analyses of meteorites, we have direct information on the content of a celestial object. We can show that these celestial objects are made of the same chemical elements as rocks on Earth, and there is no fundamental distinction between the composition of celestial and terrestrial objects.

One class of meteorites is particularly interesting. Carbonaceous chondrites contain small inclusions of unusual high-temperature minerals and these must have condensed directly from the solar nebula. They are the most primitive objects in the Solar System. The dating of these inclusions gives an age of 4.565–4.568 billion years for the Solar System. The study of carbonaceous chondrites, therefore, provides us direct hints about the chemical composition of the early Solar System.

## 17.4   External Bombardment of the Early Earth

When Galileo first observed the Moon with his telescope in 1610, he found that the lunar surface was not smooth. The nearly circular features on the Moon found by Galileo led Kepler to conjecture that these are artificial constructions of lunar inhabitants. Further telescopic observations of the Moon showed that the Moon is filled with hills and depressions, not unlike the geological features found on Earth. Some depressions have distinct round shapes similar to the craters formed by volcanos. It was therefore natural for scientists to think that the lunar craters are also of volcanic origin. In 1891, American geologist Grove Gilbert (1843–1918) found that the largest lunar craters have sizes much larger than any crater on Earth. While Earth's volcanic craters all have similar sizes, lunar craters come in vastly different sizes. From the radial rays found in lunar craters, Gilbert suggested that these rays represent splashes from an impact. The craters on the Moon are not volcanic but are caused by impacts from external objects.

The impact hypothesis was tested in the laboratory by both German geologist Alfred Wegener (1880–1930) and American planetary scientist Ralph Baldwin (1912–2010). They showed that the observed shapes and depths of lunar craters can be produced by the impacts of external objects with different velocities and impact angles. In the process of impact, much of the mass of the impactor may be vaporized and its energy is converted to heat, creating a crater in the aftermath. This was later confirmed by the testing of nuclear weapons (atomic and hydrogen bombs), which often leave behind craters.

The impact theory was finally accepted by the geological community in the mid-twentieth century. Moon rocks collected from the *Apollo* lunar missions in the 1970s gave further support to this hypothesis. Later space missions to Mars and Mercury have observed that the surfaces of these planets are also full of craters, suggesting that they have also been subjected to bombardment from external objects.

There are over 1700 impact craters with diameters >20 km on the Moon (Fig. 17.5). Because the Earth is a more massive body with a stronger gravitational pull, the number of impact events must be at least 10 times higher. Unlike the Moon, the Earth has undergone an extensive geological transformation and many of its earlier surface features have been erased. Only about 200 impact craters can still be identified on Earth today. The most well-known is the Canyon Diablo Crater in Arizona, USA, which has a diameter of 1 km and a depth of 170 m. Although initially considered to be volcanic in origin by Grove Gilbert, the crater is now believed to be created by an impact from a 50-m diameter object 50,000 years ago. The energy of the impact is estimated to be equivalent to 50 megatons of TNT, or about 3000 times the power of the Hiroshima atomic bomb.

Such impacts could still happen today. On February 15, 2013, a 20-m-diameter asteroid entered the Earth's atmosphere over Russia. The fireball that it created was brighter than the Sun and was visible up to 100 km away. Friction with the atmosphere caused it to explode at an altitude of ~30 km, producing damages over a wide area. It is estimated that once every 100,000 years, an asteroid with a size

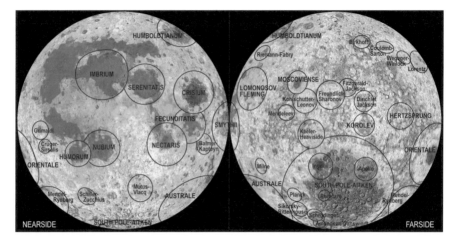

**Fig. 17.5** Map of some of the largest impact basins on the Moon with diameters >300 km. Nearside of the Moon is on the left and far side on the right. The colors represent the topographic of the lunar surface from deep blue indicating −8 km below the mean (white) to +8 km (in red). Credit: Lunar Planetary Institute, Paul D. Spudis and David A. Kring

greater than 1 km will hit the Earth, creating a crater 10–20 km in diameter. Through the release of debris into the atmosphere, a substantial amount of sunlight could be blocked, resulting in a cooling of the surface of the entire Earth. Impacts from larger objects could result in the major extinction of living species (Sect. 21.4).

## 17.5   Formation of the Moon

The Moon is the second brightest celestial object in the sky. As with other celestial objects, the Moon was assumed to have been always present and everlasting and there was no need to ask the question of where the Moon came from. After the nineteenth century when philosophical thinking has shifted to an evolutionary universe, the origin of the Moon became a fair scientific question. The first ideas were either that the Moon was a foreign object captured by the Earth or that it had been broken off from the Earth after the Earth was formed out of the same Sun-orbiting material. From the fact that the Moon has a much lower density than the Earth, we can infer that the Moon does not have an iron core as the Earth does. The current thinking is that the Moon was formed out of materials that were stripped from the surface of the Earth upon impact by a Mars-size planetesimal. This giant-impact theory neatly explains the lower density of the Moon. This event was assumed to have happened 4.5 billion years ago, during the period of heavy bombardment.

Besides providing illumination at night, the Moon exerts other effects on the Earth and on human lives. It has been known since ancient times that the Moon

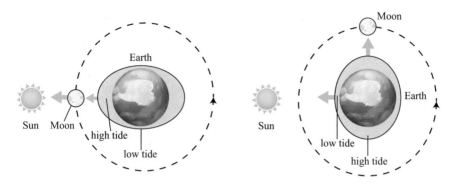

**Fig. 17.6** Cause of the tides. High tides occur on Earth in the direction facing the Moon, as well as in the opposite side of the Earth. At new Moon (left) or full Moon, the effects of tides are maximized. When the Sun and the Moon are 90° apart (first or third quarter Moon, right), the effects of tides are minimized. Diagram is not to scale

affects the coming and going of the tides. Ocean tides are caused by gravitational pulls from the Sun and the Moon, with the Moon having a larger effect than the Sun. Because the Earth rotates once a day, the times of high tides will change as different parts of the Earth turn to face the Moon. A bulge of the oceans also occurs at the opposite side of the Earth. The cycle of tides, therefore, follows the orbital motion of the Moon around the Earth and the phase cycle of the Moon. When the Moon lines up with the Sun (which occurs at new Moon and full Moon), the effects of the tides are at maximum. At quarter Moon when the Moon is at 90° from the Sun, the effects of the tides are at a minimum (Fig. 17.6).

Seasons are caused by the tilt of the rotation axis of the Earth relative to its orbital plane around the Sun. If this angle of tilt (obliquity of the ecliptic) were to change, the variation would have major effects on the climate. Gravitational pulls from the Sun and Jupiter could cause variation of the obliquity. Fortunately, the presence of the Moon steadies the angle and has maintained the climate stability of the Earth.

The Moon also influences the length of a day. After the impact event that created the Moon, the Earth was rotating with a period of ~5 hours. As the result of tidal interaction between the Moon and the Earth, the Moon is slowly moving away from the Earth. With modern laser ranging techniques, the Moon's receding rate is measured to be 4 cm per year. Tidal effects from this increasing separation also gradually slowed down the rotational period of the Earth to its present period of 24 hours. Evidence for this slowdown can be found in fossils. From the daily growth bands of corals, scientists can count the number of days in a year. From these counts, we can determine that the number of days in a year was larger in the past. Because the Earth has been rotating relatively steadily for several hundred million years, animals and plants must have evolved to adjust to the present day/night cycle. The Moon therefore also influenced the evolution of life on Earth.

Is the presence of the Moon a coincidence? Among the four terrestrial planets, Mercury and Venus have no moons, and Mars has only two small moons. The fact that the Earth has a large moon may have benefited the evolution of life on Earth.

## 17.6   Formation of the Ocean and the Atmosphere

The early Earth was a very large, hot rock without any water on its surface. Today, about three-quarters of the surface of the Earth is covered with water. Where did the oceans come from? Among the first ideas was that oceans were condensed from volcanic gases containing steam and carbon dioxide. However, we now know that the Earth suffered from a giant impact with an object the size of Mars, which ripped out a large part of the surface layer of the Earth to form the Moon. This impact event must have also removed all the water on the Earth's surface at that time. We therefore need to have an alternative way to replenish Earth with water after the formation of the Moon. The most popular theory at present is that terrestrial water was delivered by comets and asteroids that hit the Earth during the heavy bombardment phase between 4.1 and 3.8 billion years ago (Fig. 17.7).

Based on the ice content of comets and asteroids and the estimated bombardment rate, these objects have enough (more than 10 times over) capacity to have provided all the surface water that we have today. The presence of liquid water is widely believed to be a necessary condition for the development of life (Sect. 24.1). There is little doubt that many of the early life forms were found in the sea. Fossils of bacteria and algae have been found to date back to 3.3 billion years ago. So, we know that oceans have been present for at least 3.3 billion years.

The early atmosphere probably formed from gases that escaped from the crust through volcanoes as the Earth cooled down. Later impacts from comets and asteroids also contributed to the mass of the atmosphere. As cometary water vapor

**Fig. 17.7** Heavy bombardment during the early history of the Earth. This artist's depiction of the early Earth shows frequent impacts from asteroids and active volcanoes, as well as a Moon that was much closer to Earth than it is now. The formation of the Moon and the present composition of the oceans and atmosphere are the result of or affected by external impacts. Drawing by Clara Wang

condensed to form oceans, solar ultraviolet light dissociated the cometary ammonia ($NH_3$) into nitrogen ($N_2$) and hydrogen ($H_2$). Because the Earth's gravity is too weak to keep the hydrogen, only the nitrogen remains in our present atmosphere. The original composition of the atmosphere is very different from today (Sect. 21.2).

## 17.7 The Complex History of the Earth

As little as 300 years ago, the academic consensus was that the Sun, the Moon, the Earth, planets, and the stars were all created at the same time. It was only through our ability to estimate the age of the Earth that we realized that there was a sequence of events leading to the formation of celestial objects. First, the stars and the Sun came into existence, then the Earth and the planets, and then the Moon was formed out of the Earth as the result of an external impact. The early Moon was much closer to Earth than it is now and appeared larger in the sky (Fig. 17.7). As the Moon receded from the Earth, the length of a day gradually became longer. The oceans and atmosphere developed later, probably as the result of bombardments by external objects.

The identification of meteorites as having extraterrestrial origin was the first sign that broke down the barrier separating heaven and earth. By associating meteorites with the phenomenon of meteors, and meteors with comets, we have established a physical link between heavens and Earth. Our ability to directly analyze the physical properties and chemical components of meteorites gives us a hint of the makeup of celestial objects. We found that meteorites contain minerals that are in many ways like those found in rocks on Earth. However, they also contain organic materials that have no counterparts on Earth (Chap. 20). The discovery of meteorites carrying stellar material (Sect. 23.6) further extends this link to stars.

The landscape of the Earth has changed greatly from its early days. In the next several chapters, we will show that the Earth had a complex and interesting history: Continents formed and broke apart, life developed, the composition of the atmosphere changed, complex life forms emerged, the climate varied through warm and cold cycles, and living species went through several episodes of mass extinction.

### Questions to Think About
1. Why is it important to study the long-term history of the Earth?
2. Scientists feel confident that they have correctly estimated the age of the Earth. How do they know that a future method will not lead to a result indicating that the Earth is even older?
3. What is the philosophical significance of an extraterrestrial origin of meteorites?
4. Technology played an important role in the advancement of science. The invention of the telescope revolutionized the field of astronomy. What is the most important technological invention that helped our understanding of the Earth?

5. Why is it important to know the exact age of the Earth even though its presumably most important occupants, intelligent human beings, have occupied it only very recently?
6. Most of the surface of the Earth is covered by water and water plays an important part in the development of human civilization. Why are we still uncertain about the origin of the oceans?
7. What would life on Earth be like if the Earth rotated with a much shorter or longer period?
8. Describe what our lives would be like if the angle of obliquity were 90° instead of the present 23.5°.

# Chapter 18
# Common Ancestors

Where did humans come from? The common belief up to the eighteenth century was that humans, together with different species of animals and plants, were created individually and simultaneously. By the early nineteenth century, geological excavations have uncovered fossils that belong to organisms that are unlike those found on the present Earth. Fossils are traces of organisms that lived in the past, and they can be in the form of bones, shells, tracks, or impressions. Some fossils appeared to belong to species that no longer exist. In 1796, the French naturalist Georges Cuvier (1769–1832) compared the skeletal remains of elephants to the fossils of mammoths and concluded that mammoths belong to an extinct species. Another early example was the 1812 discovery of remnants of a gigantic deer (the "Irish Elk") that has no living counterpart and must be a species that has gone extinct. Extinct animals are not limited to mammals. There were also extinct large reptiles; the most famous are the dinosaurs. The study of fossil records of living organisms to understand the history of life on Earth later developed into the field of paleontology.

The fossil record also indicated that conditions on Earth had changed greatly. In the early nineteenth century, fossils of marine life were found on high mountain tops of the Alps, showing that these mountains must once have been under the sea. The first explanation for the existence of extinct animals was that they were victims of a sudden catastrophic event, for example, the flood described in the Bible at the time of Noah. The idea that the Earth was shaped by sudden, violent events was called catastrophism, which allows for a young age of the Earth.

An alternate theory was proposed by Scottish geologist James Hutton (1726–1797), who suggested that the geological features that we observe today are the result of natural processes over a long period of time. This assumption is later termed uniformitarianism which assumes that the natural processes that operate at present also worked in the past. This hypothesis was further expanded by Scottish geologist Charles Lyell (1797–1875), who maintained that slow processes could also lead to the observed geological and biological changes.

These discoveries provided the first impetus to ask the question of the evolution of life. If the geology of the Earth had changed over time, did life also evolve over

© Springer Nature Switzerland AG 2021
S. Kwok, *Our Place in the Universe - II*,
https://doi.org/10.1007/978-3-030-80260-8_18

geological time? If there had been life forms in the past that are no longer present, were they created together with present life forms, or were they their precursors? If an age can be assigned to old fossils, would it be possible to track the changes in life forms over the history of the Earth?

## 18.1   The Evolution of Living Species

In 1831, a young English naturalist Charles Darwin (1809–1882), joined an expedition on the ship HMS Beagle (Fig. 18.1). Among the books he carried on board was a copy of Lyell's *Principles of Geology*, published a year earlier. Over the five-year voyage, Darwin made extensive observations of wildlife and collected many samples of animals and plants. Heavily influenced by Lyell's ideas of gradual change, he formulated his theory of biological evolution using data he collected on the voyage. Although he had the theory developed in 1839, he did not publish it because he wanted to accumulate more evidence to support it. It was only after he received in 1858 a copy of a paper on a similar theory, *On the tendency of varieties to depart infinitely from the original type* by Alfred Russel Wallace, that he decided to document his evidence in the book *On the origin of species by means of natural selection* in 1859.

Natural selection is Darwin's proposed mechanism for evolution. He assumes that inherent in each species is a mix of genetic makeups. When a species produces more offspring than the environment can support, those with certain traits or genetic characteristics that can survive better in the environment are more likely to live long enough to reproduce. These genetic characteristics will be passed along to the next generation while other characteristics may die out. Over many generations, the species may change enough to evolve into another species. It is important to note

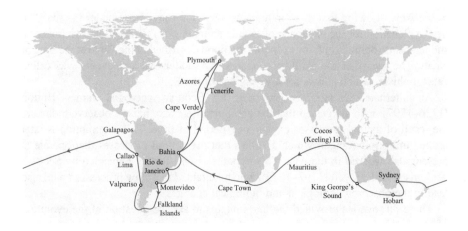

**Fig. 18.1**  Voyage of the Beagle from December 27, 1831, to October 2, 1836

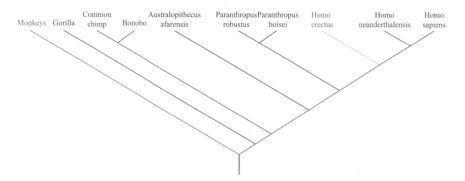

**Fig. 18.2** Gorillas, chimpanzees, and humans branched off the tree of life at different times in the past. The five branches on the right are hominids who walk on two legs, whereas monkeys, gorillas, and chimpanzees walk on four legs. The human and chimpanzee branches separated about 5–8 million years ago. *Australopithecus afarensis* lived between 4 and 3 million years ago, *Paranthropus robustus* 2 and 1.2 million years ago. *Homo erectus* appeared about 2 million years ago and became extinct about 143,000 years ago. *Homo neanderthalensis* appeared about 400,000 years ago and became extinct 40,000 years ago. The Neanderthals may have interacted with *homo sapiens,* which appeared about 350,000 years ago

that the process of evolution only affects a population of individuals, but not an individual itself. A giraffe does not grow a longer neck as the result of eating leaves on tall trees, but a long-necked giraffe in a population of giraffes may have better access to food and therefore is more likely to reproduce, leading to a future generation with longer necks.

The physical basis for the process of natural selection was provided by the work of Austrian monk Gregor Mendel (1822–1884) who gave a quantitative account of how genetic information is passed from parent to offspring. The molecular basis of genetics was provided by the understanding of the structure of the deoxyribonucleic acid (DNA) molecule as formulated by American biologist James Watson (1928–) and British biophysicist Francis Crick (1916–2004) in 1953. In reproduction, strands of DNA from each parent are passed to the child. This is how genetic information is passed from one generation to another. However, sometimes errors occur, resulting in mutations. Molecular mutation is the driving force behind the process of natural selection.

The significance of Darwin's work is not so much the idea that species can evolve, which had previously been proposed by French naturalist Jean-Baptiste de Lamarck (1744–1829) in 1809, but his detailed observations of evolution and his theory of natural selection as the driving force of evolution. Species do not evolve by inheritance of acquired characters, as envisioned by Lamarck along a one-dimensional ladder from simple to complex, with humans at the very top. Instead, natural selection produces a "tree of life" where species are related through a common joint on the tree. Humans did not evolve from apes; instead apes, chimpanzees, and humans represent different branches of the tree of life, having split off from other main branches at different times in the past (Fig. 18.2). From fossil records, we now know that 99% of all the species that have lived on Earth have gone extinct. Humans were not among the first to be created, but instead came on the scene very late. If we scale the 4.6-billion-year history of the Earth to a 24-h day, humans appeared only near the last few seconds.

## 18.2   Life Beyond What We Can See

Life on Earth is not limited to organisms that we can see with our eyes. In 1671, a Dutch amateur scientist Anton van Leeuwenhoek (1632–1723) used a microscope to observe a large variety of samples from everyday life, including water, animal and plant tissues, skin, hair, and blood. Through these observations, Van Leeuwenhoek discovered a whole new world of living things that was unknown before.

The most significant of his discoveries was his sighting of tiny creatures that he called "animalcules." These are single-cell organisms that we now call bacteria. Bacteria have sizes of a few micrometers (μm) and come in a variety of shapes such as spheres and rods. They are widely present on Earth in immense quantities. A drop of freshwater, for example, can contain over one million bacterial cells. A gram of soil can contain 40 million bacterial cells.

Despite this amazing discovery, it took over a century before van Leeuwenhoek's discoveries were extensively followed up scientifically. In 1838, German naturalist Christian Ehrenberg (1795–1876) published the book *Die Infusionsthierchen als vollkommene Organismen* (*The infusoria as whole organisms*) where he described thousands of microorganisms that he observed in water, dust, and soil by microscope. Through a series of experiments, French biologist Louis Pasteur (1822–1895) showed that microorganisms are responsible for the souring of milk and the fermentation of sugar into alcohol. Pasteur promoted the idea that many diseases are caused by bacteria. Before Pasteur, it was commonly believed that communicable diseases such as syphilis was a penalty for sexual immorality, and cholera was punishment for the underclasses of society. The assignment of the causes of diseases often carried economic or moral overtones. The germ theory of disease was confirmed through the work of German physician Robert Koch (1843–1910). He showed that the diseases anthrax (in 1875), tuberculosis (in 1882), and cholera (in 1883) were all caused by bacteria. Subsequent work found that many common diseases such as typhoid, pneumonia, gonorrhea, meningitis, leprosy, tetanus, plague, syphilis, whooping cough, and strep throat are all linked to bacterial infections.

These successes of the germ theory changed our view of the origin of diseases. Instead of viewing diseases as punishment for our sins or result of witchcraft, we learned that many diseases have physical origins. By identifying the carriers of these diseases, scientists can finally prevent and cure these ailments. Finding the physical basis of disease was the first step that changed medicine from an empirical practice to a rigorous science.[1]

## 18.3   A New Realm of Life

Microorganisms have now been found to survive in diverse habitats. In addition to common places such as land and sea, microorganisms exist in large quantities (estimated to be in the trillions) inside of human and other living bodies. They can be found also in unlikely places such as deserts, polar ice, subterranean ocean caves, hydrothermal vents, and even in troposphere clouds. Microorganisms have been found to be present under high temperatures, high acidity, extreme dryness, high pressures, high radiation, high salinity, absence of oxygen, and heavy metal concentrations. Examples include microbes found in hot springs where the water temperature is above the boiling point of 100 degrees Celsius, beyond the temperature at which scientists used to believe that life could exist. There are also living microbes under high pressure in deep seabed rocks. Collectively, they are referred to as "extremophiles."

Traditionally, we think of life as being confined to a thin layer above the solid crust of the Earth, either on land or in the oceans. With the discovery of extremophiles, we learned that life can be quite versatile and can exist under a much wider range of physical conditions. It is conceivable that a large number of bacteria and archaea exist in the deep interior of the Earth and the Earth has a deep biosphere. The total number of bacteria and archaea alive today is estimated to be $5 \cdot 10^{30}$, with a total mass exceeding that of animals and plants on Earth. The invisible realm of life is just as important as our familiar forms of life.

## 18.4   The Tree of Life

Traditional biological classification of life forms (taxonomy) is based on appearances, structures, and functions. In our everyday life, it is obvious that there are two kinds of living things: plants and animals. After the discovery of microorganisms, German zoologist Ernst Haeckel (1834–1919) added the Protista, representing most microorganisms, to animals and plants and called these three groups "kingdoms." In 1969, American ecologist Robert Whittaker (1920–1980) further expanded the number of kingdoms into five: a tree of life with four branches (animals, plants, Protista, fungi) with monera (single-cell organisms with no nuclear membrane, an example being bacteria) at the root of the tree.

---

[1] Beyond bacterial infections, we later learned that diseases can also have hereditary, nutritional, environmental, or viral origins.

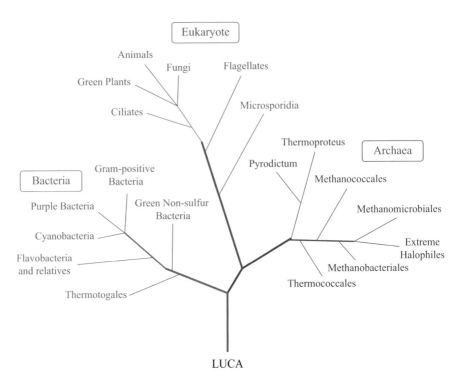

**Fig. 18.3** The tree of life based on ribosomal RNA sequence data. The last universal common ancestor (LUCA) is the root of the tree of life. Figure adapted from Woese, C. R., Kandler, O., & Wheelis, M. L. 1990, Towards a natural system of organisms: Proposal for the domains Archaea, Bacteria, and Eucarya, *Proceedings of the National Academy of Sciences*, **87**, 4576

The invention of the technique of DNA sequencing allowed a new way to classify life forms. In 1977, the American biologist Carl Woese (1928–2012) found that the RNA makeup of extremophiles is distinct from that of bacteria. Woese proposed that this group (which we now call "archaea," meaning "ancient things" in Greek) represents a new main branch in the tree of life. Beyond the domains of eukaryotes (consisting of familiar life forms such as plants, animals, fungi, and protists) and bacteria, archaea represent the third domain on the tree of life (Fig. 18.3). The use of molecular techniques is quantitative and represents a major advance in the classification of living organisms from the previous qualitative methods of taxonomy.

DNA sequencing also allows the mapping of the complete genetic makeup of a species. By comparing the number of mutations that must have happened between two related species and assuming a constant rate of mutation, one can estimate how much time is needed for the two species to diverge from each other. This technique is called the "molecular clock." From the fact that humans and chimpanzees have 98.5% of their genes in common, one can deduce that humans and chimpanzees diverged from a common ancestor about 5–8 million years ago (Fig. 18.2).

While the shapes and sizes of the branches of the tree of life are still under debate, we assume that they all share a common root at the bottom of the tree. The root is called the last universal common ancestor and is assumed to be the origin of all life forms.

## 18.5   Social Implications of Darwinism

Reactions to the theory of evolution from the scientific and the religious communities were as furious as Darwin had feared. English biologist Richard Owen (1804–1892), a prominent figure in the scientific establishment, wrote a hostile review of *On the Origin of Species*. However, Darwin was personally shielded from much of the direct criticism because his ideas were publicly defended and advocated by the English biologist Thomas Huxley (1825–1895). This cumulated in the famous debate between Huxley and Bishop of Oxford Samuel Wilberforce (1805–1873) on June 30, 1860, at the Oxford University Museum. This debate brought public awareness to the scientific issue of evolution, and generated discussions on whether the Biblical account in the Book of Genesis should be interpreted literally.

Upon the publication of *On the Origin of Species*, it was obvious that the theory of evolution was in contradiction to the account of Creation in the Bible, specifically that man was created in the image of God. The central issue of controversy was the origin of man. Were humans specially created by God or were they end products of a long process of natural evolution? The issue became emotional when humans were compared to, or suggested to have descended from, chimpanzees or apes. Many learned people at the time found this concept offensive. The place of humans in the scheme of evolution was discussed in the books *Evidence as to Man's place in Nature* by Huxley in 1863 and *Descent of Man* by Darwin in 1871. While Owen maintained that humans are fundamentally different from mammals, Huxley showed that there are many similarities between the brains of humans and apes. Darwin's *Descent of Man* put together evidence to show that humans are animals, which was an extremely provoking idea at the time. Despite all the intellectual accomplishments of humans, we came from very humble origins.

Beyond the theological implications, Darwin also introduced the concept of temporal change in our view of the world. In the views of Copernicus and Newton, the Universe was constant and everlasting. While planets revolve around the Sun, the Universe was assumed to be the same as it has always been. After Lyell's depiction of the Earth as an evolving planet and Darwin's theory of evolution of biological species, our world is no longer static. All biological species have an origin, and have been evolving, with new species emerging and old species going extinct. The process of change is driven by genetic mutation and natural selection. If this is the case, is the Universe evolving too? If all living things can be traced to a common ancestor, could all the objects in the Universe also have a common origin?

**Questions to Think About**

1. Bacteria were discovered in the seventeenth century, but it was not until the mid-nineteenth century that they were recognized as causing diseases. Why did it take so long?
2. The germ theory was the first step in assigning a physical cause to disease. Do all human diseases have a physical origin?
3. The invention of the microscope led to the discovery of new forms of life. Could future technological advancement lead to further new forms of life on Earth?
4. Is the theory of evolution in contradiction with the Biblical account of the Creation? How do modern Christians reconcile these differences?
5. What is the evidence for the existence of a deep biosphere?
6. The domain of archaea in the tree of life was discovered only a few decades ago. Could there exist other domains in life that are still undetected?
7. Bruno was persecuted for his unconventional ideas. Are there other examples of persecution of scientists because of their ideas in the last 100 years? Do you think similar persecution for academic thoughts could happen today?

# Chapter 19
# Origin of Life

The question of the origin of life was probably asked as soon as humans developed self-consciousness. Our ancestors realized very early that plants, animals, and humans, are distinct from non-living entities on Earth such as rocks, water, and air. Where did life come from? The universally common explanation among early civilizations was supernatural intervention. Almost all early cultures and ancient civilizations believed that they owed their existence to a supreme being, who also created the Earth and other celestial bodies.

Although the distinction between living and non-living seems obvious, the realm of life was broadened with the discovery of microorganisms (Sect. 18.2). If these small microscopic creatures are indeed living, then we must account for where they came from. Did God also create microorganisms in addition to plants and animals? One popular theory was that microorganisms developed directly from the non-living, in a process called spontaneous generation. While it is difficult to test the hypothesis of divine creation, a specific theory can be tested experimentally. The experimental study of the hypothesis of spontaneous creation represents the first scientific exploration of the question of origin of life.

## 19.1  Spontaneous Generation

The hypothesis of spontaneous creation, which states that life arises from non-living matter, can be traced back to ancient times. After Aristotle, ancient Greeks believed that everything is created from primary substances of earth, water, air, and fire. The idea that plants, worms, and insects can spontaneously emerge from mud and decaying meat was popular well into the seventeenth century. This theory was put to the test when the Italian physician Francesco Redi (1626–1698) noticed that maggots in meat come from eggs deposited by flies. When he covered the meat with a cloth, maggots never developed. This experiment, therefore, cast doubt on the common belief that worms originate spontaneously from decaying meat.

© Springer Nature Switzerland AG 2021
S. Kwok, *Our Place in the Universe - II*,
https://doi.org/10.1007/978-3-030-80260-8_19

The debate continued into the eighteenth century between English biologist John Needham (1713–1781) and Italian priest Lazarro Spallanzani (1729–1799). Needham was also a Catholic priest, who later became a fellow of the Royal Society. Needham supported spontaneous generation by performing experiments to show that microorganisms can grow in a broth after briefly boiling it and leaving it in an open flask. This conclusion was contradicted by Spallanzani who showed that by boiling the broth longer (over an hour) and sealing the flask, microorganisms never appear. He maintained that no spontaneous generation occurs if one adequately sterilizes the original sample and prevents air from contaminating it. In response, Needham argued that by sealing, one destroys the "vegetative force" that is required for spontaneous generation.

Spontaneous generation was built into the system of the history of the Earth developed by Georges Louis Leclerc, Comte de Buffon. Buffon was the most influential naturalist in eighteenth-century France, having published a 22-volume encyclopedia of nature, *Histoire naturelle*, in which he had a detailed description of the formation and evolution of the Earth. In Buffon's estimation, the Earth was formed 75,000 years ago, after which it continued to cool until its surface was solidified. When the temperature of the Earth was low enough, living organisms were produced by spontaneous generation. The origin of life is integrated into the history of the Earth and spontaneous generation is the mechanism.

In the nineteenth century, the French naturalist Félix Poucher (1800–1872) performed many experiments that led him to conclude that microorganisms can spontaneously emerge from liquids. Poucher's results were summarized in his book *Hétérogénie ou traité de la génération spontanée* published in 1859. Being an advocate of Darwin's theory of evolution, Poucher believed that spontaneous generation was more consistent with the theory of evolution than other theories on the origin of life.

This question was finally settled by an 1859 experiment designed by Louis Pasteur, who showed that the emergence of microorganisms is due to external contamination. He demonstrated this by first sterilizing beef broth by boiling it in a flask with a long bending neck (Fig. 19.1). The neck traps dust particles so contaminants cannot reach the flask. The broth will remain clear. However, if the neck is broken off, the broth becomes cloudy, indicating bacteria contamination as the result of exposure to air. Because air is still allowed to come in contact with the broth, Pasteur avoided the criticism that sealing from air deprives the broth of the necessary ingredient for spontaneous generation. Pasteur's pioneering experiment is the beginning of our modern belief that life only comes from life on Earth today.

Pasteur tested the hypothesis of spontaneous generation by performing two nearly identical experiments with only one differing variable. The curved neck allows air to flow through, but traps bacteria-carrying dust in the neck. The difference in outcome in the two experiments clearly shows that the bacteria are introduced from the outside, and there is no spontaneous generation after sterilization. This is a classic example of the scientific method at work. The long debate on spontaneous generation was finally over.

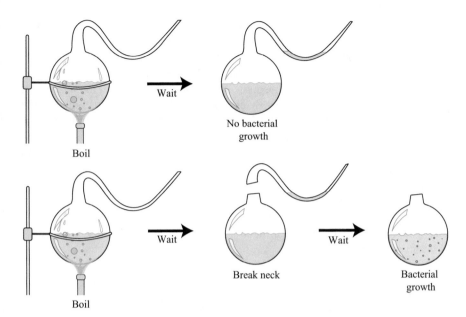

**Fig. 19.1** Pasteur's experiment to disprove spontaneous generation. The curved neck traps contaminants, allowing the broth in the flask to stay clear (top). However, when the neck is broken off, the broth quickly becomes cloudy as the result of bacterial growth (bottom)

It is interesting that Pasteur's work was viewed favorably by the Church, as the view that life only comes from life was more in line with the view of the Church at the time.

## 19.2   Panspermia: Life from Elsewhere

The modern hypothesis of panspermia refers to the idea that life is common in the Universe and the seeds of life are transported across the Universe. Life will develop in any place where conditions are favorable. However, the idea that life on Earth could have originated from heaven and was delivered to Earth can be traced to Greek philosopher Anaxagoras (~500 BC–428 BC) as early as 2500 years ago. The theory of panspermia received a major revival in the nineteenth century. In 1894, Lord Kelvin gave a presidential address to the British Association in which he criticized Darwin's theory of evolution, as the age of the Earth (which Kelvin estimated to be 20–40 million years, Chap. 17) is too short for life to have developed. Because Kelvin was such an influential figure in British science community, Darwin chose to retract his original estimate of 300 million years for the age of the Earth (derived from erosion and weathering of rock) in later editions of *The Origin of Species*.

In order to incorporate the origin of life into his low estimate of the Earth's age, Kelvin suggested that life on Earth was imported from elsewhere, specifically through the transport of life-bearing meteoric stones in space. Panspermia was further developed by German physiologist Hermann von Helmholtz (1821–1894) and Swedish chemist Svante Arrhenius (1859–1927). They suggested that simple life forms (e.g., bacteria), frozen in spores, could spread from star to star in long journeys through the interstellar medium. In the twentieth century, the concept of panspermia was promoted by British astronomers Fred Hoyle (1915–2001) and Chandra Wickramasinghe (1939–), who argued that the Galaxy is 10 billion years old and it must contain millions of life-bearing planets in other solar systems. They also cited examples of bacteria revived from bees embedded in amber for 25–40 million years as evidence of the viability of long-distance transport of microscopic life in interstellar space.

The main problem is that the panspermia hypothesis does not solve the problem of the origin of life; it simply shifts the problem to somewhere else. It may offer a solution for the origin of life on Earth, but it does not solve the problem of the origin of life in the Universe.

## 19.3  Distinction Between Living and Non-living

The crucial question of the origin of life lies in the definition of life and how living organisms are distinct from the non-living. On the superficial level, we identify living things by their ability to move, change, grow, and reproduce. While both living and non-living things are made of the same basic chemical elements, their molecular contents are not the same. By the early nineteenth century, chemists recognized certain groups of molecules, such as proteins, fat, and carbohydrates, as essential nutrients for life. This was followed by the isolation of various organic molecules as basic building blocks of life. These included amino acids such as asparagine (isolated from asparagus in 1806), cysteine (extracted in 1810 from urinary calculi), leucine (1819, from fermenting cheese), and glycine (1820, from gelatin).

Nucleic acids are substances in the nucleus of cells that we now recognize to be responsible for the storage, transmission, and processing of genetic information. There are two kinds of nucleic acids: deoxyribonucleic acid (DNA) and ribonucleic acid (RNA). DNA was isolated by Swiss biologist Friedrich Miescher (1844–1895) in 1869 from yeast nuclei. As carriers of genetic information that is passed from one generation of living organisms to the next, DNA and RNA are the most important biomolecules. The helical backbones (one for RNA, two for DNA) are connected by pairs of nucleobases: adenine, cytosine, guanine, thymine (only in DNA), and uracil (only in RNA) (Fig. 19.2). These nucleic acid bases molecules were also chemically isolated in the laboratory: guanine in 1882, thymine in 1883, adenine in 1886, cytosine in 1894, and uracil in 1900. Deoxyribose, the sugar molecule that forms the backbone of DNA, was isolated in 1909.

These molecules are extracted from living organisms and are not found in the non-living. It was therefore believed that organic molecules possess a certain

**Fig. 19.2** A schematic
drawing of the double helix
structure of the DNA
molecule. The DNA
molecule consists of two
helical strands connected by
pairs of nucleobases:
adenine (A), thymine (T),
cytosine (C), and guanine
(G). Adenine is only paired
with guanine and cytosine
only pairs with guanine

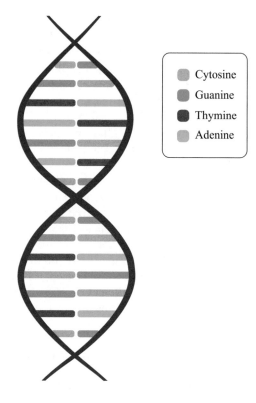

Cytosine
Guanine
Thymine
Adenine

invisible agent called "vitality" that distinguishes them from inorganic molecules. This concept is analogous to the concept of "soul" that was believed to separate humans from animals. Although the physical basis of vitality was not known (it was speculated as electrical in nature), it was nevertheless thought to be an essential ingredient for life. As summarized by chemist William Prout (1785–1850): "(there exists) in all living organized bodies some power or agency, whose operation is altogether different from the operation of the common agencies of matter, and on which the peculiarities of organized bodies depend." The reality of the concept of vitality was deeply entrenched in the scientific thinking of the nineteenth century.

The belief in vitality was gradually eroded with advancement in the laboratory synthesis of organics. In 1828, German chemist Friedrich Wöhler (1800–1882) synthesized urea, an organic compound isolated from urine, by heating an inorganic salt, ammonium cyanate. This was followed by German chemist Adolph Strecker's (1822–1871) laboratory synthesis of the amino acid alanine from a mixture of acetaldehyde, ammonia, and hydrogen cyanide in 1850, and Russian chemist Aleksandr Mikhailovich Butlerov's (1828–1886) synthesis of sugars from formaldehyde in 1861.

While a vital force in living yeast cells was thought to be responsible for changing sugar into alcohol through fermentation, German chemist Eduard Büchner (1860–1917) showed in 1897 that yeast extracts can do the same without the benefit of living cells. This marked the beginning of the realization that it is a class of biomolecules called enzymes, not the "vital force," that is responsible for fermentation. Enzymes have since been shown to be the catalysts that accelerate chemical reactions in biological systems. In 1926, American chemist James Sumner (1887–1955) found that urease, an enzyme that catalyzes the hydrolysis of urea into carbon dioxide ($CO_2$) and ammonia ($NH_3$), is a protein. Shortly after, it was found that several other crystallized digestive enzymes are also proteins. The basis of one of the key elements of life—enhancing the rates of chemical reactions efficiently and selectively—was reduced to the study of the structures and functions of protein molecules.

On the genetic aspects of life, the nucleobases that connect the strands of the DNA molecule were also synthesized artificially. The first nucleobase, adenine, was synthesized (from HCN and $NH_3$) in 1960. This was followed by the synthesis of guanine in 1967 and cytosine in 1968.

Since then, our definition of organic matter has evolved from something that possesses a special non-physical "vital force" to a group of molecules and compounds based on the chemical element carbon. The element carbon is unique in its versatility in forming different chemical bonds. Not only can carbon connect with other carbon atoms to form different structures, it can also combine with other elements such as hydrogen, oxygen, nitrogen, sulfur, and phosphorus to form a great variety of molecular forms. This special quality of connecting in a variety of ways allows many classes of organic molecules to be formed. A theoretical estimate puts the number of possible organic molecules from $10^{33}$ to as high as $10^{180}$. This group of molecules forms the basis of living organisms. Although once extremely popular, the concept of vitality has completely disappeared from scientific literature and discussion.

## 19.4   Abiogenesis: A Chemical Origin of Life

The idea that life could be developed from non-living matter via chemical means was promoted by Thomas Huxley. Motivated by the laboratory synthesis of organics, the Russian biochemist Alexander Ivanovitch Oparin (1894–1980) began in 1924 to explore the idea that life could originate from simple ingredients in the primitive Earth. Specifically, he proposed that life could have originated from a pool of inorganic materials using purely physical and chemical processes. The right mix of simple inorganic molecules in a suitable environment, injected with external energy, could in time lead to life. A similar idea was promoted by John Burdon Sanderson Haldane (1892–1964) in England. The theory of the chemical origin of life is now referred to as the Oparin-Haldane hypothesis.

The Oparin-Haldane hypothesis had some followings in the Soviet Union but found essentially no support in western scientific circles. The turn-around in opinion happened in the mid-twentieth century because of one single experiment. In 1953, American chemists Stanley Miller (1930–2007) and Harold Urey (1893–1981)

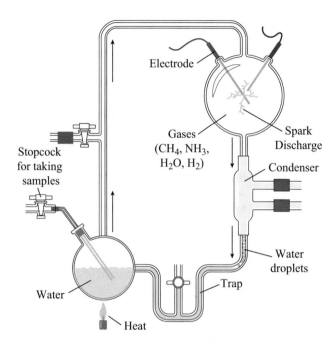

**Fig. 19.3** The Miller-Urey experiment. The experiment consists of a simple flask (upper right) containing a mixture of methane, ammonia, water, and hydrogen. An electric spark is introduced. The chemical reaction products collected include amino acids and other complex organics, showing that biomolecules could be synthesized naturally under conditions of the early Earth

conducted an experiment that simulated conditions on the early Earth. By mixing hydrogen, water, ammonia, and methane in a flask, they found that a large variety of organic compounds, including sugars and amino acids, could be formed after the flask was subjected to electric discharges for several days (Fig. 19.3). The implication was that the basic ingredients of life could form naturally on the primordial Earth.

The Miller-Urey experiment had a major impact on the scientific community. After 1953, the theory of life emerging spontaneously from simple molecules in a primordial soup became the dominant theory of the origin of life. As the original chemical mix of Miller-Urey's experiment is no longer believed to correspond to the actual atmospheric conditions of the early Earth, many other simulations using different ingredients or energy injections have since been performed. The basic conclusions of the Miller-Urey experiment remain valid.

## 19.5   Philosophy Guiding Science

Oparin's work was very well received in his home country the Soviet Union, because the idea of life originating from non-living matter fits well with the philosophy of dialectical materialism of Friedrich Engels (1820–1895). For his work, Oparin received the honors of Hero of Socialist Labor in 1969, the Lenin Prize in 1974, and five Orders of Lenin. As it happened, Haldane was also an active Marxist and a member of the Communist Party in England. The Irish physicist John Desmond Bernal (1901–1971), who was one of the early pioneers in research on the origin of life, also had strong sympathy for Soviet ideology. It is unclear whether the Marxist philosophy had any effect on the initiation of the scientific work of these scientists, but there is no doubt that Oparin's work was hailed as an example of the success of Marxist philosophy guiding science in the Soviet Union.

Neither the work of Oparin and Haldane was taken seriously in the West, probably partly based on philosophical grounds. Only after the Miller-Urey experiment in 1953 did the Oparin-Haldane hypothesis of the origin of life become acceptable in western scientific circles. Although the European and North American science communities do not subscribe to Marxist philosophy, they have now generally embraced a physical worldview of the Universe, where everything, including life, can be reduced to physical constituents operating under physical laws. Western scientists may not think that we operate under the dogmatic conditions of the Soviet Union, but we may nevertheless be unconsciously following certain philosophical principles.

On the question of cosmology, the Soviet Union also favored a steady-state model over an evolutionary model of the Universe (Sect. 16.4). This was based on Engels' belief that the Universe contains innumerable worlds in infinite space existing eternally and undergoing perpetual change. Soviet scientists did not interpret the redshift of spectral lines observed in galaxies as evidence for the expansion of the Universe (Chap. 16); instead they tried to seek alternate explanations such as the "tired photon" theory that assumes light loses energy over long distances. The Soviet opposition to the Big Bang hypothesis subsided only after the discovery of cosmic background radiation.

From the 1930s to the 1950s, there was a movement in the Soviet Union that adhered to the belief that acquired characteristics can be inherited. This idea has its origins in the work of Lamarck (Sect. 18.1) and was viewed as compatible with the dialectical materialism philosophy of Hegel. The champion of this theory was Trofim Lysenko (1898–1976), who claimed that the application of his genetic theory to agriculture could greatly increase crop yields. The adoption of his theory on farming led to disastrous results, causing widespread famine in the Soviet Union and China. The genetic theory of Lysenko was in direct contradiction to the genetic theory of Gregor Mendel (Sect. 18.1), which was already well established by the early twentieth century. However, the Mendelian theory of genetics was discredited in the Soviet Union on philosophical grounds, and biologists adhering to the Mendelian theory were dismissed or imprisoned.

Because the influence of Lysenkoism was limited to the Soviet Union and China, this genetics controversy is not as well known in the West as the heliocentric and evolution debates in the seventeenth and nineteenth centuries. All three are examples of scientific theories being judged by the religious or political dogma of the time. We must bear in mind that every generation has its own community of beliefs, and any scientific theory that does not conform to the prevailing opinions will be viewed suspiciously. One should not be so confident that we are now free of any bias, nor that we are any wiser than societies in the times of Copernicus and Darwin.

## 19.6   Remaining Questions

While we do not believe that spontaneous generation can occur now over short time scales, the current scientific opinion is that the transition from organic non-living matter to life could have occurred during the early history of the Earth over long time scales. We now recognize that complex biomolecules can be synthesized naturally from inorganic matter, but the jump to life has yet to be demonstrated. Although most scientists believe that living systems emerged on primordial Earth by natural means, we must remember that the Oparin-Haldane hypothesis remains unproven. The idea that life can emerge only through divine intervention or intelligent design, based on the argument that natural processes are too difficult or unlikely, remains popular in certain religious circles. Panspermia is also believed by some scientists to be responsible for the origin of life on Earth.

The question of the origin of life is a real one as we do have one example of life existing in the Universe. Is the origin of life an inevitable event or an extremely improbable single-step chance event? In the coming chapters (Chaps. 20 and 23), we will learn that the basic ingredients for life are routinely synthesized by stars and the basic building block biomolecules are widely present in the Solar System. Given the ubiquitous presence of the basic ingredients, could different forms of life also develop elsewhere as they did on Earth? Or does the Earth possess certain unique circumstances that allowed that to happen? If life on Earth is unique, did it arise as an accidental outcome from a cosmic/geological event?

**Questions to Think About**
1. Can we ever be sure how life on Earth originated? Is it necessary that we have a single theory of the origin of life or could life arise through multiple paths?
2. Buffon believed that the Earth was 75,000 years old, and life began from spontaneous generation. Two hundred fifty years later, our view has changed to that the Earth is 4.6 billion years old, life started by chemical synthesis, and species evolved through the process of natural selection. Do we now have the final word on these questions?
3. Can the question of the origin of life be answered purely by scientific means? Is there any other way (philosophical, religious) that this question can be discussed?

4. Is everything that exists physical matter, or results from physical processes? If life originated from non-living matter, how about consciousness?
5. If "soul" is not real, is there a fundamental difference between humans and animals? What are the ethical implications of this question?
6. What is the relationship between philosophy and science? Which one is more fundamental? Should philosophy be guiding science or the other way around?
7. Is the country of origin of a scientific theory an important factor for it to be accepted? Can you find examples of good scientific work not being taken seriously because of the scientist's country of origin?
8. Could another Lysenkoism emerge in our time? Are there any examples in our modern society where a scientific theory is discredited for being contrary to community beliefs?

# Chapter 20
# Complexity in the Universe

In Aristotle's view, celestial objects are simple, pure, and constant, in contrast to the terrestrial environment, which is dirty, complex, and variable. By the nineteenth century, we have learned that celestial objects are made of the same chemical elements as the Earth. These basic elements are in the form of atomic gases in stars. Stars are balls of gases held together by gravity. Galaxies are just assemblies of stars, which again are bound together by the force of gravity.

Although stars and galaxies are the basic physical constituents of the Universe, spectroscopic studies also reveal the existence of gaseous nebulae that are collections of low-density gases, mostly in an ionized state. These nebulae are bright in the visible region of the spectrum because they contain atoms that radiate strongly in the visible (Sect. 9.5). Unlike stars that radiate continuously over all colors, gaseous nebulae only radiate in individual discrete lines (Fig. 9.3). Because the amount of mass contained in each gaseous nebula is relatively small and they are not as numerous as stars, they were not considered a major component of the Galaxy.

Although factual knowledge in astronomy has expanded a great deal in the 2000 years since Aristotle, many astronomers still hold onto the view that the Universe is inherently simple. Atomic gas, mainly concentrated in stars with a small amount existing in interstellar space, has replaced ether as the perceived main ingredient of the Universe.

Part of this belief is the result of the fact that until the mid-twentieth century, astronomers could only study the Universe in the visible region of the electromagnetic spectrum. Light from stars is primarily emitted in the visible range, so naturally, they were the first objects that were studied. Gaseous nebulae like the Orion Nebula and planetary nebulae also radiate strongly in the visible through the emission lines of ions of hydrogen, oxygen, and other common elements. As a result, stars, galaxies, and gaseous nebulae were the celestial objects that were studied most extensively in the telescope era.

Are there other forms of matter in the Universe that we did not know about? Matter on Earth can exist in different forms as ions (also called the "plasma state"),

© Springer Nature Switzerland AG 2021
S. Kwok, *Our Place in the Universe - II*,
https://doi.org/10.1007/978-3-030-80260-8_20

atoms, molecules and solids.[1] However, the form of matter that astronomers have studied is mainly in the ionic form, as this is the common form of matter in stars and gaseous nebulae. Do the other forms of matter emit light that we can detect from celestial objects? When astronomers examined pictures of the Milky Way, they found dark patches among the dense field of stars. At first, these dark patches were referred to as "holes" or "voids" because they suggest an absence of stars. William Herschel referred to these dark patches as "openings in the heavens." It was not until the 1930s that astronomers realized that these dark patches are not regions without stars but are the result of solid particles (which astronomers call "dust"[2]) in the interstellar space obscuring the starlight.

If there is an extensive distribution of these interstellar solid particles in the Milky Way, they can dim the starlight that we observe. The effect of interstellar extinction is the reason behind Shapley's misjudgment of the distances of globular clusters and the size of the Milky Way (Sect. 12.4). The high concentration of solid particles in the plane of the Milky Way is also the reason why galaxies are mainly detected outside of the Galactic plane.

Solid particles absorb and scatter light of different colors differently, with blue light being affected more than red light. If there is a layer or cloud of solid particles between us and the star, the color of the star will appear to be redder than it should be. Because the spectral class of a star gives us direct information about its temperature (and therefore color, Sect. 9.7), a deviation from its supposed color would mean that light from that star is made redder by solid particles along our line to sight to the star. The measurement of the degree of stellar reddening can provide us with a quantitative estimate of the quantity of solids in the Galaxy.

Astronomers generally regard interstellar solid particles as a nuisance, as they affect their ability to measure the brightness of stars and galaxies accurately. It was not until the development of infrared astronomy in the mid-twentieth century that we learned about the different kinds of solids in space and the significance of the solid-state component of the Universe.

## 20.1  Molecules and Solids in the Interstellar Medium

By the mid-twentieth century, chemists have already learned in the laboratory that molecules can emit and absorb infrared light when they vibrate, and emit and absorb microwave radiation when they rotate. Examples of three simple types of rotation

---

[1] Another state of matter common on Earth is liquid. Like solids, liquids are a form of condensed matter and consist of molecules held together by intermolecular forces. Liquid requires outside pressure to maintain and cannot exist in the vacuum of space. Since liquids exist only within a narrow range of combinations of temperature and pressure, they are not common in the Universe.

[2] Astronomers use the term "dust" to refer to small interstellar solid particles. I try to avoid using this term because of possible confusion with the "dust" of our our everyday lives, which has a different composition.

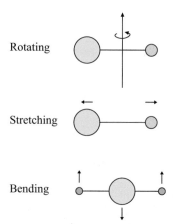

Rotating

Stretching

Bending

**Fig. 20.1** Schematic diagram illustrating the rotation, stretching, and bending motion of molecules. According to quantum theory, when molecules rotate, they can rotate only with certain discrete energies. When they change their rotational energy states from one to another, a photon with energy equal to the energy difference of the two states is emitted. The detection of the photon of this precise energy allows us to unambiguously identify the molecule from which it is emitted

and vibration are shown in Fig. 20.1. For a molecule with two atoms (e.g., CO), the molecule can rotate along an axis perpendicular to the bond joining the two atoms (top panel of Fig. 20.1), or it can stretch along the bond joining the two atoms through lengthening and shortening of the bond (middle panel of Fig. 20.1). With three or more atoms in a molecule, more vibrational motions are possible. An example illustrating a bending motion is for the water molecule ($H_2O$) (bottom panel of Fig. 20.1), where the bonds linking the hydrogen atom to the oxygen atom can change in angle.

Extensive laboratory studies of spectra of molecules were done by the German-Canadian chemist Gerhard Herzberg (1904–1999), who was also keen about the detection of molecules in space. However, the prevailing view of the astronomy community until the late 1960s was that the density of interstellar space is too low for molecules to form as molecular reactions require frequent collisions between atoms. They also believed that the strong ultraviolet light emitted by stars would destroy any molecules if they are actually formed.

From laboratory work, American physicist Charles Townes (1915–2015) found that some molecules can emit radiation in the microwave region of the spectrum. When sensitive microwave detectors became available in the 1960s, Townes mounted these detectors on telescopes to search for molecules in space. Among the first molecules he detected were water ($H_2O$) and ammonia ($NH_3$). In 1970, radio detectors of frequencies as high as 100 GHz were developed, and the molecule carbon monoxide (CO) was detected through its rotational transition at the frequency of 115 GHz (or 2.6 mm in wavelength). The wide presence of CO in the Milky Way shows that molecules are common in the Galaxy and the molecular component represents a significant fraction of the total mass of the Galaxy.

As of 2020, over 200 interstellar molecules have been detected through their rotational transitions in the microwave, or their vibrational transitions in the infrared. This list of interstellar molecules includes molecules of common elements such as H, He, C, O, N, S, and with sizes ranging from two to dozens of atoms. Also detected are molecules that are not found on Earth. For example, when a bright line at the frequency of 89 GHz was detected in 1970, the carrier of this line was named X-ogen, because this line does not correspond to any known molecular line measured in the laboratory. It was only later from theoretical and laboratory studies that this line was identified by American chemist William Klemperer (1927–2017) as a rotational line from the molecular ion $HCO^+$.

Most interestingly, many of the interstellar molecules are organic molecules. The detected species cover all classes of organic molecules, including hydrocarbons (e.g., methane $CH_4$, acetylene $C_2H_2$, ethylene $C_2H_4$), alcohols (e.g., methanol $CH_3OH$, ethanol $C_2H_5OH$, vinyl alcohol $H_2C=CHOH$), acids (e.g., formic acid HCOOH, acetic acid $CH_3COOH$), aldehydes (e.g., formaldehyde $H_2CO$, acetaldehyde $CH_3CHO$, propenal $CH_2 = CHCHO$, propenal $CH_3CH_2CHO$), ketones (e.g., ethenone $H_2C=CO$, acetone, $CH_3COCH_3$), amines (e.g., methylamine $CH_3NH_2$, cyanamide $NH_2CN$, formamide $NH_2CHO$), ethers (e.g., dimethyl ether $CH_3OCH_3$, ethyl methyl ether $CH_3OC_2H_5$), etc. Although not fully understood theoretically, Nature is able to synthesize molecules under unfavorable extreme low-density conditions of interstellar space.

Modern radio receivers are so sensitive, and the number of molecular rotational lines are so numerous, that when we sample a section of the spectrum in the millimeter wavelength region, the entire spectrum can be completely filled with lines. An example is shown in Fig. 20.2 where we can see lines from many organic molecules, as well as several lines of unknown origin (labeled "U"). From the spectral scans of various sources, it is estimated that over 10% of all the astronomically detected lines are unidentified. These results show that interstellar chemistry is much richer than terrestrial chemistry and we can still learn a lot about chemistry from astronomical observations.

Molecules are everywhere in the Universe. Molecules are found in comets, planets, and planetary satellites in the Solar System, in stellar winds from stars, in interstellar clouds, and in galaxies. Using molecular lines such as CO as a tracer, we can map the distribution of molecules in the Milky Way Galaxy. Molecules have been detected in galaxies billions of light-years away, showing that molecules were formed very early in the Universe.

## 20.2  Minerals in Space

The surface of the Earth is covered with rocks and sand, which are solids made of elements such as carbon, oxygen, silicon, aluminum, magnesium, and iron. For example, a common form of sand is silica (silicon dioxide, or $SiO_2$). Geologists use the term "minerals" to refer to pure chemical compounds that are the main

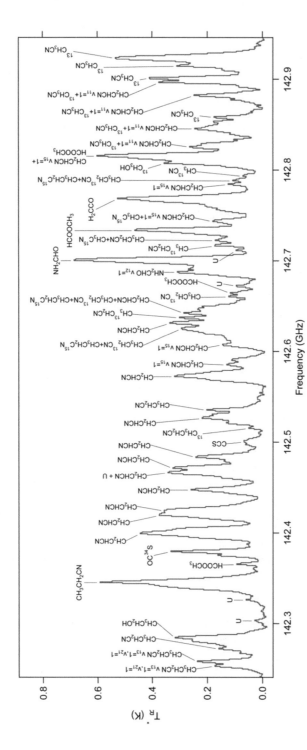

**Fig. 20.2** An example of a spectrum of the Galactic Center in the millimeter-wave (wavelength ~ 2 mm) region. All features are rotational lines of molecules with the identified molecules labeled. The symbol "U" signifies lines that are not assigned to any known molecule. Figure from Ziurys et al. 2016, *Astrobiology*, 16, 997

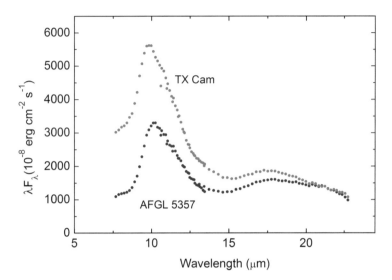

**Fig. 20.3** Infrared spectra of two red giants as observed by the *Infrared Astronomical Satellite* Low-Resolution Spectrometer. The features around 10 and 18 μm are due to amorphous silicate solids

constituents of rocks. Minerals with regular, periodic structures are called crystalline minerals, whereas those with irregular structures are referred to as amorphous.

Besides common rocks, some precious stones (gems) such as diamond, ruby, and sapphire are also classified as minerals. Diamond is a form of pure carbon, and ruby and sapphire are aluminum oxides mixed with chromium, iron, titanium, vanadium, magnesium and other impurities. Their distinct colors are the result of their reflection, refraction, and dispersion properties.

Laboratory spectroscopy of minerals shows that they have distinct spectral signatures in the infrared, which are due to vibrational motions of the mineral's lattice structure. The first mineral commonly found in old stars is amorphous silicates. In the cool atmosphere of old stars, micron-size silicate solids are found to condense in large quantities, often obscuring the atmospheres of the stars. These solids are ejected by old stars via stellar winds and they can aggregate in the interstellar medium in the form of clouds. The identification of different minerals in space using the technique of infrared spectroscopy is called astromineralogy (Fig. 20.3).

Solids are everywhere in our Galaxy, as evidenced by their ability to obscure starlight on the Galactic plane. Heated by background starlight, these solids also manifest themselves through their self-radiation in the infrared. In the mid-infrared wavelengths, the emission of the Milky Way Galaxy is dominated by radiation from micrometer-sized solid particles. While many galaxies emit most of their energy in the form of starlight in the visible range, some galaxies emit more light in the infrared, implying that they contain a large number of solid-state particles.

## 20.3   The Discovery of Extraterrestrial Organics

Although living organisms may at first seem to be the major reservoir of organics on Earth, in fact, the great majority of organic material on Earth is in the form of kerogen, a macro-organic substance found in sedimentary rocks. The inventory of carbon in kerogen on Earth is estimated to be $1.5 \cdot 10^{16}$ ton, compared to $10^{12}$ ton of carbon in all living organisms. Kerogen is amorphous in structure and is made of random arrays of aromatic carbon sites and aliphatic chains with functional groups. The term "aromatics" was first used in the mid-nineteenth century to refer to chemical substances with notable aromas, but now the term is used to refer to a class of molecules consisting of rings of 6 carbon atoms. The simplest example is benzene, which consists of six carbon atoms arranged in the form of a ring with six hydrogen atoms each attaching to one of the carbon atoms (Fig. 20.4). In contrast, aliphatic compounds consist of carbon atoms arranged in the form of a chain. Kerogen originated from remnants of past life, such as algae, land plants, and animals. Under high temperature and pressure, kerogen gradually dissociated into petroleum and natural gas. Except for small amounts of methane in deep-sea hydrothermal vents, almost all organics on Earth are biological in origin.

The biological origin of organics on Earth led to a common belief in the twentieth century that organic compounds are the exclusive domain of the Earth. In the conventional picture of the Solar System, planets, asteroids, comets and planetary satellites were formed from a well-mixed primordial nebula of chemically and isotopically uniform composition. The primordial solar nebula was believed to be initially composed of only atomic elements synthesized by previous generations of stars, and current Solar System objects later condensed out of this homogeneous gaseous nebula. Gas, ice, metals, and minerals were assumed to be the primary constituents of planetary bodies.

As early as 1834, the Swedish chemist Jöns Jacob Berzelius (1779–1848) had already found hints of the presence of organic materials in the meteorite that landed near Orgueil in southern France on May 14, 1864. It was not until a hundred years later that American chemist Bartholomew Nagy (1927–) and his colleagues found

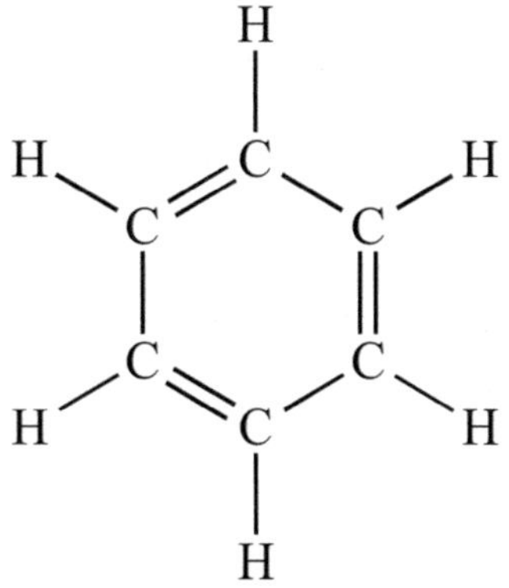

**Fig. 20.4**  The chemical structure of benzene

concrete evidence of hydrocarbons resembling products of life in the Orgueil meteorite. Because the originating source of the meteorite is probably an asteroid, which is believed to be too small and dry to harbor life, Irish physicist John Desmond Bernal (1901–1971) suggested that the Orgueil meteorite could have been launched from the early Earth and arrived back after a round-trip journey of millions of years in interplanetary space.

Further evidence for organics was found in the meteorite that fell near the town of Murchison in Victoria, Australia on September 2, 1969. Chemical analysis of the Murchison meteorite led to the discovery of complex organic compounds of both aromatic (ring-like) and aliphatic (chain-like) structures. Both aromatic and aliphatic structures are basic to biomolecules: amino acids and DNA nucleobases contain aromatic rings, and fat and sugar have aliphatic chains. Further analysis showed that many common biochemical substances, including amino acids, fatty acids, purines, pyrimidines, and sugars are found in the Murchison meteorite.

Organics are not found in all meteorites, but only in a rare class of meteorites called carbonaceous chondrites, which are the most pristine of all meteorites. With an array of laboratory techniques at our disposal, extraterrestrial organic compounds can be detected and identified with a great degree of certainty. Using different solvents, various components of the meteorite can be extracted and analyzed. Modern analysis has found that almost all basic biologically relevant organic molecules are present in carbonaceous meteorites. Organics identified in the soluble component of carbonaceous chondrites include carboxylic acids, sulfonic and phosphonic acids, amino acids, aromatic hydrocarbons, heterocyclic compounds, aliphatic hydrocarbons, amines and amides, alcohols, aldehydes, ketones, and sugar-related compounds. Altogether over 14,000 compounds with millions of diverse isomeric structures have been found.

Most interestingly, amino acids not found in the Earth's biosphere have been identified in carbonaceous chondrites. Through the work of Italian-American chemist Sandra Pizzarello (1933–) and others, many more amino acids have been identified in meteorites than the 20 that are used in our terrestrial biochemistry. Unusual nucleobases beyond the five used in terrestrial biochemistry (adenine, cytosine, guanine, thymine, and uracil) are also found. These results suggest that the organic molecules found in meteorites are not related to terrestrial life.

A large fraction (70–90%) of organic carbon in carbonaceous chondrites is in the form of a complex, insoluble, macromolecular material often referred to as insoluble organic matter. The chemical structure of this insoluble organic matter consists of small islands of aromatic rings connected by short aliphatic chains.

Another source of extraterrestrial material we can find on Earth is micrometeorites. Micrometeorites of sizes 20 μm to 1 mm are constantly falling on the Earth and can be collected in clean, isolated environments such as Greenland lake sediments and Antarctic ice and snow. Some Antarctic micrometeorites have been found to contain a large fraction of organics, including amino acids. The presence of complex organics in micrometeorites suggests that extraterrestrial organics can survive atmospheric passage and impact, and therefore has implications for the external delivery theory of the origin of life (Sect. 23.6).

With modern spacecraft, we now have the capability of flying instruments to Solar System objects to make in-situ measurements, as well as returning samples to Earth for further analysis. Before the twenty-first century, astronomers commonly assumed that comets were "dirty snowballs." From remote telescope observations, astronomers learned that the gaseous volatile component of comets contains a large variety of gas-phase molecules. Space missions to comets have found significant amounts of organics in cometary nuclei. The mass spectrometer aboard the Rosetta's *Philae* lander has detected an array of organic compounds on the surface of the Comet 67P/Churyumov-Gerasimenko. Among the molecules detected in the sample returned from Comet 81P/Wide 2 by the *Stardust* mission is the amino acid glycine. The nuclei of comets are also found to contain macromolecular compounds like the insoluble organic matter in meteorites.

From results obtained by the *Cassini* mission to Saturn, Saturn's moon Titan was found to have an atmosphere filled with organic haze. Radar imaging observations from *Cassini* have found hundreds of lakes and seas filled with liquid methane. The total amount of organic carbon on Titan is estimated to be 360,000 giga tons in its atmosphere, 16,000–160,000 giga tons in lakes, and 160,000–640,000 giga tons in sand dunes, which is much larger than the total amount of fossil fuels on Earth.

Using high-flying aircraft, interplanetary dust particles can be collected in the Earth's stratosphere. Laboratory analysis of interplanetary dust particles (sizes 1–30 μm) shows definite evidence of complex organics. Because interplanetary dust particles are believed to have originated from comets and asteroids, this gives us indirect evidence that similar organics are present in comets and asteroids.

## 20.4  Abiotic Synthesis of Organics

While the organic molecules found in meteorites resemble biomolecules found in terrestrial life, they are not biological in origin. The decreasing molecular abundance with increasing carbon numbers within the same class of compounds suggests that they are not breakdown products of living organisms but were built up from small to large through abiotic synthesis. While only 20 amino acids are used in terrestrial biochemistry, about 100 amino acids have been found in meteorites. All terrestrial living organisms use left-handed amino acids, while both right- and left-handed amino acids are found in meteorites.

The most direct evidence of abiotic synthesis of organics is in fact found in stars. When the mid-infrared part of the spectrum was first opened for astronomical observations through the installation of telescopes on high-flying aircraft, a family of strong emission bands was discovered in the planetary nebula NGC 7027 (Fig. 20.5). Because these astronomical bands do not resemble any known laboratory atomic lines, they were called "unidentified infrared emission bands." When American astronomer Roger Knacke, Canadian physicist Walter Duley, and British astronomer David Williams suggested these bands were vibrational bands of organic

**Fig. 20.5** The unidentified
infrared emissions bands
were first detected in the
planetary nebula NGC 7027
in the constellation of
Cygnus. This is a color
composite image based on
narrow-band images
obtained by the *Hubble
Space Telescope*

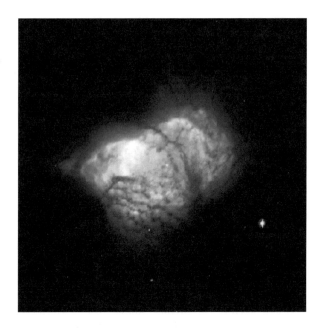

compounds, the astronomical community was skeptical as no one had expected
organics could be present in stars.

These unidentified infrared emission bands are now found everywhere in the
Universe, from star-forming clouds, planetary nebulae, and diffuse gas between
stars, to galaxies that are bright in the infrared. In some galaxies, as much as 20%
of the total energy output of the galaxy is emitted in these bands. These bands are
seen in galaxies as far as 10 billion light-years away, suggesting that very early in the
history of the Universe, soon after the first stellar synthesis of the element carbon,
organics were already being synthesized by stars in large quantities.

When molecules undergo different kinds of vibration (Fig. 20.1), they emit light
at specific infrared wavelengths. By comparison with laboratory measurements of
organics, the unidentified infrared emission bands most likely originate from complex organics with aromatic and aliphatic structures. The discovery of organics in
stars suggests that abiotic synthesis can occur efficiently even in a very low-density
environment. The fact that stars can form and distribute organics throughout the
Galaxy has significant implications for the origin of life (Sect. 23.6).

## 20.5   Unsolved Mysteries

Spectroscopic observations in the millimeter- and submillimeter-wave regions have
identified over 200 interstellar molecules. Yet thousands of spectral features believed
to be molecular in origin have yet to be assigned to individual molecules (Fig. 20.2).
The most likely candidates are molecular radicals, which have very short lifetimes in

the terrestrial environment but can be stable in low-density interstellar space for extended periods of time.

In 1814, Joseph Fraunhofer discovered hundreds of dark lines in the solar spectrum. Fifty years later, these lines were identified by Kirchhoff and Bunsen as due to absorption by atoms in the solar atmosphere. This was the first evidence that stars have the same physical constituents as the Earth (Sect. 9.2). Two hundred years later, we are facing a similar problem. Over 500 dark lines in the optical region have been discovered in the interstellar medium (called "diffuse interstellar bands"). Over the last 100 years, a large quantity of high-quality spectra has been recorded, but no theoretical interpretation has been proven successful. The diffuse interstellar bands are generally assumed to be due to electronic transitions of molecules, but the exact carrier molecules for the bands are not known.

In the red part of the visible, there is a phenomenon called extended red emission, which is seen throughout the Galaxy. The current theory is that this emission is the result of the photoluminescence of some unknown solid-state material. In order to account for energy emitted in the extended red emission in the Galaxy, a significant fraction of the diffuse starlight in the Galaxy must have been absorbed by this material. Again, given the brightness of emission, its carrier must be made of common elements.

In the ultraviolet part of the spectrum, there is a strong interstellar absorption feature at the wavelength 220 nm, which has remained unidentified since its discovery in the early 1960s. The wavelength of this feature is close to the electronic transition of carbonaceous compounds, but its exact carrier is not known.

Just as the new element helium was first discovered in the Sun (Sect. 9.4), astronomical observations have continued to lead us to new understanding of the nature of matter. The existence of spectral mysteries that have remained unsolved over many decades shows us that there is still much we can learn in physics and chemistry.

## 20.6 Extending the Frontiers of Exploration

Humans have been systematically observing the sky with their own eyes since the beginning of recorded history. The invention of the telescope in the seventeenth century is widely credited as the technological advance that revolutionized the science of astronomy. The use of the telescope allows us to see fainter celestial objects, and therefore expands our view of the Universe beyond the ~6000 stars that we could see with our naked eyes. The invention of photographic plates in the nineteenth century to record astronomical images and spectra represents another step in improving sensitivity in observations. This was advanced further in the twentieth century through the employment of digital cameras such as charged coupled devices (CCDs) on telescopes.

All the astronomical observations up to the early twentieth century have been done in visible light. Human eyes developed visions in the visible as the result of evolution. Because the Sun radiates mostly in the visible, our eyes are tuned to this narrow part in order to take maximum advantage of sunlight.

Because visible light represents only a very small fraction of the entire electromagnetic spectrum (Fig. 14.4), our view of the Universe was therefore very limited. The extension of astronomical observations beyond visible light was the next technological revolution. This was achieved in the twentieth century through the development of devices that can detect electromagnetic waves outside of the visible, first in radio and then in infrared, ultraviolet, and X-ray. However, the Earth's atmosphere is transparent only to visible and radio waves and obstructs X-ray, ultraviolet, and most of the infrared and high-frequency radio waves emitted by celestial objects from reaching telescopes on the ground. This problem was overcome by placing telescopes on Earth-orbiting satellites outside the Earth's atmosphere. Finally, we can see the Universe in its full color.

It was the opening of the spectral windows to the Universe that led to the discoveries of other forms of matter, such as molecules and solids in space. Our ability to send spacecraft to Solar System objects has allowed us to perform experiments beyond terrestrial laboratories. Not only we can perform in-situ experiments on planets and their satellites, we can retrieve and return samples from comets, asteroids, and the Moon for further analysis on Earth.

There is no doubt that technological advancements play a large role in expanding our view of the Universe. One of the philosophical implications as the result of these developments is our realization that the materials that make up celestial objects and the Earth are very similar. Galileo's telescopic observations of the Moon showed that the Moon has geological features like the Earth and was hardly the smooth and clean celestial object that was previously envisioned. Spectroscopic observations of the Sun and the stars show that stars are made of the same basic elements as the Earth. Not only do stars and galaxies have atoms like those on Earth, interstellar space is filled with molecules and minerals, many in common with those on Earth and many that are also unknown on Earth. Our belief is that interstellar molecules still obey the same rules of physics and chemistry as on Earth, the only difference being the environmental conditions in space are different, allowing unstable and exotic molecules to be formed.

Physics and chemistry may be universal throughout the Universe, yet the Earth is unique as it has the only example of biology. Is the Earth special? Earth was once considered to be the only source of organic compounds and almost the entire reservoir of organics on Earth is biological in origin. This view was challenged only when organics of abiotic origin were discovered in meteorites, comets, and other Solar System objects. Furthermore, we have strong evidence that complex organics are being synthesized by stars, and signatures of these organics are seen in galaxies as early as 10 billion years ago. It is becoming exceedingly difficult to claim that the Earth is unique or special in any way.

**Questions to Think About**

1. What lessons one can draw from the fact that many of the interstellar molecules detected have no terrestrial counterparts?

2. The mystery of dark lines in the solar system discovered by Fraunhofer was solved in 50 years. With much more powerful telescopes and laboratory instruments at our disposal, why has the problem of diffuse interstellar bands not been solved in over 100 years? What kind of scientific breakthrough do you think we will need to solve this problem?
3. We now know that complex organic matter is prevalent all over the Universe. Does that mean that life is also common in the Universe?

# Chapter 21
# Evolution of the Earth Through the Ages

Until the late eighteenth century, it was commonly believed that our world has been the same since the time of biblical Creation. By the nineteenth century, the view of a constant and everlasting Universe has become untenable and scientific opinion has shifted to a paradigm of change and evolution. The geological structure of the Earth is no longer viewed as static, but has undergone erosion, formation, and transformation. After Darwin, life is recognized as a process of nonstop evolution, with old species dying off and new species being created over the history of the Earth. This new perspective was driven by the study of biological fossils and layers of rock formation that gave scientists a record of how the Earth was at different times in the past.

After geologists accepted the concept of change, they began to wonder about the causes of these changes. How did the great mountain ranges such as the European Alps, Asian Himalayas, North American Rockies, and South American Andes, come about? Fossils of sea creatures are found in high mountains, suggesting that sea levels may also not be constant. If mountains can rise and oceans can fall, can the locations of the continents change too? Because biological evolution began in the sea, did these geological changes have an effect on biological evolution?

Did geological and biological evolutions happen separately and independently, or are they inter-related? Scientists realized that not only had the physical structure of the Earth changed, the physical conditions of the Earth had also not remained the same. Empirical evidence showed that the Earth had undergone different periods of warm and cold, which also had effects on the living conditions on Earth.

The discovery of plant fossils in the Carboniferous (coal bearing) Period showed that the climate of northern Europe must have been tropical 300 million years ago. This suggests that the climate of the Earth has not been static and has changed over the history of the Earth, or the locations of the continents have shifted over time.

It was not until the mid-twentieth century that we realized that the positions of the continents on Earth may not have always remained the same. The land that we live on, apparently so fixed and permanent, has been shifting under our feet. Because biological evolution happened during a great geological change on Earth, if the land

© Springer Nature Switzerland AG 2021
S. Kwok, *Our Place in the Universe - II*,
https://doi.org/10.1007/978-3-030-80260-8_21

and sea areas on Earth were not static, the shifting of continents may have affected biological evolution.

## 21.1   The Continents Are Moving

For thousands of years, when humans suffered from the devastating consequence of volcanos and earthquakes, they attributed these events to acts of god(s) who punished them for their sins. After the nineteenth century, when records of volcanos and earthquakes were better documented, scientists noticed that they do not happen in random places but are concentrated in certain areas of the world. The active regions lie along a band around the rims of the Pacific Ocean, which is now commonly referred to as the "ring of fire" (Fig. 21.1). Why do natural calamities cluster around this narrow region? This was a major mystery that demands a physical explanation.

The surface of the Earth is covered by a combination of land and sea. It took hundreds of years for geographers and explorers to map out the continents and oceans of the Earth. The fifteenth and sixteenth centuries represented the golden age of European exploration and the development of cartography allowed the production of maps of increasing accuracy. As early as 1620, English philosopher Francis

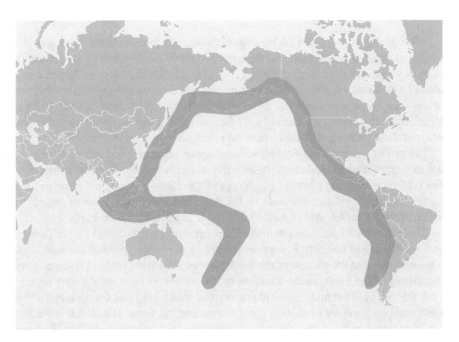

**Fig. 21.1** The ring of fire. Most of the world's earthquakes and volcanic eruptions occur around the ring of fire (colored in red)

Bacon (1561–1626) noticed that the new continent of South America and Africa on opposite sides of the Atlantic Ocean fit well together like a jigsaw puzzle. As mapping techniques improved, this coincidence became even more evident. In 1912, a German meteorologist Alfred Wegener (1880–1930), noticed that fossils from different parts of the world are remarkably similar. For example, fossils of the small reptile Mesosaurus were found only in South Africa and Brazil. If the animal had been able to cross the sea, then it should be found in other parts of the world as well. The fact that it is not suggests that the continents Africa and South America may have been close to each other in the past.

Similarities in age and structure of rocks are also found in South Africa and southeast Brazil. Because these rocks are controlled by local conditions, they might have been once close together. Coal is believed to form from dead vegetation a long time ago under warm, wet conditions. But coal is found in Antarctica and Britain, where the present climate could not have produced coal. Either these places once had a much warmer and wetter climate, or Antarctica and Britain were once much nearer the equator. All these facts point to the possibility that the continents may not be always fixed in their present locations and may have moved across the surface of the Earth through time.

Wegner proposed that our present continents originated from a one-piece super-continent called Pangaea that broke up 200 million years ago (Fig. 21.2). His 1915 book *Die Entstehung der Kontinente und Ozeane (The Origin of Continents and Oceans)* was widely criticized by geologists of the day and his continental drift hypothesis was accepted only many years after his death. It was not until the 1960s that geologists finally agreed that continents could move, and that major geological features such as mountains were created through the movement and interaction of large, rigid tectonic plates on the Earth's crust. The theory of plate tectonics suggests that the Earth's rigid outermost layer acts like an eggshell floating on the mantle of the Earth. This crust is fractured into segments of plates. There are seven major plates (Antarctic, Indo-Australian, Eurasian, African, Pacific, South American, and North American plates) plus some minor ones (Fig. 21.3). These plates are constantly moving, causing earthquakes and volcanic eruptions. When two plates collide with other, new mountains are formed. The Himalayan mountain range was lifted when the Indian plate collided with the Asian plate 50 million years ago. When two plates slide pass each other in opposite directions, a fault line is formed. The San Andreas Fault of California was formed by the relative shifting of the Pacific and North American plates over the past 10 million years.

Our world is not static, and land is continuously shifting. The theory of plate tectonics provides a physical basis for Wegner's continental drift theory. The theory can be tested by modern global positioning system (GPS) and the actual movements of the continental plates can be measured with an accuracy of centimeters per year.

The theory of continental drift also explains the diversity of life on separate continents. Australia has so many unusual animals because Australia had been isolated from the rest of the world by vast oceans since the breakup of Pangaea. Its animals and plants no longer had contact with animals and plants from other parts

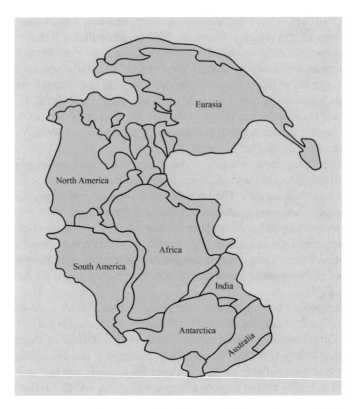

**Fig. 21.2** A conceptual map of the supercontinent Pangaea. All the continents used to be connected in one piece in the form of a supercontinent. The names of the modern continents are labeled on this map. Image credit: Wikimedia Commons

of the world. This is an example of how changes of the Earth can also affect the evolution of life.

## 21.2    The Oxygen Evolution

Not only the ground has been changing, the air in the atmosphere has not always been the same. In the eighteenth century, chemists found that our atmosphere consists mainly of two gases: oxygen and nitrogen (Chap. 14). However, the chemical composition of the Earth's atmosphere during its early days was very different. This change was driven by biology.

The earliest life forms on Earth were single-cell microorganisms without a nucleus. They are called prokaryotes. Bacteria, as well as later-discovered archaea (Sect. 18.3), are examples of prokaryotes. Prokaryotes generally have sizes smaller than a micrometer, too small to be seen by the naked eye. They survive by

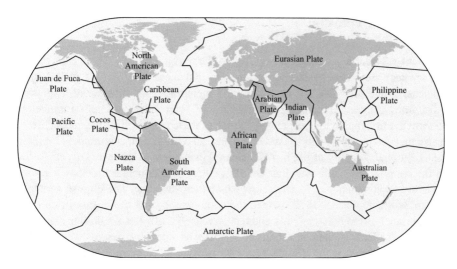

**Fig. 21.3** The Earth has seven major tectonic plates and more minor ones

consuming molecules in their environment. For bacteria, their food usually consists of organic compounds, but some archaea can consume inorganic compounds. Beginning about 3.4 billion years ago, a new kind of bacteria called blue-green bacteria (cyanobacteria) developed the ability to use sunlight to combine with carbon dioxide in the air to chemically produce carbohydrates (a class of biomolecules that includes sugar). These cyanobacteria formed layers on water surface of oceans and lakes to absorb sunlight and release a waste product of oxygen molecule ($O_2$) through a process called photosynthesis. Over time, the accumulated oxygen in the atmosphere increased, changing the oxygen content of the Earth's atmosphere to the roughly 20% that we have today.

The availability of oxygen in the atmosphere allowed other life forms to develop. Some bacteria acquired the ability to inhale oxygen to digest carbohydrates in their food to release energy. The waste product of this process (called respiration) is carbon dioxide ($CO_2$). Organisms that make use of photosynthesis and respiration achieve a balance, leading to the steady oxygen content in the atmosphere. As the degree of biodiversity increased on Earth, the evolution of life also changed the Earth.

When the atmospheric oxygen level reached 1–3% around 2.3 billion years ago, a layer of ozone ($O_3$) began to form in the stratosphere. The ozone layer was crucial in protecting life from harmful solar radiation in the far ultraviolet (wavelengths between 200 and 300 nm) and allowed life to expand onto the continents.

## 21.3   Life Explosion

At about 1.8 billion years ago, a new form of large cell emerged. These cells (called eukaryotes) are distinct from prokaryotes in that these cells have a well-defined nucleus, where their DNA is protected by a membrane. An example of single-cell eukaryotes is algae. About 600 million years ago, some eukaryotic cells got together to form the first multicelled organisms. Some of these multicelled organisms grew to a size that can be seen by the naked eye. Then, all of a sudden, at around 542 million years ago, the diversity of life forms increased in a dramatic manner. This explosion in the complexity of life was first noted in layers of sedimentary rocks where one layer has an abundance of fossils below which there are none. The English geologist Adam Sedgwick (1785–1873) named the geological period of the fossil-rich layer "Cambrian." Biologists call this sudden increase in life the Cambrian explosion (Fig. 21.4). This sudden appearance of life was cited by Sedgwick and other scientists as evidence for Biblical Creation and was used against Darwin's theory of evolution.

We now understand that this sudden transition is not because the Earth was devoid of life before. The increase in fossils may in part be due to animals forming hard skeletons that can leave imprints. The exact cause of the higher pace of evolution remains unknown. Was this surge in life complexity the result of continental drift (Sect. 21.1), the result of an increase in oxygen content in the atmosphere, global climate change (Sect. 22.1), or purely biological?

About 400–500 million years ago, animals with a backbone structure appeared. These vertebrates are the first lifeforms that later developed into the fish, amphibians, reptiles, birds, and mammals that we have today. Although all early life organisms lived in the sea, the coming and going of tides may have left some eukaryotes on shore. They became the earliest forms of land-based plants and fungi. Beginning about 400 million years ago, some animals developed the ability to live both on land and in sea. They were the first amphibians. Following in the steps of amphibians were the reptiles, which began to appear around 350 million years ago. These land-based animals had thick skins and laid eggs with shells. The largest reptiles were the

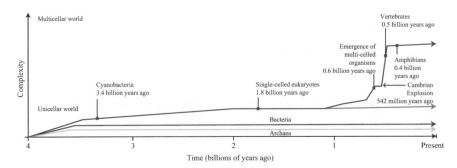

**Fig. 21.4** Change in the complexity of living organisms over the history of the Earth. Most notable is the Cambrian explosion about 550 million years ago

dinosaurs, which dominated the world from 240 to 65 million years ago, until they were completely wiped out by a major extinction event (Sect. 21.4). The only branch of dinosaurs that survived evolved into birds.

Through continuous evolution and increasing complexity, the number of species of animals and plants in existence today is higher than at any time in the history of the Earth. However, this growth of number of species is not constant and has been interrupted by episodes of extinction events.

## 21.4   Extinction Events

Biological species do not persist forever. In 1813, French naturalist Georges Cuvier (1769–1832) noted the sudden disappearance of fossil records, which could be evidence for large-scale extinction of biological species. In his book *Théorie de la Terre (Essay on the Theory of the Earth)*, Cuvier suggested that such extinctions are the result of abrupt changes in the Earth. He further suggested that there was more than one catastrophe responsible for such extinctions. However, with the rise of uniformitarianism in the nineteenth century, Cuvier's idea became unfashionable and the interest in extinction events was only revived in the late twentieth century.

From fossil records, geologists found that there were at least five major extinction events in the last 500 million years of the Earth. The first major extinction event occurred at the end of the Ordovician Period (488–444 million years ago, or Mya), followed by another event in the late Devonian Period (60–420 Mya) at 360 Mya. The massive extinction event at 251 Mya near the end of the Permian period (299–252 Mya), is the most dramatic. It is estimated that between 80–90% of all living species on Earth were wiped out. Even in the more protected marine environment, half of the species became extinct. All this happened over a relatively brief period of 10,000 to 100,000 years. The next event at 200 Mya near the end of the Triassic Period (251–199 Mya) removed 20% of all marine families and many of the large amphibians. The elimination of species and the change in the ecosystem allowed the emergence of dinosaurs as a dominant group of species in the following Jurassic Period (201–145 Mya). The most popularly known of all the extinction events is probably the one near the end of the Cretaceous Period (145–66 Mya) and the beginning of the Paleogene Period (66–23 Mya) 65 million years ago that killed off the dinosaurs (Fig. 21.5).

The exact causes of the mass extinction events are not known, with the exception of the Cretaceous-Paleogene event, which is attributed to external impact (Sect. 17.4). The site of the impact causing the Cretaceous-Paleogene mass extinction is identified as the Chicxulub Crater in Yucatan, Mexico. This crater has a size of 180 km and is buried under 1 km of sediments on the seabed. This crater is estimated to have been caused by the impact of a 10-km asteroid, releasing energy equivalent to $10^{14}$ tons of TNT. The impact generated tsunamis 100–300 m high along the surrounding coast. The ejected dust sent to the atmosphere blanketed the surface of the Earth and diminished sunlight for at least several years. Without enough sunlight

**Fig. 21.5** The five
extinction events since the
Cambrian period.
Geological time can be
divided into eons (left
column), eras (middle
column), and periods (right
column). The Hadean eon
was when the Earth and
Moon were formed, the
Archean eon began after the
first single-cell organisms
emerged, the Proterozoic
eon was when life
diversified, and the
Phanerozoic eon was when
life became easily visible.
The three eras of the
Phanerozoic eon are divided
into 11 periods. After the
Neogene period is the
Quaternary period, which
we are in now. The
extinction events are marked
by arrows on the right

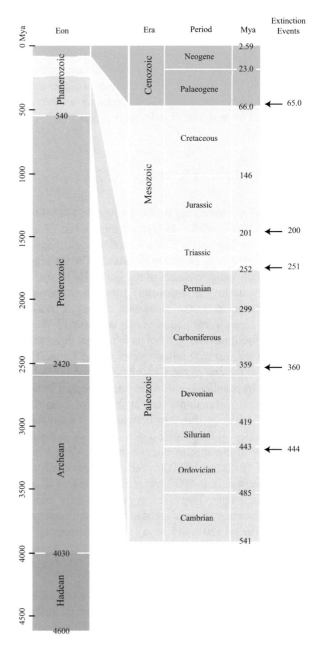

for photosynthesis, plants died. Without plants for food, most of the land-based
living species, including all dinosaurs, did not survive. An impact event of this
magnitude happens about once every 100 million years (Fig. 21.6).

**Fig. 21.6** An artistic perception of an asteroid impact. The impact creates a large crater, expels materials into the atmosphere that blocks incoming sunlight and can cause a mass extinction of life on Earth. Drawing by Clara Wang

The Permian-Triassic extinction event 250 million years ago has also been suggested to be due to an external impact, although the evidence for this is not as direct. Other possible causes that could be responsible for mass extinction include global warming or cooling, nearby nova or supernova events, continental drift, massive volcanic eruptions, or ocean level changes. However, all these environmental changes could be traced back to large-body impacts, so even though the impacts may not lead to direct, immediate extinctions, the effect they have on the environment may do so.

The link between infrequent external impacts and massive extinction also revived the concept of catastrophism (Chap. 18), as sudden events can also shape the evolution of the Earth in addition to slow, gradual processes over billion-year timescales.

## 21.5   Emergence of Humans

For 150 million years, dinosaurs were the largest animals on land. The Cretaceous-Paleogene extinction event ended the domination of reptiles on Earth. Emerging from the shadows of the dinosaurs were small mammals. Without their previous predators, these small mammals evolved to early monkeys having hands and fingers, which can be used to pick fruits from flowering trees. Instead of two eyes pointing in different directions, the eyes of early monkeys shifted closer together, allowing them to see an object with both eyes and therefore gain a perception of distance. When

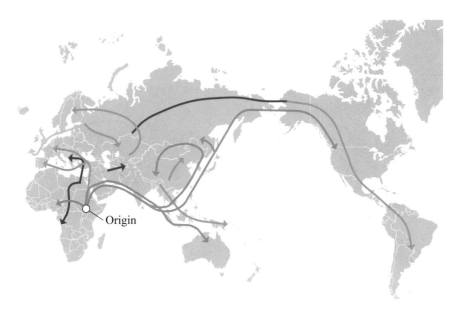

**Fig. 21.7** A schematic diagram illustrating the routes of human migration out of Africa based on genome analysis. The orange curve represents the approximate travel route of human migration starting about 50,000 years ago, the blue curve about 30,000 years ago, the purple curve about 20,000 years ago, and the pink curve about 10,000 years ago. Adapted from the diagram in *Tracking Y Chromosomes Through Time,* from the magazine *"Natuurwetenschap en techniek,"* October 2009

they spent more time on the ground, they developed the ability to walk on two hind legs, leading to early species of apes (Fig. 18.2).

Modern humans evolved in Africa about 300,000 years ago. This "out-of-Africa" theory is supported by genome analysis of different populations that shows that modern populations are branches of a genetic tree beginning in Africa. About 100,000 years ago, groups of humans began moving out of Africa (Fig. 21.7). By 60,000 years ago, they had settled in East Asia, and arrived in America no later than 13,000 years ago. Early civilizations in Mesopotamia, Egypt, China, and India developed between 5000 and 3000 years ago. No matter which race on Earth, we all come from the same ancestor.

Early humans were social animals. They developed social interactions and established community living to improve their chances of survival. We learned about the behavior and social structure of early humans through the study of remains of bones, tools, and living sites. Early humans survived by hunting, fishing, and gathering fruits and plants. They became more proficient in these activities through the development of an increasing array of tools. Because the availability of food is seasonal, humans needed to move from place to place. About 10,000 years ago, groups of humans settled together on a piece of land to systematically grow plants and domesticate animals for food. This represents the beginning of agriculture and the first step of humans shaping their environment. The gathering of people living permanently in groups was the beginning of society.

## 21.6   Lessons from the History of Life on Earth

Although the exact mechanism of how life originated on Earth is unknown (Chap. 19), we do know that life arose relatively quickly under conducive conditions. In the beginning, life forms were small and simple and remained so for billions of years (Fig. 21.8). Because of environmental change, evolution accelerated, resulting in increasing complexity. The present biodiversity is the consequence of heterogeneous environments of the Earth. Over the history of the Earth, we have undergone multiple extinctions and species explosions. Although individual species are fragile, life itself has proven to be sustainable through changes in the environment. An external impact knocked out the once dominant dinosaurs, but mammals emerged as the result, leading to the present human dominance on Earth. Even if humans were to become extinct in the future, it is likely that some form of life will remain.

Through the tree of life, we can trace the millions of present and past species to one common ancestor. Can we extrapolate the tree of life to the future? Will different new branches grow out of the present tree? It has been argued that because of humans' control and alternation of the environment, no further natural evolution is possible. Can the species of humans evolve further? What might humans be like 1000, or 10,000 years from now? Will humans create machines of high intelligence that can replace humans themselves? Will future civilizations on Earth be based on artificial machines? Are we witnessing the end of biological evolution?

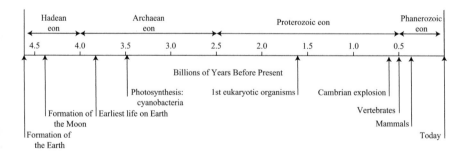

**Fig. 21.8**  Geological and biological timeline of the history of the Earth. The periods of the four geological eons (Hadean, Archean, Proterozoic, and Phanerozoic) are also shown. Significant geological and biological events are noted below the time axis. The emergence of humans is too recent to show on this scale

**Questions to Think About**
1. How can scientists be confident about things that happened millions of years ago?
2. Many scientific theories that overturned conventional ideas only became widely accepted decades after their first proposal (e.g., Wegner's theory of continental drift). How should we view the scientific theories that are popular today?
3. What is our modern assessment of the nineteenth-century debate between catastrophism and uniformitarianism?
4. The Cambrian explosion was once cited as evidence for Creation. Discuss the evidence to support such a belief.
5. History of the Earth and its living contents were closely related. Could contemporary developments of human technology affect the Earth?
6. Could evolution of life on Earth have stopped at the stage of single-cell organisms? Is increase in complexity an inevitable result of evolution?
7. We now attribute all calamities and diseases as the result of physical processes of Nature. Are there any examples of natural events that are not yet unexplained by physical causes?

# Chapter 22
# Climate Changes Through the Ages

The Earth system consists of the physical components of land, oceans, atmosphere, and biosphere. The biosphere refers to the thin layer on the crust of the Earth where living organisms are found. The concept that the Earth is a self-contained system was first proposed by Russian geologist Vladimir Vernadsky (1863–1945) in his book *Biosphere* in 1926. This idea was revived 50 years later by English environmentalist James Lovelock (1919–), who proposed the Gaia hypothesis, which states that the Earth's crust, the oceans, the atmosphere, and the biosphere are all part of a single entity. All these components are actively interacting with each other and changes in one component will lead to feedback in the other components. We now know that the early Earth acquired some water from its crust. After the emergence of life from this small pond, oxygen released from bacteria substantially increased the amount of oxygen in the atmosphere (Sect. 21.2). The subsequent evolution of life was affected by the changing physical conditions of the Earth.

As we discussed in Chap. 21, all these components have undergone significant evolution since the formation of the Earth 4.6 billion years ago. The Earth is not an isolated object and was subjected to external influences ranging from an external collision that led to the formation of the Moon, to impacts from asteroids and comets that brought water to the Earth and caused major extinction of life (Sect. 21.4). An extraterrestrial impact can cause the collapse of entire ecosystems through the disruption of food chain and destruction of habitat, leading to extinction of species. Although the scale of external impacts has greatly reduced since the early days of the Earth, the current rate of external mass delivery is estimated to be about 40,000 tons per year, based on the deposit of micrometeorites. A proper understanding of how the Earth functions as a whole depends not only on the interactions between its internal components, but also on external factors such as changes in solar radiation, solar wind, and impacts from comets and asteroids.

The main source of energy of the Earth is the Sun. The rate of energy output from the Sun has been relatively constant, which allows the Earth to be kept relatively warm throughout its history. However, it does not mean that the environmental conditions on Earth have always been the same. In fact, through various

© Springer Nature Switzerland AG 2021
S. Kwok, *Our Place in the Universe - II*,
https://doi.org/10.1007/978-3-030-80260-8_22

astronomical and geological factors, Earth's climate has undergone drastic changes. These climate changes have impacted the evolution of life on Earth.

The idea that large-scale climate changes have occurred over the history of the Earth was proposed by Swiss geologist Louis Agassiz (1807–1873), who in 1837 suggested that ancient glaciers covered large parts of Europe, Asia, and North America. His theory was based on earlier observations by others that ice in the Swiss Alps has moved over time. Evidence of ancient glaciation was also found in England, Wales, and Ireland. The term "ice age" (Eiszeit) was coined by German naturalist Karl Schimper (1803–1867) in 1837 to refer to the period of wide glacier coverage of the world. Agassiz summarized his glaciation theory in his 1840 book *Études sur les glaciers* (*Studies on Glaciers*), but his theory was not accepted by most scientists at that time because they found the idea of large-scale temperature variation incredible. The concept of global climate changes over the Earth's history only became popular in the late nineteenth century, after the death of Agassiz in 1873.

## 22.1   Climate Cycles

Using tree rings (Sect. 17.1), we can derive a record of temperature changes over the past several hundred years. By examining the air bubbles contained in ice cores extracted from great depths (many kilometers), scientists can know the Earth's past surface temperature precisely, going back almost a million years. Armed with such data, scientists can track variations of the Earth's climate over history. In the past one million years, the Earth has undergone semi-regular cold spells on hundred-thousand-year time scales (glaciation periods).

There is also evidence that there are climate cycles over longer periods called ice ages. Ice ages are defined as periods of the Earth when one or both of its polar regions are covered with ice. Because there is still year-round ice in both Arctic and Antarctic regions, technically we are still in an ice age. Based on geological records, five major ice ages can be identified: the Huronian (2.4–2.1 billion years ago), Cryogenian (720–630 million years ago), Andean-Saharan (460–420 million years ago), late Paleozoic (360–260 million years ago), and the latest Quaternary Ice Age (2.58 million years ago to present).

Under the most extreme scenarios during some of these ice ages, the whole Earth might have been completely covered with ice and this has been called Snowball Earth. It has been suggested to have occurred during the Cryogenian ice age, just before the onset of the Cambrian explosion of life (Sect. 21.3).

Within the last ice age, scientists can identify several glacial periods, the last of which occurred between 115,000 to 11,700 years ago, with a glacial maximum at about 22,000 years ago (Fig. 22.1). Because it is the most recent, it is also the best studied. It is estimated that approximately 30% of the Earth's land area was covered with ice. The affected areas included Canada, the high mountain regions on the west coast of South America, Antarctica, Greenland, northern Europe, most of Russia,

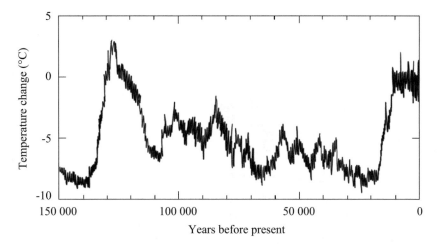

**Fig. 22.1** Temperature variation of the Earth in recent history as determined from Antarctic ice cores. Our present warm period (right peak, mid-Holocene) started about 10,000 years ago. The previous warm period (the left peak) occurred about 130,000 years ago. Data from Petit et al. 1999, *Nature*, **399**, 429

and the Himalaya highland, and Tibetan Plateau of western China. Siberia and Alaska were connected by a land bridge due to low sea levels, allowing human to cross from Asia to North America (Fig. 21.7). Even in areas not covered by ice, the climate became drier due to lower water evaporation and rainfall. Forests became grasslands and then deserts. Starting about 17,000 years ago, these effects were reversed as the Earth became warmer and wetter. The melting of ice raised sea levels, cutting off America from Asia and the British Isles from the European continent. The regrowth of vegetation allowed humans to engage in farming and ranching, laying the foundation of civilization.

We are currently in a warm interval (interglacial period). Using measured time intervals of previous interglacial periods of 40,000–100,000 years, we can expect to enter another glacial period in a few tens of thousands of years.

If there were cycles of climate change over the history of the Earth, what could be the forces driving such cycles? The cause of the ice ages is not known, although several factors probably contributed to it. Among the factors are the chemical composition of the atmosphere (Sect. 21.2), volcanos, and the motion of the tectonic plates (Sect. 21.1). For shorter interglacial periods, astronomical effects could be a factor.

The Serbian astronomer Milutin Milankovitch (1879–1958) identified three astronomical cycles that could have effects on the Earth: (i) wobble of the Earth's rotation axis (which causes the precession of the equinox, Chap. 11, Vol. 1) which has a period of 26,000 years; (ii) variation of the angle of inclination between the Earth's rotational axis and the axis of the Earth's orbital plane around the Sun (angle of obliquity, Chap. 6, Vol. 1), which has a period of 41,000 years; and (iii) deviation of the Earth's orbit around the Sun and change in the eccentricity of the orbit due to

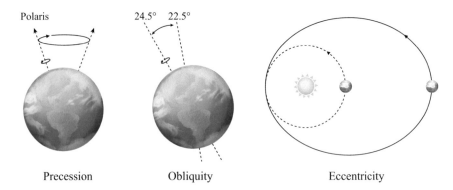

Precession                    Obliquity                           Eccentricity

**Fig. 22.2** Schematic illustrations of the three effects contributing to the Milankovitch cycles. Left: Change in the direction of the Earth's rotational axis relative to the celestial sphere. Middle: Change in the inclination angle between the Earth's rotational axis and its orbital plane. The expected range of the variation is from ~22.5° to 24.5° (Appendix G, Vol. 1). Right: Change in the eccentricity of the orbit of the Earth. The current eccentricity of the Earth's orbit is 0.017 and decreasing. The drawings are not to scale and the effects are exaggerated

planetary perturbation (particularly by Jupiter and Saturn), with periods of 100,000 years and 400,000 years. These are collectively known as the Milankovitch cycles (Fig. 22.2).

The seasons are the result of a finite angle of inclination between the Earth's rotational axis and its orbital axis. A larger angle of obliquity means more extreme seasons. At present, the Earth is closest to the Sun (at perihelion) near the time of winter solstice in the North Hemisphere, so the effects of the Earth-Sun distance and the angle of solar radiation work against each other, making Northern Hemisphere seasons less extreme. It is the opposite in the Southern Hemisphere, where the distance and angle effects add together, making the seasons more extreme. As the result of precession, a reverse situation will happen in 13,000 years. At that time, the Earth's north pole will be tilted towards the Sun at perihelion, making the Northern Hemisphere summer warmer.

When the Earth's orbit is more eccentric, there is a larger variation in the distance between the Earth and the Sun, resulting in an increase in the radiation the Earth receives from the Sun, and therefore larger temperature variations. According to Kepler's second law, a more elliptic orbit would also result in a larger variation in the Earth's orbital velocity, and therefore affect the relative lengths of the seasons (Chap. 11, Vol. 1).

As in the case of Mercury, gravitational perturbations by Jupiter and Saturn will also cause the orbit of the Earth around the Sun to shift in orientation relative to the stars (Fig. 6.4). This variation has a cyclic period of ~112,000 years. This shift of orbital orientation will affect the beginnings and the relative lengths of the seasons.

The Milankovitch cycles suggest that there are natural causes for variations of the Earth's climate due to changes in solar radiation received by the Earth as the result of these effects. However, these natural effects do not seem to explain the extent of climate change over the history of the Earth. So, there must be other effects that contribute to the climate cycles of the Earth. For example, natural causes may have induced positive feedback to local atmospheric/oceanic processes that lead to an exaggerated climate change.

## 22.2  The Warm Earth

Temperatures of planetary bodies are determined by the balance of absorption of sunlight and self-radiation. At the distance of the Earth from the Sun, the temperature is expected to be around $-18\,°C$, about the temperature of the Moon. However, the average global temperature of the Earth is about $15\,°C$, $33\,°C$ higher than expected. This anomaly can be traced to the presence of certain molecules such as water vapor ($H_2O$), ozone ($O_3$), carbon dioxide ($CO_2$), and methane ($CH_4$) in the atmosphere. These molecules can absorb infrared light and therefore block the release of radiation from the Earth to space, resulting in a warming of the Earth's atmosphere. This is known as the "greenhouse effect" and these molecules are collectively known as greenhouse gases in the Earth's atmosphere. This is named after the use of glass in a greenhouse to allow visible sunlight to come in but prevents re-radiated infrared radiation from plants inside from escaping, resulting in a warmer environment inside the greenhouse (Fig. 22.3).

The presence of greenhouse gases is critical in allowing the presence of liquid water on Earth. Water has a freezing point of $0\,°C$. Without greenhouse gases in the atmosphere, the Earth would be at the same temperature as the Moon. Earth would have no liquid water and it would be doubtful that life could have developed and survived. The atmosphere and its greenhouse gases are therefore essential elements of existence of life on Earth.

## 22.3  Human-induced Climate Change

As soon as plant life occupied a significant fraction of the Earth's continents, wildfire became frequent. Forests and ground vegetation can be ignited by lightning (globally estimated to happen 1.4 billion times a year) and fueled by abundant oxygen in the atmosphere. When humans came onto the scene, they harnessed fire from wildfires, first for protection against predators and heating of habited areas, and later for cooking. Fire could have played a role in human evolution as consuming cooked food leads to a smaller and more efficient digestive system, allowing more energy for brain growth. Fire was later deployed to make pottery (allowing the boiling of water in a container) and metal making (development of bronze- and iron-based tools). When agriculture developed, fire was used as a tool to clear forest land for planting. Deforestation accelerated after the development of metal tools, because axes and saws are much more efficient in cutting down trees, leading to the wide utilization of wood for furniture, housing, and transport. In Europe, it is estimated over half of the

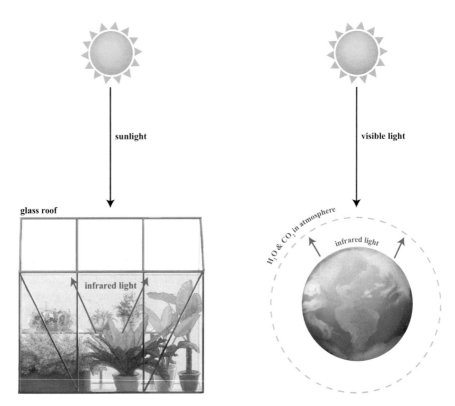

**Fig. 22.3** Greenhouse effect. Left: The glass roof of a greenhouse allows sunlight to pass through but absorbs the infrared light re-emitted by the plants inside, making the greenhouse warmer. Right: Water and carbon dioxide molecules in the Earth's atmosphere serve as a greenhouse, allowing visible sunlight to pass through but blocking the release of infrared light from the Earth to space

original forest has been lost over human history. In the United States, approximately one-quarter of the original forests have been lost since the arrival of European settlers. Because forests are major sinks of carbon dioxide, deforestation has direct consequences on climate (see below).

In order to secure water supplies for crops, humans started irrigation projects by altering natural waterways. Canals were built to divert water from one location to another. Beginning in the twentieth century, numerous dams were constructed along major rivers for control of river flow and generation of electricity.

Until the beginning of the nineteenth century, human societies mainly relied on the burning of wood, employment of animals (horses, oxen, and donkeys), and to a lesser extent, wind (windmills) and flowing water (water wheels), as sources of energy. This all changed with the invention of steam engines, which utilize the burning of coal to boil water into steam to drive a mechanical wheel. The broad adoption of steam engines allowed large-scale machine-based manufacturing in factories. The invention and deployment of internal combustion engines in vehicles and ships for transportation created huge demands for petroleum. Coal, oil, and

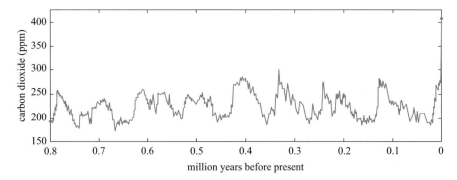

**Fig. 22.4**  Global atmospheric carbon dioxide concentrations ($CO_2$) in parts per million (ppm) for the past 800,000 years. The peaks and valleys track ice ages (low $CO_2$) and warmer interglacial periods (higher $CO_2$). During these cycles, $CO_2$ was never higher than 300 ppm. In 2020, it reached 417 ppm (blue dot on the right edge of the plot). On the geologic time scale, the recent increase looks virtually instantaneous. Plot is based on EPICA Dome C data (Lüthi, D., et al., 2008, *Nature*, Vol. 453, pp. 379–382) provided by NOAA NCEI Paleoclimatology Program

natural gas are all used for heating homes. Because coal, oil, and natural gas are all derived from remnants of past lives, they are collectively known as fossil fuels.

One of the byproducts of the burning of fossil fuels is carbon dioxide ($CO_2$). With an increasing amount of fossil fuels being burned, the amount of carbon dioxide deposited into the atmosphere skyrocketed (Fig. 22.4). This man-made release of carbon dioxide outstrips the amount that can be naturally absorbed by plants or deposited into carbonate minerals.

Soot from smoke produced from burning is also deposited in the atmosphere, which in large concentrations can also affect the climate. The scenario of "nuclear winter" warns that the large-scale release of soot into the atmosphere as the result of nuclear bombs may block sunlight for long periods and wipe out life on Earth. The same mechanism has been invoked to explain the mass extinction after an asteroid impact on Earth (Sect. 21.4).

## 22.4   Effects of Climate on Society

The warming of the Earth that began about 13,000 years ago corresponded to the beginning of rapid human social development. Permanent communities with large populations began to appear in the form of villages, with individuals collaborating in food acquisition, sharing of resources, and mutual protection. These collective communities led to the formalization of myth, ritual, and religion. They also made possible the passing of accumulated knowledge from one generation to the next, first orally and later by writing. The accumulation of observational and descriptive knowledge (e.g., changing sunrise positions with the seasons) laid the foundation of rational thinking and science.

The need for maximizing crop yields necessitated a better understanding and prediction of the seasons, which motivated systematic and long-term observations of the movement of the Sun. The practical need for an accurate calendar for agriculture was a strong incentive for astronomical studies (Chap. 9, vol. 1). The identification of the four special dates of the year (solstices and equinoxes) gave rise to festivals; the winter solstice was a special day for celebration in many cultures.

The advantages of agriculture also brought uncertainties. Successful crop yields rely very much on rain, the absence (drought) or excess (flood) of which could ruin the entire growing season. Weather is much less predictable than the movement of the Sun. Divine intervention was often invoked, and religious offerings were made in order to secure the proper amount of rain. Science and religion were often used in parallel in support of agriculture. The practice of astronomy and religion are intertwined; many ancient monuments (e.g., Stonehenge) were built for both astronomical and religious purposes.

Because agriculture is a much more efficient way of food production, this resulted in rapid economic, social, and cultural advancement of the human race. The availability of leisure and free time led to the development of cultural pursuits such as music, art, and sculpture. The inevitable conflicts that arise from people living near each other necessitate the establishment of governing bodies to enforce rules of behavior. This has led to an evolution of different forms of government, from tribal, oligarchy, monarchy, theocracy, to republic.

The possible effects of the onset of another ice age, or global warming as the result of human activities, have been the subject of extensive speculation. While the next glacial period is expected to be thousands of years away, some climate models predict that the onset of human-induced global warming could be much more sudden. The melting of polar ice caps could lead to a rise in sea levels, making large parts of low-lying continental areas uninhabitable. There will also be shifts in agricultural regions, resulting in large-scale human migration. The melting of Arctic ice may lead to changes in ocean currents. Although the global average temperature is expected to rise, changes in the gulf stream that currently brings warm air to Europe may result in significant cooling of the European climate. The scientific study of the cause and possible effects of climate change is therefore of great urgency.

## 22.5   Balance Between Development and Conservation

Much of the human development in the last 5000 years has been the result of human alteration of the surroundings. Forests and grasslands are transformed into farmland, and trees are cut down for energy and construction. Minerals and fossil fuels are extracted from land as resources. While humans only take up 0.01% of the biomass on Earth, the total mass of man-made materials and structures increased from 3% of total biomass in 1900 to 100% in 2020. These changes in the Earth's landscape have also led to the destruction of habitat of animals and plants, both on land and in the sea. Changing acidity of the oceans as the result of increasing carbon dioxide content perturbs the living conditions of sea-based animals (e.g., corals). Increased human demand for food leads to overfishing and depletion of fish stocks. Although many biological species have gone extinct over the history of the Earth, the rate of extinction is accelerated recently due to human activities. Could human activities lead to another mass extinction of living species?

Increasing human population and improving living conditions have led to an increase in the demand for energy. The burning of fossil fuels has altered the chemical composition of the atmosphere. There is a finite reservoir of fossil fuels in the Earth's crust and there is a chance that continued demand may eventually deplete supply. These changes have led to the debate between economic development and conservation and preservation of the environment. How to continue to develop while keeping natural resources sustainable is a challenge.

What is the long-term solution to human development? Is continuing technology development the answer? To what extent should we utilize technology to reshape the Earth's environment? If we run out of fossil fuels, can we deploy space-based solar shields to capture a larger fraction of the energy of the Sun? If we exhaust all natural resources on Earth, should we attempt to exploit resources in other planetary bodies? These are ethical issues that go beyond scientific and technological advancements.

**Questions to Think About**
1. Why were scientists in the nineteenth century reluctant to accept that there had been large-scale climate changes over the Earth's history?
2. The emergence of human civilization coincided with the end of the last glacial period. Was it a coincidence? What would be the state of civilization today if the last glacial period ended 10,000 years later than it did?
3. There is some controversy over how immediate the effect of human-induced climate change is. Cite some evidence in support of each side of the argument.
4. In the past several thousand years, humans have greatly transformed the state of the Earth, e.g., through the conversion of forest into farmland. Should such practice continue?
5. Human burning of fossil fuels is a major source of carbon dioxide input into the atmosphere. Should we greatly reduce our use of fossil fuels? What are the practical alternative sources of energy?
6. If there is another ice age coming, why should we be concerned about global warming?
7. The human population has multiplied greatly in the last several thousand years. Is there a limit to human population growth? Is there a maximum size population that the Earth can support? List some of the limiting factors on the scenario of indefinite population growth.
8. If there had been large-scale climate changes over the history of the Earth, how can we be certain that the present climate changes are induced by human activities?
9. Science and religion coexisted in ancient times. Can they coexist today?

# Chapter 23
# The Link Between Stars and Life on Earth

In Aristotle's cosmology, stars are eternal. By the nineteenth century, astronomers realized that stars can be sorted into different groups based on their spectral properties. The spectral classification of stars (Sect. 9.7) led to the idea that stars are born differently and evolve differently, leading to a paradigm shift from a static view to evolution. We now believe that stars do not live forever, and they have a life cycle of birth and death just like humans and animals. From the age of the Earth and the Solar System, we know that the Sun must be at least 4.6 billion years old, and it has been shining almost constantly at the present luminosity for much of this time. By comparing the present rate of nuclear fuel consumption and the expected total supply of fuel, one can estimate the lifetime of the Sun. Through the construction of models of the structure of the Sun and calculations of its rate of energy generation, astronomers estimate that the Sun will have a total lifetime of about 10 billion years. In other words, the Sun will have about another 5 billion years to live.

Observationally, we can predict the future of the Sun by observing other stars. Assuming that the Sun is a typical star, we can get a glimpse of the future of the Sun by observing other stars in the Milky Way Galaxy. Because stars were born at different times in the past, a sample of stars should represent stars in different stages of their lives, just as when we observe a sample of the human population, we will find a mix of the young and the old. By taking snapshots of the physical appearances of the young and old, one can develop a theory on how humans grow, age, and die. Similarly, we can put together a scenario on how stars evolve over time by observing a large sample of stars. The modern theories of stellar evolution were developed with this approach.

© Springer Nature Switzerland AG 2021
S. Kwok, *Our Place in the Universe - II*,
https://doi.org/10.1007/978-3-030-80260-8_23

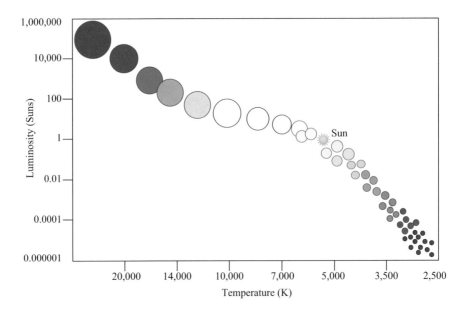

**Fig. 23.1**  Stars on the main sequence. The Sun is located near the middle of the main sequence. Both the vertical and horizontal axes are on a logarithmic scale. The colors and sizes of the stars are symbolic representations of colors and sizes of the stars and are not actual or to scale

## 23.1  Lifetimes of Stars

From spectroscopic observations of stars, astronomers can determine the luminosity, gravity, temperature, and masses of stars (Sect. 9.7). Most stars are found to obey a strong luminosity-temperature relationship, with hotter stars having higher luminosity. When the distribution of stars is plotted on a diagram of luminosity vs temperature, they tend to lie on a well-defined band, which is known as the "main sequence" (Fig. 23.1). The fact that stars are confined to a band on the luminosity-temperature diagram appears to suggest that stars evolve from hot to cold, moving along this sequence from the top-right to the bottom left.[1] This turned out not to be the case. Instead, a star is born onto a specific point on the main sequence and this point is determined by the mass of the star. A massive star is hot and luminous, and a low-mass star is cool and less luminous. The star Vega is about twice the mass of the Sun, and it has a temperature of ~10,000 K (compared to ~6000 K for the Sun) and is 40 times more luminous than the Sun.

A star can remain stable in temperature and luminosity for a long time while generating energy from the nuclear fusion of hydrogen to helium in its core (Sect. 15.2). The Sun can do this for about 10 billion years. As a massive star needs to

---

[1] The belief that stars evolve from hot to cold is the reason why astronomers call hot stars "early type" and cool stars "late type." This convention is still in use today.

maintain a higher luminosity, it is burning its fuel at a higher rate and therefore has a shorter lifetime on the main sequence. For Vega, this translates to a main-sequence lifetime of less than 1 billion years. For a star 10 times the mass of the Sun, its lifetime can be as short as 10 million years. Because biological life requires billions of years to evolve, it is unlikely that there is life on planets around massive stars.

After a star uses up the supply of hydrogen in its core, its envelope will expand to become a red giant, while its core contracts. A red giant draws its energy from fusing hydrogen in a layer above the core. In the case of the Sun, this will occur about 5 billion years from now. The surface of the Sun will expand to the orbit of Venus and the Sun will be several hundred times brighter than now. When observed from Earth, the Sun will cover a large part of the sky and will take hours to rise and set, instead of the two minutes that it does now.

After further contraction of the core, higher temperature will allow helium to be ignited in the core, followed by further expansion of the envelope. This results in an even redder, brighter, and a larger red giant. The existence of life on Earth is dependent on a stable Sun, as a slight change in the radiation output of the Sun will cause major environmental changes on Earth with dramatic effects on the delicate balance required for life. The heat from this future Sun as a red giant will evaporate the oceans and destroy almost all life forms on Earth.

## 23.2   Death of the Sun

The Sun will stay a red giant for several hundred million years. The ash of helium burning will accumulate in the core, while helium will burn in a shell surrounding the core.[2] The core of the star is made up of oxygen and carbon, with nuclear fusion taking place in two separate layers, an inner layer fusing helium into carbon and an outer layer fusing hydrogen into helium. An outer hydrogen envelope occupies most of the volume of the star (Fig. 23.2). This stage of "double-shell burning" is called the asymptotic giant branch with an expected lifetime of about one million years.

Asymptotic giant branch stars are among the largest stars known. When the Sun becomes an asymptotic giant branch star, its surface will approach the orbit of Mars. Its luminosity also increases to several thousand times that of the present Sun. A stellar wind develops which gradually depletes the hydrogen envelope and exposes the hot core. The remnants of the stellar wind become a planetary nebula (Fig. 23.3), and the exposed core becomes the central star of the nebula. As all the hydrogen of the central star is burned up, it dims and contracts to become a white dwarf. A schematic diagram illustrating the evolution of a sun-like star is shown in Fig. 23.4.

---

[2] Astronomers use terms like "hydrogen burning" or "helium burning," but these are not chemical reactions but nuclear fusion reactions.

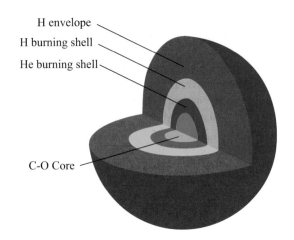

**Fig. 23.2** A schematic diagram illustrating the structure of an asymptotic giant branch star. This drawing is not to scale. The actual size of the core and the two burning shells are much smaller than the hydrogen envelope.
Figure adapted from Kwok, S. 2001, *Cosmic Butterflies*, Cambridge University Press

H envelope

H burning shell

He burning shell

C-O Core

## 23.3   The Final Fate of Stars

Although stars are nothing but balls of gas, they manage to maintain a spherical shape and a stable size for billions of years. The stability is made possible by a delicate balance between the star's tendency to collapse from gravitational force and to expand from pressure from the hot gas inside. White dwarfs, however, have such high densities that the gravitational force is too strong to be resisted by gas pressure alone. Instead, a new force called "electron degenerate pressure" takes over. In extreme high densities when electrons are squeezed against each other, they are forced to move at higher and higher speeds, according to the quantum statistical theory developed by Italian American physicist Enrico Fermi (1901–1954) and English physicist Paul Dirac (1902–1984). Just as the movements of molecules in the air exert a pressure on the walls of a room, these fast-moving electrons exert a pressure to resist the gravitational attraction. This force is a purely quantum mechanical phenomenon and is not observable in our everyday lives.

In 1926, English physicist Ralph Howard Fowler (1889–1944) applied the just-published theory of Fermi and Dirac to the structure of stars. He found that the electron degenerate pressure could indeed support a star at high densities. In 1930, the 19-year-old Indian student Subrahmanyan Chandrasekhar realized that at the densities typical of white dwarfs, these electrons move near the speed of light. He then applied Einstein's special theory of relativity to the structure of white dwarfs and demonstrated that the maximum mass for white dwarfs is 1.4 times the mass of the Sun, a value which has since been named the Chandrasekhar limit.

The existence of the Chandrasekhar limit suggests that a star born with a mass higher than 1.4 times the mass of the Sun could not become a white dwarf. What could be the end state of such stars? In 1939, American physicist Robert Oppenheimer (1904–1967) and Russian-Canadian physicist George Volkoff (1914–2000) found that another stable state is theoretically possible. At densities higher than those in white dwarfs, electrons and protons can combine to form neutrons. Neutrons

**Fig. 23.3** NGC 6302 in the constellation of Scorpius is one of about 4000 known planetary nebulae in the Milky Way galaxy. This is a composite color image processed from multi-band images obtained by the *Hubble Space Telescope*. Image credit: NASA, ESA, and the Hubble SM4 ERO Team

have quantum properties like those of electrons, and an equivalent force called "neutron degenerate pressure" can keep a star from collapsing. Such stars are called neutron stars. While a white dwarf with a mass of one solar mass is the size of the Earth, a neutron star of the same mass has a size of only 30 km. The actual existence of neutron stars was confirmed with the discovery of pulsars, which are a manifestation of rotating neutron stars.

In the early 1970s, it was commonly believed that the final state of a star at the end of its life is dependent on its mass. Stars born with a mass less than 1.4 solar masses will become a white dwarf, whereas a star two or three times the mass of the Sun will end as a neutron star. For stars more massive than that, there is no known stable state and they are assumed to collapse to an infinitesimal size. Such objects are called

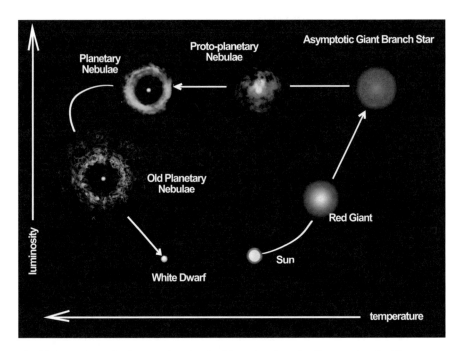

**Fig. 23.4** A schematic diagram showing the life cycle of a solar-like star. A solar-like star can brighten to several thousand times its present luminosity. After a late-life loss of mass via stellar wind, the star loses its outer envelope, exposes its core, and evolves to higher temperatures until its supply of hydrogen fuel runs out. It then decreases in luminosity and ends as a white dwarf. The luminosity scale (vertical axis) is approximately logarithmic. Figure adapted from Kwok, S. 2001, *Cosmic Butterflies*, Cambridge University Press

black holes because beyond certain densities, the gravitational attraction is so large that even light cannot escape. By definition, black holes cannot be seen, and their existence can be inferred only from the gravitational effects on their stellar compan- ions, or from the X-ray emissions that radiate as matter falls into a black hole.

We have seen that white dwarfs evolved from central stars of planetary nebulae after the envelopes of their red giant progenitors are ejected. Similarly, neutron stars and black holes are the remnants of the cores of their progenitor stars after the stellar envelopes are ejected in the form of supernovae. If every star heavier than 1.4 solar masses becomes a supernova, then supernovae must be a very frequent occurrence; at least several per century are expected. However, the last supernova seen in the Milky Way was the one seen by Johannes Kepler in 1604. So observationally, we have not seen a supernova in our Galaxy for over 400 years. There is clearly a discrepancy between theoretical expectations and observations.

This mystery was solved through the discovery of the stellar mass loss. Near the end of the asymptotic giant branch phase, a star develops a very strong stellar wind, about a billion times stronger than the solar wind observed in the present Sun. This massive wind can remove most of the mass of the star in a matter of hundreds of thousands of years. A star born with a mass as high as eight times the mass of the Sun can reduce its weight to below the Chandrasekhar limit, therefore avoiding the fate of becoming a supernova. Planetary nebulae and white dwarfs, not supernovae and neutron stars or black holes, are the most common end states of stars.

## 23.4 Stellar Synthesis of Complex Organics

It is interesting to note that the element carbon, which is the basis of all organic compounds and life, is not made until the late life of a star. During the asymptotic giant branch phase of stellar evolution, carbon is made from the fusion of three helium nuclei. These carbon nuclei are brought to the surface through the process of convection and ejected into interstellar space by a stellar wind.

Infrared and millimeter-wave spectroscopic observations of old stars in the 1970s have shown that they are prolific molecular factories. Over 80 gas-phase molecules have been detected through their rotational transitions in the circumstellar envelope of asymptotic giant branch stars. Minerals including silicates, silicon carbide, and refractory oxides have been detected by infrared spectroscopy (Sect. 20.2). During the evolutionary stage between the asymptotic giant branch and planetary nebulae, molecules of increasing complexity are being synthesized in sequence. Starting with simple molecules such as $C_2$, $C_3$, and CN, chains (HCN, $HC_3N$, $HC_5N$), rings ($C_3H_2$), and acetylene ($C_2H_2$) are formed in the stellar winds of asymptotic giant branch stars. Acetylene becomes the building blocks of the first aromatic molecule benzene (Fig. 20.4), from which aromatic compounds are formed. Over periods of just thousands of years, complex organics are made in the circumstellar environment (Sect. 20.4).

The ability of stars to form complex organics on such a rapid time scale was totally unexpected. In terrestrial laboratories, chemical reactions only proceed under high-density conditions. But the densities in planetary nebulae are so low that they are near total vacuum conditions on Earth. How stars manage to create complex organics under such conditions is still a mystery.

## 23.5 Birth of New Stars and Planetary Systems from Stellar Debris

Planetary nebulae expand and gradually disperse themselves into the interstellar medium. They are only visible for about 30,000 years. After that, the nebulae are too faint to be seen and the only remnant of the star that can be observed is the core. The cores of central stars of planetary nebulae further evolved into white dwarfs (Fig. 23.4). Their brightness continues to fade as their supply of nuclear fuel runs out. White dwarf stars are therefore the common end of typical stars in our Galaxy.

When the shining moment of a star ends, the Galaxy becomes its graveyard. At the same time, the atoms, molecules, and solids ejected earlier in the asymptotic giant branch and planetary nebulae phases may aggregate together elsewhere in the Galaxy and become the birth sites of another generation of stars.

Out of the whirlpools of these interstellar clouds, new stars are born. When gravitational self-contraction concentrates enough mass for a nuclear reaction to ignite, the central part of these gaseous nebulae becomes a new star or a cluster of stars. The remaining gas and solids in the cloud aggregate into planetary bodies, comets, and asteroids around the new star (Fig. 10.1).

## 23.6  External Delivery of Organics to Earth and Their Effects on the Origin of Life

Stars of different masses are not born in equal numbers. The star formation rate is heavily skewed toward the low end. There are far more stars of low mass than stars of high mass in the Galaxy. Given the observed rates of mass loss, astronomers estimate that only stars born with masses higher than eight times the Sun will become supernovae, and these stars constitute only 5% of all stars in the Galaxy. Because the remaining 95% of stars go through the asymptotic giant branch and planetary nebulae phases of stellar evolution, large amounts of minerals and organic solids must have been produced by ordinary stars over the history of our Milky Way Galaxy. These stellar products are ejected into the interstellar medium and spread throughout the Galaxy. Direct evidence for stellar manufactured solids to have traveled across the Galaxy to arrive at the Solar System and Earth can be found in the form of pre-solar grains preserved in meteorites (Fig. 23.5). It is truly remarkable that we can actually hold a piece of stellar material in our hands.

Pre-solar grains are grains of diamonds, silicon carbide, corundum, and spinel that have isotopic abundances not typical of the Solar System but consistent with their origin in asymptotic giant branch stars. These solids are condensed from gases in stellar winds and ejected into interstellar space (Sect. 20.2). This process is similar to how organic solids are formed in carbon-rich planetary nebulae (Sect. 23.4). Because organic solids are extremely sturdy, they are therefore likely to be able to transverse across interstellar space even under a high ultraviolet radiation background. Minerals and organic solids from old stars can chemically enrich interstellar clouds. When new planetary systems are formed from these clouds, they may already have embedded in them complex products from previous generations of stars.

From geological evidence and observations of the Moon and the planets, we now know that the Earth is not a closed system. The Earth was subjected to extensive external bombardments in the past and continues to be bombarded at present, although at a lower rate (Sect. 17.4). Even now, on average every year one meteoroid of 5 to 10 m in size enters the Earth's atmosphere. In every one million years, we can

**Fig. 23.5** Electron microscope image of a pre-solar silicon carbide (SiC) grain. The size of the grain can be seen in comparison to a bar of length 1 micrometer, shown in the lower left. Image credit: E. Zinner

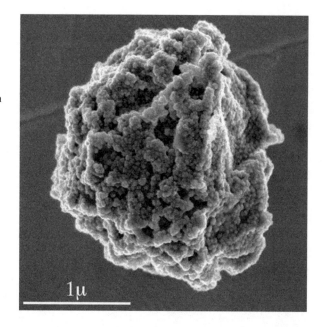

expect a km-size object will hit the Earth. The impact events in the past are likely to have caused extensive perturbations to our physical and biological systems (Sect. 21.4).

From the craters on the Moon, we can reconstruct the rate of bombardments. Comets are the primary agents of bombardment. From the crater record on the Moon, one can estimate that the Moon was hit by comets with sizes ranging from 5 to 500 km. The same bombardments must have also struck the Earth. The larger gravity of the Earth compared to the Moon means that Earth must have suffered a larger rate of external bombardment and retained most of the fallen materials. From Chap. 20, we learned that comets contain not just ice and minerals, but also organic compounds. In 1992, American astronomers Christopher Chyba and Carl Sagan (1934–1996) estimated that during the period of giant bombardment in the first one billion years, comets and other external bodies must have brought in a substantial amount of organics to Earth. The possible influence of these prebiotic molecules on the development of life on Earth must therefore be considered seriously (Fig. 23.6).

The heavy bombardment of the Earth is believed to have occurred about 3.8 to 4.1 billion years ago, and remarkably, life emerged on Earth shortly after. Is it a coincidence or did Earth really do it all on its own in spite of the tremendous help that has come from the outside? If indeed the Earth has received life-forming materials such as amino acids and nuclear bases from the outside, why does life on Earth only use 20 amino acids in its proteins? Also, why are only five nucleobase pairs used in our DNA and RNA? Laboratory experiments have shown that the other amino acids are perfectly capable of making different proteins, and other nucleobase

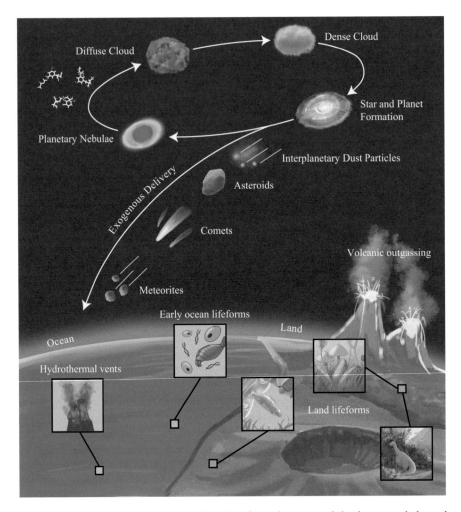

**Fig. 23.6** A cartoon illustrating the manufacturing of organic compounds in planetary nebulae and their ejection into interstellar space. The primordial Solar System, formed from interstellar clouds, could have inherited these stellar materials. These organics were carried by asteroids and comets which rained onto the early Earth, leading to the creation of life. Drawing by Clara Wang

pairs can function well in enzymes. The American astrochemist Lucy Ziurys suggests that life happened so quickly on Earth that it did not have time to try all chemical possibilities. Life just evolved based on whatever chemicals happened to be handily around at the time. There was no design, no careful planning, no detailed experimentation; life just began using the externally supplied ingredients available.

If this is true, life could have taken paths resulting in biological organisms that are totally different from what we are familiar with. If indeed the basic ingredients of life are widely distributed in the Galaxy, life on another planet in the Galaxy could develop using entirely different sets of amino acids and nucleobases and proceed along different paths of biochemistry, resulting in living organisms that are very different from our own. It would be chauvinistic of us to assume that life elsewhere would remotely resemble a life that we are familiar with on Earth.

## 23.7 The Future of the Human Species

In contrast to the ancient belief that stars are eternal, we now believe that stars have finite lifetimes. Stars go through cycles of birth and death, and the future fate of the Sun can be reasonably predicted based on observations of other stars in the Galaxy. When the Sun expands and engulfs the Earth, the existence of the Earth will end. While we think we can predict the future of the Sun and the Earth, the future of the human species is much more uncertain. Although life on Earth began billions of years ago, human civilizations have a history of only a few thousand years. How long will *homo sapiens* last before it self-destructs or is wiped out by an ecological disaster? We have seen many examples of species extinction over the history of the Earth (Sect. 21.4). A typical mammal species lasts for about 2.5 million years. Will we last 2.5 million years? What factors will contribute to our continued survival and eventual demise?

If humans indeed become extinct, what will happen? Life is resilient; just as mammals became dominant after the extinction of dinosaurs 65 million years ago, some other forms of life will surely emerge to take over the Earth. Life will go on.

**Questions to Think About**
1. The spectral classification of stars led to the development of theories on stellar evolution in the twentieth century. What effects did stellar evolution have on our perception of cosmological evolution?
2. Pre-solar grains in meteorites are made in stars. What is the significance that we can hold a piece of stellar material in our hands?
3. What are the philosophical implications if life on Earth indeed developed using organic compounds delivered by stars?
4. Human civilization occupies only a very small fraction of the history of the Earth. In our short (hundred thousand years) presence, we have already made major impacts on the condition of the Earth. How might humans affect the future of the Earth?

# Chapter 24
# Life on Other Worlds

The speculation that there could be extraterrestrial life began soon after Copernicus demoted the Earth to the status of just one of the planets. Galileo's 1610 telescopic observation of the lunar surface led Kepler to suggest that the circular features seen on the Moon might be artificial constructions by lunar inhabitants. John Wilkins (1614–1672), Master of Trinity College, Cambridge and Bishop of Chester, wrote in his 1638 book *The Discovery of a World in the Moon* that the Moon may be inhabited by intelligent beings, and speculated on future journeys to the Moon and encounters with a lunar civilization. William Herschel was a firm believer in life on the Moon. In 1780, he wrote "... the almost absolute certainty, of her (the Moon) being inhabited" and observed features on the Moon were the "works of the lunarians and may be called their towns." The idea that the Moon is populated by life remained popular well into the nineteenth century. An 1835 news report in the *New York Sun*, wrongly attributed to observations by John Herschel, reported the observation of forests, fields, beaches, and even winged creatures on the Moon. The Scottish astronomer Thomas Dick (1774–1857), in his 1838 book *Celestial Scenery*, estimated that there were over 22 trillion inhabitants on planets of the Solar System.

Among the planets, Mars drew the most attention. From early observations of Mars, astronomers concluded that it had a topography very similar to that of the Earth, with the darker regions identified as seas and the lighter regions as continents. In his book *Other Worlds Than Ours: The Plurality of Worlds Studied Under The Light of Recent Scientific Researches* in 1870, British astronomer Richard Proctor (1837–1888) suggested Mars had an atmosphere, vegetation on the surface, and seasons similar to those on Earth. From the maps of Mars, French astronomer Camille Flammarion (1842–1925) suggested that Mars had the climate and geography to be a living world. The American astronomer Percival Lowell (1855–1916) reported finding artificial canals on the Martian surface, and his work generated huge public interest in Martian life for several decades. The idea that Mars harbors life, even intelligent life, remained popular throughout the nineteenth century.

As the result of planetary explorations that began in the mid-twentieth century, all the planets and planetary satellites have been imaged at close range, and we no

© Springer Nature Switzerland AG 2021
S. Kwok, *Our Place in the Universe - II*,
https://doi.org/10.1007/978-3-030-80260-8_24

longer believe that there are large life forms on the Moon, on Mars, or anywhere else in the Solar System. As of 2020, the only examples of life are what we have on Earth. However, we cannot rule out the existence of microscopic life and scientists are actively working to seek evidence for microscopic life in the Solar System. To this day, the search for extraterrestrial life remains a topic of high public interest.

The existence of intelligent life in other stellar systems remains a possibility. As we have learned in this book, there is really nothing special about the Earth. We are one of four rocky planets in our Solar System, with a structure that is very similar to those of our closest neighbors, Venus and Mars. The source of our energy is the Sun, which is a very ordinary star—one of more than 100 billion stars in our Milky Way Galaxy. Many stars in the Galaxy have planetary systems around them. After the first discovery of an extrasolar planet orbiting another star in 1995, thousands have been identified (Sect. 10.2). It is therefore likely that there exist extrasolar planets that have conditions suitable for life to develop.

## 24.1  Conditions for Life

We know from fossil records that life has been present on Earth for at least 3.7 billion years, but we are not sure exactly how it emerged. Life probably began in a "primordial soup" where suitable ingredients in a favorable environment gathered into self-replicating organisms. Until recently, many scientists believed that life on Earth developed in isolation, beginning with the simplest ingredients such as ammonia, methane, and water. In the last chapter, we learned that old stars produce complex organic compounds in abundance, and these products have been distributed all over the Galaxy, including our early Solar System. For about 300 million years during the early days of the Earth, our Earth was heavily bombarded by debris from the primordial solar nebula. These bombardments brought heavy doses of organic matter, which probably facilitated the origin of life.

If external bombardments brought ingredients favorable to the formation of life to Earth, they probably did the same to Mars. Mars is a planet with a rocky surface and an atmosphere: could there be life on Mars at present or at some time in its past? In our Solar System, there are also planetary satellites that possess a solid surface, an atmosphere, and liquid oceans. Their geological resemblance to Earth has raised the expectation that life could exist there.

Humans need to consume water and breathe in oxygen, and we can survive only within a narrow range of temperatures. Many scientists believe that the most probable reason why only Earth harbors life is the presence of liquid water on Earth's surface. Life is believed to have begun in the ocean and liquid water provided the ideal environment for life to emerge. If this is true, then planets can harbor life only within a limited distance from their parent stars. If they are too close, then water on the planets' surface would evaporate. If they are too far, water will condense into ice. The conditions for the existence of water, therefore, define the "habitable zone."

However, we must be cautious not to put too many constraints on life's existence based on our own needs. As we learned from Chap. 18, life takes many forms. We now know that microorganisms are resilient and can flourish in high temperatures, high acidity, and extreme dryness (Sect. 18.3). Bacteria have been found in the driest deserts of Chile, the frozen lands of Antarctica, highly acidic water bodies, and hot hydrothermal vents deep under the sea. The discovery of extremophiles shows that terrestrial life thrives in extreme conditions. It also challenges our traditional beliefs about what constitutes favorable conditions for life to emerge and flourish. The versatility of life and the possible existence of a deep biosphere on Earth suggests that simple life forms such as bacteria and archaea could exist on Mars, on planetary satellites such as Titan and Europa, and maybe even on asteroids and comets.

## 24.2   Search for Life in the Solar System

If life on Earth indeed originated from simple inorganic molecules as proposed by the Oparin–Haldane hypothesis, could the same process be at work in an extraterrestrial body? The most direct way to test this hypothesis is to search for evidence of life in objects in the Solar System, such as planets and planetary satellites. Since Earl Slipher's (1883–1964) early observations at Lowell Observatory in the early twentieth century, the seasonal variations of the dark regions of Mars have been cited as evidence of vegetation. This speculation seemed to have been confirmed by American astronomer William Sinton's (1925–2004) detection in 1959 of infrared absorption features in the dark regions of Mars that are similar to spectral signatures of organic matter on Earth. It later turned out that these infrared features were due to deuterium water molecules in the Earth's atmosphere, and the idea of plants on Mars was proven wrong. As of 2020, there is no evidence of plant life on any of the Solar System planets except the Earth.

Various landers and rovers have been sent to explore the surface of Mars in search of conditions that may have supported life on Mars in the past. Pictures taken by roaming rovers show that Mars' landscape is very similar to deserts on Earth. These missions have also analyzed the contents of Mars' atmosphere, sampled the soil on the surface, and drilled underground for signs of life. They have revealed evidence of liquid water 3.5 billion years ago and the presence of organic matter on the surface. These findings show that Mars had conditions suitable for the emergence of life. Mars has a thin atmosphere and used to have water flowing on its surface. If life could develop on Earth, it could have done the same on Mars.

Saturn's satellite Titan is the second planetary object that is known to host a liquid ocean or lake on its surface (Fig. 24.1). The widespread presence of organic matter in Titan's atmosphere and surface (Sect. 20.3) makes this planetary satellite a prime candidate to search for life. Could these hydrocarbon lakes harbor life on Titan?

**Fig. 24.1** An artist's rendition of the surface of Titan based on data obtained from the *Cassini* mission. Credit: European Space Agency

## 24.3   Search for Signs of Life on Other Planets in the Galaxy

The chemical composition of the Earth's atmosphere has been significantly altered by the presence of life. The prebiotic atmosphere was dominated by $N_2$ and $CO_2$. From 2.4 billion years ago, an increasing amount of oxygen was produced through photosynthesis by cyanobacteria, algae, and plants (Sect. 21.2). Ozone ($O_3$) was produced by reactions between $O_2$ and sunlight. If we observe the Earth from outside, we will find spectral features of $O_2$ and $O_3$, which are the results of life on Earth (Fig. 24.2). We can get an idea of the Earth's spectrum as observed from space by observing the Moon's night side during a time close to the new Moon. While the night side of the Moon is not lit by sunlight, it does receive light from the Earth. The reflected sunlight from the Earth is reflected back from the Moon to the Earth in the form of earthshine (Fig. 24.3). Depending on which part of the Earth is facing the Moon, the reflectivity of the Earth's oceans, land, or clouds can be determined.

If we use the Earth's spectrum as a template to obtain spectroscopic observations of extrasolar planets, we may discover planets that are similar to the Earth. When an extrasolar planet passes in front of its parent star, signatures of the planet's atmosphere can be detected in the stellar spectrum (Fig. 24.4). The European Space Agency's *Atmospheric Remote-sensing Infrared Exoplanet Large-survey (ARIEL)*

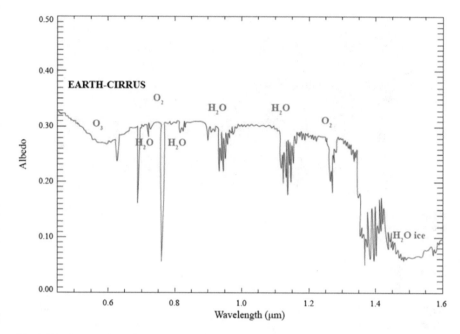

**Fig. 24.2** The spectrum of the Earth's atmosphere and clouds carry signatures of life. The vertical axis (albedo) is the reflectivity of the Earth as measured by earthshine

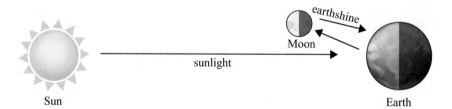

**Fig. 24.3** Earthshine. The night side of the Moon is not totally dark as it receives light from the dayside of the Earth. The light reflected from the night side of the Moon is called earthshine. Earthshine can be observed on a crescent Moon a few days before and after the new Moon

mission plans to observe the atmospheres of 1000 extrasolar planets using this method. While this method can identify Earth-like planets, their detection does not represent a conclusive proof for extraterrestrial life.

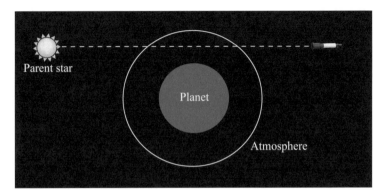

**Fig. 24.4** A schematic drawing illustrating how the atmospheric spectrum of an extrasolar planet can be observed during planetary transit. The drawing is not to scale

## 24.4  Search for Extraterrestrial Intelligence

Beyond the search for life in the Solar System, astronomers have also attempted to search for extraterrestrial intelligence (SETI) through remote observations. Our Galaxy is 10 billion years old and consists of 100 billion stars. There are probably more than 100 billion planets in our Galaxy. It is therefore not unreasonable to think that this Galaxy contains other intelligent life forms, and we are not alone or unique. If alien civilizations are characterized by their use of energy, degree of technological development, extent of communication and outreach, their existence can be discovered through a search for evidence of utilization of energy, artificial structures, or communication.

There are two ways to connect with extraterrestrial intelligent civilizations: We can either actively broadcast our existence by sending out signals or passively observe signs of extraterrestrial intelligence. However, both efforts are limited by our current state of technology. The idea of actively broadcasting the existence of life on Earth started in the nineteenth century when there was a widely held belief that there could be intelligent beings on the Moon and Mars. In 1820, Carl Friedrich Gauss (1777–1855) proposed constructing a giant right triangle with three squares on each side to illustrate human's understanding of the Pythagorean theorem. Different fields of vegetation would be planted on the triangle and the three squares (Fig. 24.5). This figure would be constructed on the Siberian tundra with a size large enough to be visible from Mars. Upon its sighting, the Mars civilization would recognize that there is an advanced civilization on Earth.

In 1840, Johann Joseph von Littrow (1781–1840), director of the Vienna Observatory, proposed that a large circular or square canal be dug in the Sahara Desert. This canal would then be filled with kerosene and lit, making a giant sign visible from the Moon or Mars, objects that were popularly believed to have intelligent inhabitants (Fig. 24.6). If the shapes of the circle and square are perfect, these could be seen by aliens as evidence of artificial construction.

**Fig. 24.5** Gauss's proposal
for extraterrestrial
communication. A
schematic drawing of
Gauss's proposal to
construct a geometric sign
on Earth for communication
with extraterrestrial
intelligent beings on Mars

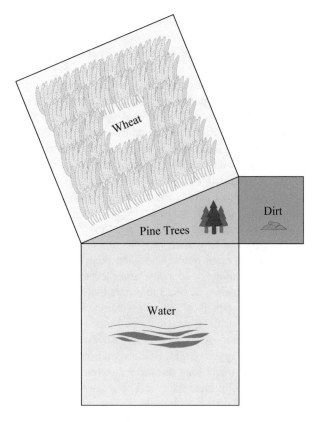

**Fig. 24.6** A proposal by
von Littrow to send signals
to intelligent extraterrestrials
on Mars

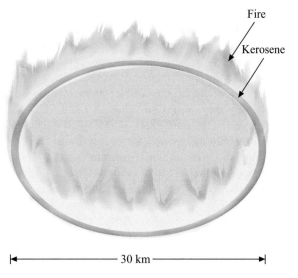

The experiment by Gauss assumes that mathematics is universal. There is some evidence to support this view; for example, the Pythagorean theorem was independently proved by both Chinese and Greek mathematicians. However, since mathematics is strictly a construction of the human mind (Sect. 8.6), we cannot be certain that alien civilizations would develop the same kind of mathematics as we did.

After the development of radio technology in the twentieth century, we have been broadcasting radio, television, and cellular phone signals on a large scale. These signals could easily be picked up by extraterrestrial civilizations and deciphered as artificial signals from an intelligent society. Due to the finite speed of light, our radio signals have reached stellar systems only about a hundred light-years from Earth.

The best-known example of a passive search for artificial structure by extraterrestrial intelligent beings was the observation of canals on Mars. Using his 18-inch telescope in Flagstaff, Arizona, Percival Lowell found evidence of canals of artificial construction on Mars. Through a series of lectures, articles, and books (*Mars* in 1895, *Mars and Its Canals* in 1906, and *Mars as the Abode of Life* in 1908), Lowell excited the general public about the possibility of intelligent life on Mars. Subsequent observations of Mars by larger telescopes and images taken from spacecrafts orbiting Mars showed that the canals that Lowell saw were just natural geological features or products of his imagination. We no longer believe that there is any evidence for artificial structures on Mars.

Since the mid-twentieth century, astronomers have devoted a lot of effort to using radio technology for SETI. Radio signals have the advantage over optical signals in that they can travel long distances through interstellar space without being affected by extinction by interstellar solid particles. SETI researchers have surveyed thousands of nearby stars with radio telescopes in the search for artificial signals. These signals being sought could be in the form of very short pulses, pulses of periodic intervals, or other forms that are not consistent with natural processes. In spite of decades of efforts, no confirmed artificial signal has been detected.

However, we should keep in mind that the radio technique is only one step beyond the primitive techniques that were proposed in the nineteenth century. If alien civilizations are thousands of years or even millions of years more advanced than us, it would be impossible for us to imagine what kind of communication technology they would be using. A radio SETI assumes that the targets of our searches are at the same exact technological stage as us, which is extremely unlikely.

The increasing demand for energy by an alien civilization may result in their construction of an artificial structure that surrounds their parent star in order to capture the entire radiative output of the star. Such a structure (called a Dyson sphere) will have a size of ~1 AU and can convert sunlight into thermal energy. It will radiate like a blackbody and its infrared radiation may be detectable by remote telescopic observations. The difficulty lies in distinguishing such artificial structures from natural circumstellar envelopes ejected from stars by stellar winds (Sect. 23.3).

## 24.5   Direct Contact with Alien Life

Every year the Earth is rained on by 40,000 tons of micrometeorites, which are fragments of comets and asteroids (Sect. 20.3). Meteorites of lunar and Martian origins have been found on Earth. To date, there is no evidence that any of these extraterrestrial objects carry living, frozen, or fossilized microorganisms. If there is, it would be evidence for panspermia (Sect. 19.2). In addition to passively receiving extraterrestrial materials, we are also actively bringing back samples from other Solar System objects. We have already obtained samples from the Moon (the *Apollo* programs), comets (the *Stardust* mission), and asteroids (the *Hayabusa* missions) and will soon attempt sample-return missions from Mars. Should any of these samples contain alien microorganisms, their effects on the Earth's biosphere are unpredictable.

There is a possible analogy from history on the effect of cross-contamination. Within a few decades after Columbus landed in the Caribbean in 1492, most members of the native Taino population were wiped out as they did not have immunity to smallpox, influenza, and other viruses carried by the Europeans. Shortly after Hernán Cortés's Spanish forces landed in Veracruz, Mexico in 1520, half of the population in the Aztec capital Tenochtitlán succumbed to diseases.

If we send returning probes or humans to other potentially habitable Solar System bodies, we could be bringing back alien life forms that may pose a danger to life on Earth. The British astronomers Fred Hoyle and Chandra Wickramasinghe had promoted the idea that comets could be the cause of pandemics, such as the influenza outbreak of 1917–1919. Much of the meteoritic materials that fall on Earth are fragments of comets, and comets are now known to contain complex organics and biomolecules such as amino acids (Sect. 20.3). Although we cannot rule out the possibility of frozen bacteria or viruses in meteoritic materials, it is unlikely that extraterrestrial viruses could cause pandemics. The cellular structures of terrestrial and extraterrestrial life forms are likely to be so different that it would be impossible for extraterrestrial viruses to survive and reproduce in a terrestrial host. Bacteria, being organisms capable of independent living, are more likely to subsist on Earth.

If we are successful in finding extraterrestrial microorganism, either in a sample return mission or exploration of another planetary body, and are able to decipher the genetic materials of such microorganisms, it is possible that they will belong to a new domain of life beyond the three that we have on Earth (Sect. 18.4). They may even not share our common root—our last universal common ancestor.

## 24.6   Did Aliens Visit the Earth?

If advanced intelligent civilizations are indeed common in the Galaxy, would they not have already visited us? An analogy that is often made is that when the European civilizations developed seafaring capabilities, they had the human desire, religious zeal, and economic need to explore the world, leading to contact with native civilizations in America, Asia, and Australia. If alien civilizations have the capability of space travel, would they not have come in person, or have sent a robotic probe to Earth? There have been thousands of reports of unidentified flying objects (UFOs) —light-emitting objects moving in the sky in an erratic fashion. The common interpretation is that these are spacecraft from visiting alien civilizations. While many UFO reports have been explained as planets (usually Venus), airplanes, balloons, meteors, fireballs, clouds, artificial satellites, and optical atmospheric phenomena, some remain unexplained. Even if these sightings are real, we must ask the question: If an alien civilization capable of space travel came hundreds of light-years to reach us, why would they travel in a primitive mechanical device? Why would their spacecraft emit visible light, when they have the entire electro-magnetic spectrum available to them?

We should remember that we are a very young and primitive civilization. Humans have been around for only 300,000 years, and we have been a technological society capable of distant communication for only a hundred years. Given the 10-billion-year age of our Galaxy, probability suggests that if there are alien civilizations in this Galaxy, they will be millions of years older and technologically far more advanced than us. Just as the civilizations of one hundred years ago could not imagine the technologies we have today, we similarly cannot imagine technologies one hundred years from now, let alone a thousand or million years. If an alien civilization is visiting us, it is certain that it would not be using twentieth-century Earth technology, which is what a light-emitting flying saucer is. If extraterrestrials visit our planet, they will travel in forms that are entirely unknown and possibly invisible to us.

Evidence for past alien visits could also be found in the form of archeological artifacts. Indeed, some landscape markings and artificial structures have been suggested as put here by aliens. Although none of these claims have been confirmed, it is possible that alien civilizations did visit the Earth without us knowing it. Just imagine that if we could travel back in time to England one hundred years ago and leave behind a computer storage device (e.g., a data disc) containing texts of thousands of books, images of thousands of pictures, and recordings of thousands of pieces of music. Would the English population see the disc as anything other than a piece of shiny metal? It is possible that even if alien civilizations had left behind artifacts during one of their past visits, we may not be able to recognize or decipher them.

We also need to consider the possibility that alien civilizations may be also much further up the evolutionary tree than us. We can talk to dogs with confidence that some of our messages are understood. But can we communicate with ants? Our degree of comprehension and consciousness may be very limited when viewed in a larger context. A more highly-evolved extraterrestrial species may have no desire to talk to us. They could be already observing us, the same way we view animals in a zoo or bacteria in a petri dish. Such a scenario may be impossible for us to test: How would bacteria be aware that they are being studied?

Is there a way that we can discover objects that we can be certain are of extraterrestrial origin? No one knows what an alien artifact may look like, but it is not impossible to tell. Let me offer an example. A higher dimension of space seems to be totally abstract and unreachable. But if we see a sphere gradually growing larger and larger to a maximum size, then gradually shrinking in size and disappearing, we can be relatively certain that we are witnessing a four-dimensional sphere passing through our three-dimensional space. This is based on the analogy of a sphere passing a plane as it would be observed by a person living in a two-dimensional world. This example suggests that if there is a mysterious phenomenon that is beyond our understanding, we may be able to conclude that it is not natural. To put this in more concrete and simple terms, if you find an object with a structure that is clearly the result of design and not naturally made and it is definitely not of human construction, this may be considered an extraterrestrial artifact.

## 24.7 Different Paths of Scientific Development among Alien Civilizations

Astronomy was the first science to be developed by human civilization. One would naturally assume that it is the same for other alien civilizations. We must remember that the Earth is blessed with an atmosphere that is transparent to visible light where stars emit most of their energy. As a result, the wonders of the Universe were easily accessible to early humans. It is not difficult to imagine planets that are completely covered by clouds so that no celestial objects other than their parent star are visible. Venus is an example of such a planet. Venus and the Earth are both terrestrial planets with solid surfaces and are similar in many ways. Even our early atmospheres are similar. One difference is that the condensation of water into oceans absorbed much of the carbon dioxide and sulfur dioxide in the early Earth's atmosphere, and the emergence of living organisms 3.5 billion years ago enriched the atmosphere with oxygen (Sect. 21.2). Venus's atmosphere consists of mostly carbon dioxide and clouds of sulfuric acid which make its atmosphere totally opaque in visible wavelengths. As a result, no visible astronomical observations can be made from the surface of Venus.

Could an intelligent species evolve from a planet with a thick atmosphere? The greenhouse effect of a thick atmosphere creates a warm environment on the surface, which could be advantageous, in particular for a planet located farther from its parent star than our Earth is from ours. Such a species may observe the apparent motion of their sun and feel the effects of seasons (if their planet's rotation axis is inclined with respect to its orbital plane), but they will not have access to the observations of planets and stars to develop a sophisticated theory of cosmology. Such a society would develop other sciences such as biology, chemistry, and physics, but not astronomy, at least not until they develop the technology of space flight to get above their atmosphere for a view of the Universe.

In more extreme scenarios of underwater or underground civilizations, there would be even less motivations for astronomical development. Such a non-astronomical civilization may be advanced in other ways but may be less inclined to interstellar communication.

## 24.8   The Social Implication of Discovery of Extraterrestrial Life

If life is indeed discovered on Mars, what would be the potential significance? We would know that life is not unique, and Earth is not special or privileged. If extraterrestrial life is different from organisms on Earth or operates under different rules of biochemistry, our understanding of life would broaden. These differences may make our present differences between races seem trivial.

The detection of extraterrestrial life would also bring home the message that the rise of humans is a result of opportune events, and life could be quite common in the Universe. Hopefully, this discovery would give us a new sense of community. We are a class of living organisms who, through accidental evolution, are presently cohabitating on a small, fragile planet. If life existed on Mars in the past but is now gone, why did life go extinct? What circumstances led to the demise of life on Mars but allowed life on Earth to thrive? What lessons can we learn from Mars to ensure the continued survival of organisms on Earth?

## 24.9   Ethical Issues of Planetary Exploration and Engineering

Even if no life is detected on planetary bodies such as Mars, the exploration and colonization of other worlds by humans still raise ethical, political, and economic issues. Is it within our right to visit, and possibly contaminate, other worlds? Should we be allowed to exploit the resources of the Moon and other planetary bodies? Will our visits to asteroids, planets, and planetary satellites disrupt the environment of these objects to an irreversible extent?

Colonization of Mars may require us to bring Earth-based plants and other species to the planet to create a life-supporting system. What are the potential consequences? What kinds of safeguards or precautions should we exercise before engaging in these activities? Given the current high public interest in human habitation of the Moon and Mars, we need to have a thorough discussion of these issues before embarking on such endeavors.

On a longer timescale, should humans be engaged in planetary engineering? If our energy consumption continues to increase as the result of population growth and improvement in the standard of living, we will reach a point that we exhaust all resources of this planet. Should the next step be the exploitation of resources in other planetary bodies in the Solar System? With the development of future technologies, it is conceivable that we could alter the atmosphere of Venus to make the planet habitable. Chemical or biological means can be introduced to convert carbon dioxide in the Venus atmosphere to oxygen, so that the surface temperature can be lowered, and oxygen made available for human respiration. Humans have made large-scale changes to the landscape of the Earth through agriculture and urban development; is planetary engineering inherently different from the way that we have transformed planet Earth in the last several centuries? As an analogy to planetary colonization, there was large-scale migration from Europe to America, populating the American continent at the expense of the native population. These are ethical questions that we need to explore before we engage in planetary engineering or colonization.

## 24.10 Nonbiological Alien Life

Are we being chauvinistic to demand that extraterrestrial life resemble our own? From the study of meteorites, we know that abiotic synthesis can produce more amino acids than the 20 that we use in proteins, and more nucleobases than the five that we use in our DNA and RNA (Sect. 20.4). Can different proteins be produced using the other amino acids? Can different sets of genetic code be constructed using different nucleobases?

Popular science fiction literature often suggests that other elements, such as silicon, can be used to replace carbon as the basic element for life. This is unlikely to be true. Because of the versatility of carbon atoms in forming chemical bonds, the potential families of organic compounds are very large. Just using hydrogen, carbon, nitrogen, oxygen, and a few other common elements, $10^{33}$–$10^{180}$ different organic molecules can theoretically be constructed. The possible biochemical pathways leading to different forms of life are close to infinite.

It is therefore doubtful that extraterrestrial intelligent life will look like us; in fact, it may not look like any other forms of terrestrial life at all. Even if we do have a "close encounter," it is not certain that we can recognize our counterparts from space.

Finally, we may speculate on a nonbiological form of an intelligent alien world. Continued development of machine-based intelligence may result in a society without humans. We already know that machines can learn. It is therefore not impossible that machines can also evolve and improve on their own. Machines can build better machines. Self-reproducing machines of intelligence can essentially live forever, unrestrained by the finiteness of human lifetimes. A machine-based species could downgrade or replace the human species on this planet.

The needs and demands for human subsistence imply that there is a limit to human space travel. A machine-based civilization will not suffer from such limitations and can engage in long-duration space travel and colonization. We can imagine human colonization of Mars, but it seems impossible for humans to travel over light-year distances to colonize the planetary system of another star. If interplanetary colonization is indeed common, does it mean that such colonizations are mainly carried out by machines? If this is the case, does it mean that most of the extraterrestrial intelligent civilizations are nonbiologically based?

**Questions to Think About**

1. Should we look for intelligent civilizations in other stars? Will contact with other intelligent life be beneficial to us?
2. How should we, as a human race, react if we discover life forms on other planets that are behind us in terms of technical developments? What if they are way ahead of us?
3. Will we ever be aware of or encounter other alien civilizations? Is any effort in SETI futile?
4. Both van Maanen and Lowell saw things in their observations that were not real. Did subjective wishful thinking play a role in their observations? What role does psychology play in science?
5. Kepler and Lowell interpreted features on the Moon and Mars, respectively, as artificial constructions by aliens. Both turned out to be wrong. How can we be sure that any signals that we detect in the future are truly from extraterrestrial intelligent civilizations?
6. The idea that life is prevalent in the Solar System was popular in the nineteenth century, which turned out to be wrong. Are there any popular scientific beliefs we hold today that may turn out to be false in the future?
7. Some scientists believe that extraterrestrial civilizations are common in the Universe and some think that intelligent life is unique on Earth. What are the arguments in support of these very different views?
8. Does the existence of extraterrestrial life, in particular intelligent life, cause any conflict with existing religious doctrines?
9. How would the discovery of extraterrestrial life or contact with alien civilizations affect our current social structure?
10. What kind of ethical guidelines should be in place before we colonize another planet?

# Chapter 25
# Our Place in the Universe

Since the early days of human civilization, our ancestors have been keen observers of the sky. They noted that although the apparent motions of the Sun, the Moon, the stars, and the planets are complicated, they all have repeatable patterns that can be summarized and predicted. They also learned that these observational patterns can be described by mathematical representations, from which cosmological models were developed. These models were revised and refined with improving observational capabilities, resulting in a continual evolution of the view of our place in the Universe.

At first, it seemed there was a clear distinction and separation between heaven and Earth. The heavens are pure, peaceful, and everlasting, while the Earth is dirty, complex, and in constant turmoil. We later learned that the heavens and the Earth are more connected than we thought. The Sun and the stars are made of the same chemical elements as the Earth. Except for hydrogen, helium, and a few light elements created during the Big Bang, all other elements were made by nuclear reactions inside stars. These stellar materials are distributed throughout the Milky Way Galaxy and are the primordial ingredients for the formation of later generations of stars, solar systems, planets, and life. We are all stellar materials.

The development of astronomy in the last 3000 years has changed many of our long-held perceptions of ourselves in the Universe. As the result of scientific progress, we no longer believe that the Earth is at the center of the universe; the heavens are constant; the Earth is young; the Earth is unchanging; all living creatures on Earth were created together and have remained unchanged over time; and mankind occupies a unique and central place in the Universe. These changes of perception occurred on three fronts: the spatial scale, the temporal scale, and the relative scale.

© Springer Nature Switzerland AG 2021
S. Kwok, *Our Place in the Universe - II*,
https://doi.org/10.1007/978-3-030-80260-8_25

## 25.1   Changing Spatial Scale of the Universe

Ancient people's knowledge of the world was limited to places that they could travel to, and places to which they heard other people had traveled. For the early Western civilizations, their world covered the land areas surrounding the Mediterranean Sea, from present-day Spain to Turkey in the north, Iran to the east, the Atlantic to the west, and North Africa from Egypt to Morocco in the south. The total area is about 10 million square km. When Eratosthenes in the third century B.C. calculated the circumference of the Earth to be 40,000 km, the surface of the Earth was estimated to be 500 million square km. This means that the known world represented only 2% of the entire world. This was the first realization that the world could be much larger than previously believed.

Ancient astronomers were also aware that there are celestial objects beyond the Earth. The Sun, the Moon, and the stars must be very distant compared to distances in everyday life. When Aristarchus calculated the distance to the Moon to be 70 times the radius of the Earth (Chap. 8, Vol. 1), and the distance to the Sun to be 1400 times the radius of the Earth (Chap. 12, Vol. 1), they had the first quantitative estimate of the size of the celestial world.

In addition to the Sun and the Moon, there are also the stars. The stars were believed to be located on the surface of the celestial sphere, beyond the orbits of the planets. From the size of the epicycles of the planets, Ptolemy estimated the size of the celestial sphere to be 20,000 times the radius of the Earth (Chap. 15, Vol. 1).

From the fact that an Earth observer's horizon bisects exactly the celestial sphere within 0.1 degrees, the celestial sphere must be larger than the size of the Earth by a factor of $1/0.1°$ ($= 573$) in the two-sphere Universe model. After Copernicus, the fact that no such difference can be detected as the Earth orbits around the Sun means that the size of the celestial sphere must be 573 times the Earth-Sun distance. Assuming Aristarchus' value for the Earth-Sun distance, this means that the celestial sphere is 800,000 times the radius of the Earth (Sect. 18.1, Vol. 1). Since Aristarchus underestimated the Earth-Sun distance by a factor of 17, the size of the celestial sphere is at least 10 million times the radius of the Earth, using the modern value of the AU.

In the heliocentric model, the fixed stars do not rotate together and there is no need for them to all lie on the surface of a celestial sphere. Stars are freely distributed in three-dimensional space and each star may have different distances from Earth. With the measurement of stellar parallaxes, we finally have a good estimate for the distance for nearby stars. The distance to the nearest star Alpha Centauri is 4.4 light-years, or 280,000 AU. At the beginning of the twentieth century, Milky Way was known to be the stellar system of which our Sun is a member. The diameter of the Milky Way is about 200,000 light-years, or about 10 billion AU. The distance to the Andromeda Galaxy, one of our nearest galactic neighbors, is about 2.5 million light-years. The Virgo Cluster of galaxies is about 60 million light-years away. If we use the redshift of spectral lines in galaxies as a guide to distance, there are galaxies billions of light-years away. The history of astronomy has been a progressive increase in our estimate of the size of the Universe.[1]

## 25.2 On the Temporal Scale

Until the eighteenth century, the time interval of our perceived existence was a short one. It was bounded by the Creation a few thousand years in the past and the second coming of Christ in the (not-so-distant) future. Our view of the world was also a static one. The stars were assumed to be eternal, and they are as they have always been. Our present Earth was no different from the way it was at its beginning. All living organisms were created together at the time of Creation. This view was first called into question by Kant and Laplace, who raised the idea that the Solar System may not be eternal but had an origin. The concept of time was introduced into our understanding of the Universe. The abandonment of the static nature of the Universe and the search for evidence of evolution continued over the last 200 years. Our current belief is that the Universe is constantly evolving; stars form and die, and the Universe is expanding. The landscape and climate of the Earth have changed greatly since the Earth's formation. Living organisms evolved from simple to complex over billions of years and all living organisms can be placed on various branches of the tree of life. Living species are not permanent: Old living species die off and new species emerge. All living organisms, past and present, can be traced to a universal common ancestor.

Part of this revolution is the result of our improved ability to date past events, in particular through radioactive dating. It is interesting to note how recent our modern view of the age of the Earth is. Our current estimate was arrived at only a few decades ago. In a short period of 300 years (from 1650 to 1950), our view of the age of the Earth has changed from several thousand to several billion years, an increase by a factor over a million. From the analysis of the most pristine meteorites, our estimate of the age of the Solar System is 4.6 billion years. Our parent star, the Sun, was born in the Milky Way galaxy less than 5 billion years ago and can be expected to live for another 5 billion years.

The first sign of life appeared about 3.8 billion years ago and animals appeared about 800 million years ago. Humans (*homo sapiens*) appeared about 300,000 years ago. If we condense the 4.6 billion years of earth history into one year, then the first unicellular organism appeared in late March, animals in late October, and humans about 30 minutes before the end of the year. Humans are latecomers, and our existence represents a small fraction of the Earth's history.

Our ability to measure the age of the Universe is based on measurements of distances and receding velocities of galaxies. Accurate stellar models also put

---

[1]Conversions between different distance units (km, $R_E$, AU, light year, parsec) are given in Appendix II.

constraints on the age of the Universe as stars cannot be older than the Universe. Current estimates put the Universe at about 14 billion years old.

Because of observations of temporal evolution, we can speculate on the question of origins. What is the origin of life, the Earth, and the Universe? What were the original ingredients, environment, and processes that led to the creation of life? How was the Earth formed? What was responsible for the expansion of the Universe and the Big Bang? We have partial answers to these questions and our quest for an understanding of origins is still ongoing.

## 25.3   On the Relative Scale

Our view of humanity's place in the Universe has also evolved over time. After the integration of the cosmological model of Ptolemy with Christianity, we believed that humans were specially created by a supernatural being, placed at the center of the Universe, served by the illuminating Sun and the Moon, and entertained by a sphere of shining stars. The plants and animals were there to serve us and for our consumption. We were unique and we were special. After Copernicus, the center of the Universe was moved from the Earth to the Sun. We later learned that the Sun is just an ordinary star and is only one of about one hundred billion stars in our Milky Way Galaxy, which is also just one of one hundred billion galaxies in the Universe. The galaxies are distributed on a three-dimensional surface that is expanding in the fourth dimension. There is no center of expansion and all galaxies are equivalent.

The historical scientific progress that resulted in the removal of the Earth from the center of the Solar System and the center of the Galaxy has led to the formulation of a generalized principle (often called the Copernican Principle) that states that the Earth holds no special position in the Universe. If this principle is true, then the Earth cannot be the only place where life exists, and life must be common around other stars, and in other galaxies.

The Earth that we live on is also not unique. For over two millennia, humans have maintained the belief that the Earth was special. The Earth is a place with water in lakes and rivers for us to drink, the air in the atmosphere to breathe, grasslands to raise livestock, land to grow crops, and forests to collect wood to build fire and shelters. For these privileges, humans must endure occasional storms, floods, volcanos, earthquakes, and other calamities. This world was given to us for a purpose. We now realize that these amenities are not charities from God but are the result of geological and biological processes. The changing weather and natural disasters are not punishments handed to us but are the result of the natural processes of the Earth. Typhoons and hurricanes are the results of physical interactions between ocean and atmosphere, and volcanos and earthquakes are consequences of plate tectonics.

Space explorations have discovered that other planets and planetary satellites in our Solar System also possess atmospheres, oceans, and mountains. Planets are common around other stars in our Galaxy, and many are rocky like the Earth. The

Earth is nothing but a tiny speck of dust in the vicinity of an ordinary star that happened to have the right conditions to harbor life.

## 25.4   Constitution of Matter

Aristotle proposed that everything on Earth is based on four elements: earth, water, air, and fire. From developments in chemistry, we learned that air is made of a mixture of molecules, water is a molecule composed of oxygen and hydrogen, rock and sand are constituted of chemical compounds, and fire is a combustion phenomenon created through the consumption of oxygen in the air. Molecules such as oxygen in the air, water in the ocean, and carbonates in rocks, can be broken down into atoms, which are entities with protons and neutrons in a nucleus surrounded by clouds of electrons. Every substance on Earth is built from less than 100 chemical elements, which are defined by the number of protons in their atomic nuclei. Stars, instead of being made of a celestial element called "ether," as envisioned by Aristotle, are made from the same list of elements on Earth. As far as we know, there is not an element that uniquely belongs to stars. From the analysis of spectra of galaxies, we have reason to believe that distant galaxies are also made of the same atoms as we have on Earth. Although based on the same chemical elements, there may exist molecules and solids elsewhere in the Universe that are not common on Earth because of the different physical environments. In summary, there is no basic difference in the composition of the Earth, planets, stars, and galaxies.

From our understanding of nuclear physics, astronomers proposed a scenario that the first few elements (hydrogen, helium, and lithium) were synthesized during the Big Bang. After stars were formed from these early elements, the other chemical elements were built up through nuclear reactions inside of stars. Our present relative abundance of chemical elements is the result of nucleosynthesis over several generations of stellar evolution. The chemical elements that we have on Earth are remnants of stellar ejecta into the interstellar medium.

## 25.5   The Role of Humans in the Universe

Under the Christian doctrine in the Middle Ages, humans have a unique place in the Universe. Our bodies are made of matter, but we also have a soul that distinguishes us from animals. We have free will that allows us to choose between good and evil, and the choice to accept Jesus and have an afterlife in Heaven or to descend to Hell. We live on Earth with the privilege of observing the wonders of celestial objects in heaven, but also suffer uncertainties and calamities. God provides us with plants and animals to enjoy as food, but also with diseases for our punishment.

Our interpretations are now different. The atoms in our bodies are inherited from the stars and are no different from the atoms that make up rocks and trees. Our body functions under the rules of biochemistry utilizing molecules that can be artificially synthesized in the laboratory. These organic molecules are also known to be present in meteorites and comets and are routinely synthesized by stars. The theory of evolution puts us on one of the branches of the tree of life, with no intrinsic difference from other species on the tree. When our lives end in this world, our bodies decay as do those of other animals and plants, without a separate soul escaping from the physical body.

Organic compounds, the basic ingredients of life, are widely present in the Universe. Although we have not yet detected any sign of extraterrestrial life, there is widespread belief among scientists that because the building blocks of life are ubiquitously present in the Universe, life could be too. Optimists claim that we will soon detect evidence of extraterrestrial life, probably first within our Solar System on a planetary body such as Mars or Titan. The assumed widespread presence of intelligent life in the Universe suggests that someday we will be contacted or visited by intelligent extraterrestrial beings.

However, there is also uncertainty about the existence of free will. Can future developments in artificial intelligence reproduce human consciousness? Machines can make strategic decisions if the rules of the game (e.g., chess) are known, but can they make ethical choices? Will the species of *homo sapiens* last forever, or will we evolve into another biological species or be replaced by machines?

## 25.6    Looking into the Future by Learning from the Past

After Eratosthenes, Greek scholars must have wondered what is in the other 98% of the world. Over the following centuries, Europeans pondered whether there are other people and civilizations in those unknown lands. They speculated whether these people look like them, or perhaps have wings like birds and horns like goats. Would these inhabitants of faraway lands be more advanced or more backward than they? These questions were finally answered during the age of exploration, where direct contact was made with native peoples in Asia, Africa, the Americas, the Pacific islands, and Australia. In general, the European explorers found the native populations in these regions were less technologically advanced, and as a result, were able to exploit the wealth of these regions and colonize their people.

Our ancestors' state of mind was not dissimilar to ours today wondering about extraterrestrial civilizations. With astronomical observations, we are aware of the existence of planets revolving around distant stars. Are these planets inhabited by other forms of intelligent life? When will advances in technology allow us to make contact with alien civilizations? We probably should not be chauvinistic, as extraterrestrial beings may have evolved based on different rules of biochemistry, even though their bodies may have shared the same atoms and molecules as we do. They probably will not be human-like, and may not even resemble any life forms we know. Given this uncertain outcome, shall we continue to search for extraterrestrial life? My own view is that it will not matter, and the contact will probably be initiated by our extraterrestrial counterparts, rather than by us. I hope that this encounter will be a benign one, unlike the native American encounter with the Europeans in the sixteenth century. One thing is for sure: the day that we have an encounter with an advanced extraterrestrial civilization will be a watershed moment in human history, as our lives will never be the same again.

## Questions to Think About

1. While the Greeks were aware that their known world occupied only a very small fraction of the Earth, the Chinese did not know that the Earth is round and had little idea about the size of the Earth. What effects do you think this perception had on social development in China over its history?
2. Is a steady-state Universe philosophically preferable to an evolving Universe?
3. Do you share the optimism common in the scientific community about our ability to understand the origin of the Universe and the origin of life? If not, why not?
4. In the past 200 years, we have witnessed a progressive rapid growth in man-made technology. Are there limits to the future development of technology?
5. Could our current theory of the structure and origin of the Universe be wrong? Ptolemy's geocentric theory stood for 1500 years before it was discarded. If we cannot claim we have the final word, how long might our current theory hold?
6. Is the origin of the Universe a scientific question or a theological and metaphysical one?

# Chapter 26
# The Common Links in Our Journey

Among all the living species, humans are the only ones who observe the world and ask questions about how and why. The ability to seek patterns from observations and formulate theories to explain and predict past and future events is the essence of science. In this book (and Volume 1), we have explained why astronomy is the oldest science. Recognizing that the celestial objects move with repeatable patterns, our ancestors developed quantitative models to describe these patterns. They were also motivated by the desire to seek possible connections between heaven, Earth, and humanity. The ancient discipline of astrology later became the discipline of astronomy, which systematically studies the positions, movements, brightness, and shapes of celestial objects. Astronomy further evolved into astrophysics, using our knowledge of terrestrial physics acquired in the laboratory to understand the structure of celestial objects. Spectroscopy made possible the identification of atoms and molecules in stars, interstellar space, and galaxies. These identifications suggest that the laws of physics and chemistry are applicable over large spatial scales across the Universe and back in time to billions of years ago.

Ancient observations of the sky and the development of mathematical astronomy played a major role in the adoption of the scientific method as a way of rational thinking in pursuit of truth (or at least the best approximation of the truth). Until the seventeenth century, astronomy was the only science that had accumulated enough data to allow sophisticated mathematical models to be constructed and subjected to quantitative testing of their predictions. The success of astronomy in utilizing the scientific method to seek the truth paved the way for the development of other sciences such as physics, chemistry, biology, and geology. Through complementary advances in these fields, we arrived at our current understanding of our place in the Universe.

© Springer Nature Switzerland AG 2021                                              279
S. Kwok, *Our Place in the Universe - II*,
https://doi.org/10.1007/978-3-030-80260-8_26

## 26.1   A 300-year Success Story

In contrast to the ancient belief that heaven and Earth are fundamentally different, we have learned that the Sun and the stars are made of the same chemical elements as the Earth. From the spectra of stars, nebulae, and galaxies, we see that the same spectral patterns are observed in atoms, molecules, and solids on Earth. This gives us confidence that the laws of physics and chemistry are widely applicable over the Universe. Because galaxies can be millions or billions of light-years from us, and the light we see from them was emitted millions or billions of years ago, this implies that the laws of physics and chemistry that govern the behavior of atoms, molecules, and solids are the same over large distance and temporal scales.

From the work of Newton, we learned that the laws of motion and gravitation that govern terrestrial objects are the same laws that govern the movement of the planets. Einstein's theory of gravitation, which is so successful in explaining the advance of the perihelion of Mercury, is also used to interpret the distance–velocity relationship observed in galaxies as evidence for the expansion of the Universe. This suggests that Einstein's theory applies equally well to the small scale of our Solar System and the large scale of the Universe. Our concept of the beginning and evolution of the Universe is based on the faith that physics as we know it was also valid billions of years ago.

Exploration of the surface of Mars by robotic landers has shown that Mars has been subjected to many of the same geological processes that we observe on Earth. Volcanism and external impacts are also likely to have happened in the other terrestrial planets, Mercury and Venus. With the discovery of thousands of extrasolar planets, we can begin to test whether the processes that govern the Earth's crust, atmosphere, and oceans are also present on planets in other solar systems.

At present, we have only one example of life, and that is on Earth. However, we know that complex organic compounds, which are the building blocks of life on Earth, are also widely present elsewhere in the Universe. Could these organic compounds create similar, or totally different, forms of life? Will these extraterrestrial life forms be governed by the same laws of biochemistry that we have on Earth?

In spite of the past successes, there are prominent cases of unsolved problems in astronomy. Thousands of spectral lines in the millimeter-submm region remain unidentified (Sect. 20.1). Given the wavelength region, they are most likely rotational lines of unknown molecules. In the optical region, there are hundreds of very strong interstellar absorption lines called diffuse interstellar bands that have remained unidentified for over 100 years. The strength of the lines suggests that the carrier molecules must be made of common elements such as hydrogen, carbon, and oxygen. The carrier of these lines must be a significant fraction of the reservoir of carbon in the Universe.

Newtonian physics seems to work well throughout the Universe, except many spiral galaxies seem to have rotational motions that seem to suggest they have masses much greater than the masses contained in their stars (Fig. 26.1). There are also clusters of galaxies that seem to be bounded by more masses than could be

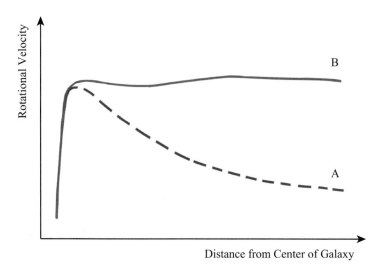

**Fig. 26.1** A schematic diagram illustrating the change in rotational velocity in a typical spiral galaxy. Curve A represents the decline in velocity expected from the visible mass of the galaxy and Curve B shows what is observed. The flatness of the observed rotational curve suggests that there is unseen mass beyond what is visible in the starlight

accounted for by the visible mass of their member galaxies. These observations led to the hypothesis that there is an unknown form of "dark matter" in the Universe that is not the same as the "normal" matter that is represented by atoms. In order to account for the observations, either Newtonian physics works differently on large scales, or there is a large amount of matter that produces gravitational effects but does not emit light.

These unsolved problems tell us that we can learn about the fundamental structure of matter and physical and chemical processes from astronomical observations.

## 26.2   The Interdisciplinary Nature of Science

One of the emerging features of twenty-first-century science is interdisciplinary studies. Previously, scientists used to work on a single discipline and dig deep into the subject with a narrow focus. It is now realized that many problems require knowledge in multiple subjects to solve. For example, the change in the Earth's climate over time is affected by the Sun; the relative movement of the Earth and the Sun; and physical and chemical interactions between the Earth's atmosphere, ocean, and the continents. These interactions also affect biological processes in the biosphere, which in turn feed back into its environmental surroundings. The social and economic behavior of humans also plays a role.

Even astronomy, a subject with a 3000-year history, is no exception. Astronomy has evolved from a purely observational science to a multidisciplinary subject, drawing knowledge from other disciplines such as physics, chemistry, geology, and biology. From knowledge of atomic physics, we learned that stars are made of the same chemical elements as those on Earth. From nuclear physics, we learned that stars generate power through nuclear fusion. From chemistry, we learned that molecules, even complex organic molecules, are prevalent in space. From geology, we learned that geological processes observed on Earth may also apply to other Solar System objects. Our current effort to understand global climate change requires interdisciplinary collaboration by physicists, chemists, biologists, geologists, oceanographers, atmospheric and planetary scientists. We cannot expect to solve the problem of the origin of life without contributions from many different areas of science.

All research is now done by professional scientists and the age of amateur scientists is long gone. The volume of accumulated knowledge in science is so large that it takes many years for a student to acquire enough background to embark on research. In spite of the need for cross-disciplinary fertilization, our scientific training has become increasingly specialized. Candidates for a Ph.D. degree are expected to perform an in-depth study of a narrow problem. How do we reconcile the requirements for specialized training to the need for interdisciplinary research at the frontier of science?

## 26.3   The Path to Discovery

The general public has the perception that scientific research rigorously follows a fixed recipe. This cannot be further from the truth. In fact, science is a creative exercise very similar to the original composition of music and works of art. The major distinctions are that the tool we use is mathematics and that the objects of our creation –scientific theories– are subjected to tests. Scientific discoveries have often been presented as the result of a eureka moment of a single individual. In fact, scientific discoveries are more accurately described as the cumulation of a process in which many people take part and the exact moment of discovery may not be easy to define. Thomas Kuhn, in his book *The Structure of Scientific Revolutions* uses the discovery of oxygen as an example. Although Lavoisier suspected that there was something wrong with the then-popular phlogiston theory of burning and was convinced that something in the air played a part in the burning of matter, the atmospheric agent of burning (oxygen) was not isolated or identified for many years.

When a new idea is proposed, it is often not immediately accepted. We now consider Dalton's atomic theory (Sect. 14.1) as a milestone in the development of chemistry, but it was not perceived as such at the time. Initially, Dalton (himself not a chemist) did not get a good reception from chemists. The problem was that in the early nineteenth century, chemists could not distinguish between compounds and mixtures (e.g., salt in water, oxygen and nitrogen in air, glass), and the law of proportionality required by the atomic theory does not apply to mixtures. Because his theory did not explain everything, it could not be correct, so chemists believed at the time.

However, Dalton's atomic theory could explain the chemical reactions leading to the formation of chemical compounds so neatly that we now define compounds as those which can be explained by the atomic theory and put chemical mixtures into a separate category of substances. The fact that Dalton's theory did not fit all experimental facts did not diminish the success of the theory, it allowed us to better recognize what constitutes a chemical reaction and what does not.

Major discoveries are quite frequently not what originally motivated the research. Copernicus set out to eliminate the concept of equant from the construction of planetary models but ended up with a theory whose implications go beyond planetary motions. The impact of the heliocentric theory on social structure far exceeded what Copernicus had imagined (Chap. 18, Vol. 1). Einstein was motivated to seek a mathematical formulation of Mach's Principle. Although unsuccessful in achieving this goal, this effort led to a new theory of gravitation.

It is interesting to note that the progress of science is not a one-way street. Old debates can be reignited as new evidence is found. The once-discarded theory of catastrophism was revived with the giant impact theory of the formation of the Moon and sudden massive extinction events caused by external impacts or volcanic eruptions.

Serendipity also plays a part in scientific discoveries. Galle discovered Neptune by luck in spite of the erroneous prediction of Le Verrier. Penzias and Wilson found anomalous signals while testing their radio antenna, unaware that they actually had discovered the cosmic background radiation left over from the Big Bang.

What we can discover may also be constrained by time and conditions. When we look back to the most influential books in science, from Ptolemy's Almagest, to Copernicus' *Six Books on the Revolutions of the Heavenly Spheres*, to Newton's *Principia* to Darwin's *On the Origin of Species*, is there a sequential relationship that leads from one to the next? Could Darwin have formulated his theory of evolution without Newton first solving the problem of planetary motions, and instilling a physical interpretation of our natural world? Ptolemy's work was built upon extensive observations of motions of celestial bodies over hundreds of years. Would an alien civilization that lives under a thick atmosphere and lacks knowledge of the celestial bodies develop the sciences of chemistry and biology before physics and astronomy? (see Appendix VIII).

## 26.4   The Human Aspects of Science

Scientific discovery is a human endeavor and requires certain human qualities such as ingenuity and effort to achieve. However, discovery is not the result of a eureka moment and the road to the understanding of Nature has many detours and failures. Perseverance is an element that is often overlooked. It took Copernicus almost 40 years to repeat Ptolemy's calculations for the heliocentric model and Kepler had to perform hundreds of pages of calculations to arrive at the conclusion that Mar's orbit is an ellipse. Darwin developed his theory of evolution through many years of careful observations of the characteristics of living species. Nothing worthwhile is easy.

Science is about the unknown. Scientists working on the frontiers of science and exploring uncharted territories often run into dead-ends and detours and make many errors and wrong guesses in the process. For hundreds of years, astronomers worked hard to add epicycles to better explain planetary motions, on the mistaken assumption that planets move in perfect circles with uniform speed. Although these assumptions seem absurd now, they were treated as sacred at the time. Some concepts that were once very popular in the seventeenth, eighteenth, and nineteenth centuries, such as ether, vitality, phlogiston, and caloric, have all now been abandoned. In the nineteenth century, astronomers spent decades searching for the planet Vulcan which turned out to be nonexistent. These failures are all part of the path of progress of science.

When the timing is ripe for discovery, often several people will come to similar conclusions at the same time. Adams and Le Verrier separately predicted the existence of Neptune. The theory of evolution by natural selection was developed independently by Darwin and Wallace. The periodic table was developed by both Mendeleev and Meyer. The theory of stellar nucleosynthesis was separately worked out by both Hoyle and Cameron. Breakthroughs in science also draw on work by many people. The distance-velocity relationship of galaxies depended on the distance determination by Hubble and redshift measurements by Slipher and Humason, with the theoretical interpretation provided by Lemaître. Just as Newton's formulation of the theory of gravitation was based on the work of Galileo and Kepler, Einstein's theory of special relativity relied heavily on the work of Poincaré and Lorentz.

This reality of discovery is often in contrast to the public perception that it is the work of a lone genius, which tends to give the credit to a single person while ignoring the contributions of others. In this aspect, science is different from arts, as music, paintings, and novels are often the work of a single person. In his book *Everything's Relative: and other Fables from Science and Technology* Tony Rothman cites many examples of scientists and inventors who were not given proper credit for their discoveries and inventions because of their nationality, social status, and other factors. The science community is not immune to social biases as in other human endeavors.

When a novel idea emerges, it must face the political, religious, and social views of the time. Both Copernicus and Darwin delayed the publication of their theories as they knew that they would be opposed by the powerful Church establishment. In the eighth century, the Chinese monk Yi Shin measured the change in altitude of the Pole Star along a 3000 km north-south baseline, but failed to conclude that the Earth was round, even though he had all the data and mathematical tools to do so (Chap. 12, Vol. 1). He probably did realize the conclusion but was unable or unwilling to announce it because of Confucian philosophical beliefs and political reality at the time.

More often than not, oppositions to new ideas are found within the academic establishment. In the words of Arthur Koestler: "The inertia of the human mind and its resistance to innovation are most clearly demonstrated not, as one might expect, by the ignorant mass—which is easily swayed once its imagination is caught—but by professionals with a vested interest in tradition and in the monopoly of learning."

## 26.5   How Science Should Be Taught?[1]

Universities can trace their origins to the institutions of higher learning in the thirteenth century. The early universities were mostly operated by the Church and the subjects of instruction included logic, mathematics, astronomy, and theology. Today's universities cover a wide range of subjects, from humanities and science to practical subjects such as engineering and medicine, and vocational training such as business and law. Among the diverse purposes of university education, science has a unique role to play as the scientific method is the only known way of getting to the truth, or at least to an approximation of the truth. A proper training in science should equip students with mastery of the scientific method and the techniques of quantitative logic, the ability to make a rational judgment, in order to rise above ignorance and prejudice common in our society. The essence of science is to instill a sense of skepticism and to keep an open mind on issues.

However, our current mode of science education is not perfect. Science textbooks often highlight successes and minimize failures, giving students a mistaken perception that the path to truth is easy and self-evident. Ancient astronomers were keenly aware that the motions of the Sun, the Moon, and the planets and stars are regular but complicated. The Sun rises at different times and from a different direction every day; the lengths of the seasons are not equal; the Sun has a daily cycle of 24 hours but the stars have a daily cycle that is 4 minutes shorter; the axis around which the stars turn is tilted from the horizon; eclipses occur repeatedly but the cycles are not simple; the planets generally move through the stars from west to east but reverse directions from time to time. All these phenomena were very difficult to explain but were successfully modeled by Ptolemy, who could also predict the positions of celestial objects hundreds of years into the future.

Textbooks give a trivialized version of science and students who are not aware of the difficulty of the problem often have the mistaken impression that Ptolemy was obviously wrong. Consequently, they fail to appreciate the greatness of Copernicus because they think the heliocentric model was obvious. In fact, the heliocentric theory was not accepted by the academic community for over 100 years as the theory contradicts common sense (if the Earth is turning and moving at high speed, why do not we feel it?). The experimental and observational proof of the heliocentric theory

---

[1] This section is based on the article Kwok, S. 2008, "Science Education in the 21st Century", *Nature Astronomy* 2, 530–533.

did not come until over 200 years after Copernicus in the form of aberration of starlight and Foucault's pendulum (Chap. 2). The theoretical issue of us not feeling the motion of the Earth was not resolved until Newton developed his laws of gravitation and motion (Chap. 3).

What are some of the problems with current science education? The way science is taught at all levels of schooling is often seen as abstract and irrelevant to real life. Students in chemistry and biology are burdened with memorization of facts, and students in physics and mathematics feel that their discipline contents are abstract, and they cannot relate these materials to the real world. Students, in general, fail to see that science is in Nature all around them, and the scientific method is widely applicable in different aspects of their lives.

In school curricula in most parts of the world, science subjects are segregated into physics, chemistry, and biology, and the connections between these fields are usually not emphasized. Memorization, rote-learning, and keyword marking in exams are well-known ills. Young people regard schooling as a game to be won but not as a road to intellectual fulfillment. We are also faced with rapid technological change. A lot of factual information is available on the internet, and artificial intelligence is making certain occupations obsolete. It is therefore much more important to give our students fundamentals that will stay with them for the rest of their lives. These essential tools include language skills such as comprehension, expression, and communication, as well as quantitative skills such as performing analysis, seeing hidden patterns, identifying variables, and formulating solutions to problems.

What is the purpose of university science education? It should not be just to train scientists or engineers but to introduce students to a scientific way of thinking that will make them better citizens. Science education benefits not only the individual but society as a whole. In a democratic country, the collective views of citizens influence the nation's directions. The inability of common people to understand and interpret graphs, statistics, and scientific data could threaten democracy. The internet, in spite of greatly facilitating access to information, has also caused the spread of false information and theories. The internet has made everyone an expert and the public is confused about who the real authorities are. The flood of information makes it necessary for the public to be able to assess, not just access, information and be able to make judgments about true or fake "facts."

While current science curricula often focus on factual knowledge content, I argue that it is more important to teach the process of science. Training should include mastering methods such as building theoretical models, constructing experiments, taking data, revising models based on data, and communicating results. Students should acquire the ability to solve problems by studying examples of previous work. In the process, they should develop free, bold, independent, and creative thinking. They should be able to make rational judgments and rise above the ignorance and prejudice that are prevalent in society.

Science students should be encouraged to develop their sense of curiosity and acquire the confidence to ask questions and challenge assumptions. Science students should be knowledgeable about our world and aware of how Nature works. They should also think analytically and quantitatively, keep an open mind, and remain independent from public opinion. Our goal should be to train students as people of intellect, not for a vocation. Graduates should be versatile enough to take on any job. Most importantly, their education should lay the groundwork for lifelong learning, as society's needs are constantly changing.

The practice of teaching should encourage students to observe their surroundings, rekindle their childhood curiosity, and learn to be aware of natural phenomena. From these observations, they should learn how to think independently, logically, and analytically. Through historical examples, they can learn how to solve problems.

## 26.6  How Science Is Done

For most of the history of science, science was either done by gentlemen of independent means, or was supported by kings, nobles, or the Church. The way that science is done underwent a dramatic revolution in the mid-twentieth century where state-sponsored research became the dominant mode of operation. After the Soviet launch of the artificial satellite Sputnik 1 in 1957, the U.S. government began to inject huge amounts of federal funds into scientific research. Large experimental and observational facilities were built, national laboratories established, and millions of scientists, technicians, and students went on government payrolls through direct funding of government laboratories or indirect funding to universities. Overhead charges on government contracts became such a large component of university budgets that no major university can operate without federal research funds. This U.S. model of state-funded research was copied by Europe and later also by Japan, Korea, and China in Asia.

Science also became much more expensive. Currently, a state-of-the-art ground-based telescope costs hundreds of millions of U.S. dollars, and the capital and operational costs of astronomical and planetary science space missions can run into billions of dollars. Large experimental facilities such as the *Large Hadron Collider* cost close to 10 billion dollars to build, and 1 billion dollars a year to operate. There is no doubt that the large investment of federal funds into manpower and infrastructure is responsible for the rapid progress in science in the late twentieth century.

This new model of doing science means that the way of doing science has changed. Scientists now often work in large teams on common projects decided and run by the team. The mode of a lone scientist working in the basement with homemade equipment has completely vanished. For a scientist to participate in research, he or she has to apply for research grants from funding agencies in order to acquire the laboratory equipment and hire the personnel needed for research. It is not uncommon to find medical research teams with more than 100 members, all supervised by a principal investigator. Astronomers have to apply for observing time for access to telescopes and particle physicists have to gain access to particle accelerators before research can be undertaken. These proposals have first to pass peer reviews, which means that unconventional ideas are unlikely to be approved. The dissemination of results is through publications in journals, which is subject to peer review. While such review processes can eliminate crackpots and unsound ideas, they also promote conformity, which is the greatest enemy to true progress in science.

In ancient times, science was driven by personal curiosity or practical needs such as agriculture, navigation, and healing. In the twenty-first century, the motivation of scientists has changed. Like being a clergyman in the Middle Ages, being a scientist can be a way to make a comfortable living. Although research funding was originally designed to be a means toward a goal (scientific discovery), it can easily be confused with being a goal in itself. Successful accumulation of funding has become signs of power and prestige, upon which honors are bestowed. Leaders of large research teams serve as managers, salesmen, and public relations agents for their teams. The demands for time for these activities, however necessary, remove many team leaders from being directly involved in scientific activities. On the other side, there is tremendous pressure on young scientists to conform to current ideas, as it is unlikely that radical ideas will receive funding. The surest way to "success" is through incremental work on existing models.

Maybe science is a random-walk process and there is no perfect system to maximize the probability of success. Almost all professional scientists argue for increased injection of funds into research, and there is no doubt that much of the progress in science in the last 50 years can be attributed to the increase in financial support for science. It is, however, healthy to take a pause from time to time to make a careful assessment of our current way of doing science.

## 26.7   The Ethics of Science

Like other human activities, science is not immune to fraud. The perpetuators of scientific hoaxes may be motivated by money, vanity, or ideology. One of the most notorious examples is the Piltdown Man. In 1912, a skeleton was discovered in Sussex, England that supposedly represented the long-sought missing link between ape and human. The discovery took the archeology community by storm and was hailed as evidence that modern man originated in Europe rather than Africa. It was not until 1953 that the skeleton was determined to be a modern human skull combined with an orangutan jawbone. Unknowingly, the scientific community was taken for a ride for 40 years.

One of the hottest debates in paleontology was whether birds descend from dinosaurs. In 1999, a 125-million-years-old fossil representing a link between dinosaurs and birds was found in China. The result was announced with much fanfare in Washington, D.C. by the National Geographic Society, which funded the research. It was later discovered that the specimen was a dinosaur tail fossil glued to a bird fossil. Fortunately, in this case, the duration of the hoax was short.

Because hoaxes represent an embarrassment of the scientific establishment, these events are quickly forgotten. Even with all the modern safeguards, we cannot rule out that there will not be future hoaxes so long as the incentives for such hoaxes exist.

As science advances at an ever-faster rate, there is an increasing gap between knowledge possessed by scientists working at the frontiers of research and that of the average layman. Narrowing this gap through public education is a desirable and noble goal. Because scientific research is supported by public funds, scientists have a duty to report their findings to the public. Through the dissemination of scientific research outcomes, the public can be in a better position to influence public policy, with public health and environmental issues being obvious examples. Does smoking cause cancer? Is nuclear power safe? Should we consume genetically modified food? How harmful is air pollution? Is human-caused climate change real? The answers to these questions rely on research and the outcomes can change the way we live our lives. Outreach and popularization of science have therefore become a part of the work of a modern scientist.

Due to the increasing competitiveness of research funding, scientists are under pressure to publicize their work and seek public attention. This unfortunately has led to the premature announcement of results, which is detrimental to the long-term public trust in science. There are several recent examples of highly publicized scientific results which later turned out not to be true. In 1989, the announcement of energy released by nuclear reactions under room temperatures (cold fusion) generated worldwide excitement, but the results were not experimentally confirmed by others. In 1996 NASA announced the discovery of fossils of bacteria in the Martian meteorite Allan Hills 84001 (ALH 84001), but the results did not stand up to further scientific scrutiny. In 2011, the *Oscillation Project with Emulsion-tRacking Apparatus* (*OPERA*) project reported that neutrinos can travel faster than light, which later was found to be due to experimental errors. In 2014, the *Background Imaging of Cosmic Extragalactic Polarization* (*BICEP2*) telescope at the South Pole announced the detection of gravitational waves but the signal later turned out to be an emission from ordinary dust. Public interest in the search for extraterrestrial life has probably motivated various announcements of detection of life on Mars and Titan based just on the detection of methane. The finding of extrasolar planets around the star TRAPPIST-1 also fueled claims of the existence of extraterrestrial life. In 2020, newspapers headlines all over the world proclaimed detection of life on Venus, based on a single spectral line of the molecule phosphine ($PH_3$), which later turned out to be mistaken. All these extravagant claims are not helpful to the credibility of science, and sow public mistrust in scientists and the scientific method.

This raises the question: what is the appropriate behavior of scientists in the public dissemination of scientific results? Scientists are human and mistakes have happened often in the history of science. However, knowingly exaggerating or misinforming the public is unethical behavior. There is a minority school of thought that any publicity is good because it promotes public awareness of science. Even if the results later turn out to be false, proponents maintain that the public has a very short memory and will not hold it against them. They also suggest that the public is interested in exotic science, so the more extravagant the claim, the better. Science needs public funding, so the end justifies the means.

Science is hard. To perform research at the frontier of science requires many years of training and there are no shortcuts. Being a scientist is not just a job, but a calling. Consequently, scientists have social responsibilities beyond their job duties. Scientists are not there to appease or entertain the public, but to inform the public of the hard truth, no matter how unpopular it may be. Scientists should be the conscience of society and must be willing to go against the current of public opinion. Scientists who act this way may have to sacrifice their own self-interest for the long-term good of humanity.

There is a trend of science is becoming increasingly politicized, where only certain ideas are considered acceptable. This does not happen just in authoritarian countries, but in democratic countries as well. A true scientist must be willing to stand up to political pressure. As scientists, we must be tolerant of different ideas, so long as the rigorous scientific method is followed.

## 26.8   Science and Technology

Much of the progress in astronomy has been driven by technology. The use of telescopes greatly expanded the observable universe. With naked-eye observations, only thousands of bright stars could be studied. But with increasingly larger telescopes, we can observe faint stars, gaseous nebulae, and distant galaxies. The installation of spectrographs on telescopes allowed the study of the chemical composition of celestial objects and made astronomy a discipline of physical sciences. The invention of photography and later digital imaging devices such as CCD cameras made possible longer exposures to capture images of fainter objects and made measurements of colors and brightness quantitative. The placement of telescopes on artificial satellites orbiting the Earth bypasses the adverse effects of the atmosphere. Since the first telescopic astronomical observations in the early seventeenth century, our ability to detect faint celestial objects has increased by over a factor of ten billion, as the result of the construction of telescopes of larger aperture, placement above the atmosphere, and improvement in light-detecting devices.[2]

Our view of the Universe has been greatly expanded not only because of increasing the sensitivity limit but also because of expanding color coverage. The invention of detecting devices outside of the visual range, first in the radio and later in infrared, ultraviolet, and X-rays, opened the electromagnetic spectrum beyond visible light for astronomical observations. Space-based telescopes can bypass the

---

[2]Our naked eyes can see stars of the sixth magnitude. A modern ground-based telescope can detect stars (or galaxies) fainter than the 26th magnitude. The *Hubble Space Telescope* can extend this limit to the 31st magnitude. This 25-magnitude difference corresponds to $25*2/5 = 10$ orders of magnitude in brightness (see Appendix III), or a factor of 10 billion.

blockage of the atmosphere and can see the Universe in its full color from radio to gamma rays.

On the theoretical side, the use of computers has allowed complicated mathematical calculations to be carried out in a fraction of the amount of time needed for analytical calculations by hand. The coupling of computing power with large data storage systems allows large data sets obtained from telescopic observations to be archived and analyzed.

The space program has also changed astronomy from being a purely observational science to experimental science. The sending of spacecraft and robotic landers to the Moon, planets, comets, asteroids, and planetary satellites allows us to experiment and directly analyze the environment of other Solar System objects. The ability to collect and return samples to Earth has opened up the entire arsenal of laboratory equipment to analyze celestial materials.

Similarly, the use of the microscope opened our eyes to the small world. We became aware of microorganisms as part of living organisms, which greatly expanded our view of life. The link between bacteria and many diseases led to great progress in medical science (Sect. 18.2). It is interesting to note that traditional practitioners are often skeptical about technological innovations. There were medical doctors who refused to look through a microscope to see bacteria for themselves, as there were contemporaries of Galileo who refused to look at stars through a telescope.

Although the effect of technology on science is obvious, the reverse is not clear, at least not at the beginning. The scientific revolution in the seventeenth century did not immediately lead to technological advances. The industrial revolution of the eighteenth century had more to do with practical inventions (such as the steam engine) than any scientific revelations. It was not until the twentieth century that scientific advances in electricity and magnetism, atomic theory, and hydrodynamics, led to practical devices such as radio, television, telephones, telegraphs, electronics, computers, and airplanes. The coupling between science and technology is so tight that now most people use the terms science and technology interchangeably, not realizing that science is a way of thinking and our view of the world, whereas technology represents the development of practical devices.

## 26.9 Science and Society

Although technology is a significant element in the development of science, it cannot be the only driving force. In this book, we have discussed the effects of science on society, e.g., how the Copernican revolution changed the social structure of Europe. We also learned that society can also affect scientific progress. The flourishing scientific development in ancient Greece virtually came to a full stop after the rise of Christianity in the Roman Empire. The 500 years between 800 and 1300 A.D. were the golden age of Islamic science. What prompted its rise and decline? The conformity imposed by Confucianism was probably the reason why science never developed in China in spite of its relative economic prosperity over two millennia. After the Renaissance, astronomy took off again in Europe, in particular after the invention of the telescope. Although the telescope was introduced to the Muslim world and China in the seventeenth century, it had practically no effect on the development of astronomy in these regions. What were the social constraints that deterred the proper utilization of this wonderful instrument in these societies? The economic power of the United States is probably one of the reasons for it becoming a dominant scientific power in the twentieth century. Did the egalitarian American society and other social factors contribute to its rise? The Soviet Union was a major military power after the second world war. Why did it fail to become a dominant scientific power? There is a lot of discussion on the rising economic power of China in the twenty-first century. Will China also become a scientific power that can rival the United States and Europe?

In Sect. 26.6, we discussed the democratization of science funding that broadened the scientific manpower base and unleashed the power of scientific research. What kind of social structure and economic model would best promote the development of science? Science is often accused of being elitist. Is elitism a necessary condition of doing science? Is it a good or bad thing for society?

## 26.10   The Hidden Assumptions Behind Modern Science

Although not often explicitly stated, most scientists subscribe to a number of principles. Principles are philosophical statements that are deemed desirable, where no counterexamples can be found, but do not require proof. The correctness of these principles is supposedly confirmed by successful examples. A list of commonly adopted principles includes the principle of reducibility, principle of universality, principle of evolution, and principle of relativity.

The principle of reducibility states that everything in the Universe can be reduced to smaller internal constituents. Galaxies are made of stars. Stars are made of gases of atoms. Atoms are made of protons, neutrons, electrons, and other elementary particles. There is no need for non-physical entities such as ether. Rocks and mud on Earth are chemical compounds, and rivers and seas are mainly made of water molecules with various minerals mixed in. Living organisms are made of cells, cells are made of molecules, molecules are made of atoms, atoms are made of elementary particles.

Not only is matter reducible, the operations and interactions of matter are also reducible. Lightning and the aurora borealis are light generated by atomic processes. Typhoons and tornados are the result of the systematic motion of molecules in the air. All biology can be reduced to chemistry and all chemistry to physics. There is no "vitality" that separates the living from the non-living. Humans and chimpanzees have very similar genetic makeup and they function using similar biochemical processes. The mutations that make possible biological evolution can be traced to molecular re-arrangements. Diseases are manifestations of chemical processes, genetic defects, or invasion by bacteria or viruses. Mental illness is due to a chemical imbalance in our brains. Our physical body defines our existence, not our "soul."

The question facing the principle of reducibility is whether there is an end to this reductionism? Is there a family of truly fundamental particles that can no longer be subdivided? If indeed there is such a family of fundamental particles, does it mean that the science of physics has come to an end? What we actually mean by "matter" is also unclear. In common language, matter is a physical entity that can be defined by mass, size, electric charge, and physical properties such as hardness, transparency, solubility, etc. If this is the case, do we consider photons and gravitons to be matter? Both have no mass. Even a familiar concept such as the electron is not easy to define. Although electrons have mass and charge, the way that they interact with other matter is not as straightforward. Most people usually visualize electrons as particles like billiard balls, but according to quantum theory, electrons have the properties of both waves and particles. Scientists believe in the existence of electrons because they are part of a mathematical model that is self-consistent and can correctly predict the spectra of atoms.

The principle of universality states that physical laws apply to everything in the Universe and over all time intervals. The theory of gravity is applicable from the smallest scale, between molecules, to the largest scale, clusters of galaxies. It is gravity that holds a star together and it is gravity that drives the revolution of stars around the center of the Milky Way. The theory that governs the structure of atoms produces the same spectra from atoms in the laboratory and in stars. The rotational transition lines from the CO molecule in a galaxy 10 billion light-years away are the same CO lines we see in the laboratory. These are strong evidence for the universality of the laws of physics and chemistry.

The principle of universality also motivates physicists to strive for further unification. Newton unified the laws of motion in heaven and on Earth. Maxwell unified the phenomena of electricity and magnetism into a single theory and recognized that light is an electromagnetic phenomenon (Sect. 14.4). By the early twentieth century, four fundamental forces were known: the gravitational force, the electromagnetic force, and the strong and weak nuclear forces. By the mid-twentieth century, a mathematical model was developed to unify electromagnetic and weak nuclear forces. Physicists believe that unification is desirable, or even inevitable. There are current efforts to develop a "grand unified theory" to unify all forces of nature.

The main uncertainty in the principle of universality is in the realm of biology. Because we have only one example of life (life on Earth), we cannot state with absolute certainty that the rules that govern extraterrestrial life are the same as those we have for Earth's living organisms. Does extraterrestrial life also undergo evolution by natural selection? We have no knowledge that this is true.

The principle of evolution states that the presently observable diverse range of objects in the Universe have all undergone continuous change, and everything has an origin. The hundred or so chemical elements did not appear all of a sudden together but were built up from small to large through nuclear processes in stars. Biological evolution began with simple organisms and it took billions of years to evolve to the complex organisms that we have today. The Universe has an origin, and it has been evolving. The principle of evolution contrasts with the principle of steady state where the Universe is the same now as it was in the past and will be the same in the future. Everything that we see now came from something else sometime in the past.

This principle also implies that everything has a beginning. The Universe began in a Big Bang, and all life evolved from a universal common ancestor. There was a specific time in the past that the Universe was born (now believed to be 13.8 billion years ago) and the first life originated (now believed to be over 3.8 billion years ago). The Solar System was formed as a spiraling gaseous nebula, from which the Sun, planets, comets, and asteroids emerged. The Earth was formed from the accretion of planetesimals 4.6 billion years ago and its surface structure, oceans, and atmosphere have been changing ever since.

The principle of relativity states that there does not exist a special frame of reference that is unique. We cannot perform an experiment to determine what state of motion we are in. Unfortunately, there is a counterexample: one can easily determine that one is in a rotating frame of reference because of the presence of inertial force. Mach suggested that there is a physical origin for the inertial force, but no physical theory has yet been formulated to explain the inertial force.

There is also the Copernican Principle which states that the Earth holds no special position in the Universe. This principle has been generalized to state that there is no place in the Universe that is unique or special. We have learned that the Universe is expanding in the fourth dimension of space, all galaxies are receding from each other, and there is no center of the Universe. According to this principle, then the Earth cannot be the only place with living organisms, and life must be common in the Universe. Coupled with the principle of evolution, this could imply that life and civilization are continuously changing, and there are life forms and civilizations that are more advanced than others as the result of evolution.

There is also the anthropic principle, which states that the physical condition of the Universe is the way it is because it is the one that allows the development of intelligent life to observe it. This principle, therefore, couples the existence of the present Universe to human awareness of it. The present construct of the Universe is not necessarily the result of design or luck, but the result of the selection effect.

One could argue that we believe in these principles because they have been demonstrated to be successful. However, we should not lose sight that these principles are fundamentally a matter of faith. Astronomers for centuries have believed that the orbits of all celestial bodies are circles, and they move at uniform speed at all times. They believed in this principle because the circle is the most beautiful of all geometric figures and uniform motion is the simplest of all motions. They can also point to the tremendous success of Ptolemy's model in explaining the motion of celestial objects. The discarding of this principle by Kepler laid the foundation of the new physical model of planetary motions by Newton. The dogma of Aristotle lasted almost 2000 years until it was dismantled by the scientific revolution. How long will our current dogma last?

At a more fundamental level, is science universal? When we contact an alien (probably more advanced) civilization, would we share the same knowledge of science or mathematics? One can argue that because we all live in the same Universe, and physics and chemistry are universal, we should share the same theories and interpretations of our understanding of the working of the Universe. We use Newton's and Einstein's theories of gravitation to explain the motion of planets, stars, and galaxies, but there is no guarantee that an alien civilization could not develop another totally different interpretation of the same observed phenomena. We know that languages are not universal. The pictorial writing of Chinese is very different from the phonetic writing of Latin languages, and yet both can express human feelings and emotions. Many of our scientific theories make use of mathematics, which like languages, is a human construct. Even if science is universal, there is no guarantee that the way alien civilization formulates their scientific theories is the same as ours.

## 26.11   Is There a Limit to Science?

Since Hipparchus adopted mathematics as a tool to describe natural phenomena, we have been following this recipe. Ever more complicated mathematical models are constructed to explain gravity (as in general relativity), quantum phenomena (Heisenberg and Schrödinger formulations of quantum mechanics), and the interaction of subatomic particles. Are there natural phenomena that cannot be described by mathematics?

It is interesting to note that the discrete nature of quantum physics allows atoms and molecules in distant celestial objects to be uniquely identified. If the interactions (absorption and emission) between light and matter are continuous rather than discrete, we will not be able to identify distant materials with the degree of certainty that we have today. It is physics on the micro-scale that allows us to understand astronomy on the macro scale. Without a quantum theory of matter and the technique of high-resolution spectroscopy, astrophysics would not have developed to today's sophistication. Is the quantum nature of matter a coincidence? There is no a priori reason why the nature of matter must be quantized. One could easily imagine a structure of atoms that allows electrons to be in any orbit, as planetary orbits are in the Solar System. The role of scientists is to describe Nature, but we cannot say (or we do not know that we can) why Nature behaves in such a way. Science can describe but cannot provide the reason why.

Our current world view is dominated by a mechanical view that galaxies, stars, planets, Earth, plants, animals, and humans are all physical objects that are made of chemical elements operating under the rules of physics and chemistry. Everything has an origin. The Universe has a beginning, stars and galaxies were formed after the Big Bang, chemical elements synthesized in stars and spread across the galaxies, planetary systems condensed out of interstellar gas and dust, and life began and evolved on planets. All these happened under terrestrial physical and chemical laws. As of the early twenty-first century, scientists believe that we have understood the origin of elements, have some clues about the origin of the Universe, and will soon find the source of the origin of life.

With ever more powerful telescopes, we can peek into the distant parts of the Universe, to galaxies billions of light-years away. Is there a limit to how far we can observe? As we look to large distances, we are also looking back in time. Certainly, we cannot look beyond the Big Bang, and in practice, we cannot look beyond the last scattering surface when light decoupled from matter 400,000 years after the Big Bang (Sect. 16.4). To what extent can we extrapolate our knowledge of terrestrial physics to the minutes, seconds, or fractions of seconds after the Big Bang? Because we cannot observe to test these theoretical predictions, are such extrapolations a matter of faith?

Can the scientific method, which has been so powerful in helping us to understand the physical world, be applied to understand the individual and collective human behavior and the working of human society? Can the physical laws that govern our body's digestions and genetic reproduction be extrapolated to understand human emotions such as happiness and love? For almost 2000 years, people's lives were heavily guided by religion. Religious doctrines dictated what we believe and how we behave and offered explanations for everything we can observe. In the last 300 years, scientists have gradually moved to the view that all natural phenomena and human diseases have a physical origin. In many parts of the world, religious rules of behavior are replaced by social laws. Upon death, many of our contemporaries believe that their souls will not go to Heaven or Hell, but instead, their bodies will decompose into soil and dirt. Humans are just a complicated chemical factory that runs like a clockwork. We have almost completely abandoned our previous faith in religious doctrines and replaced it with a new dogma of materialism and reductionism.

There are still some uncertainties in this way of thinking. Our brain can store and process information, and so can a computer. Given sensory input and repeated trials, computers can also learn from its experience. A fast computer with a smart algorithm can outperform humans in a variety of tasks, including games such as chess and Go. Artificial intelligence can also compose music that laymen cannot distinguish from human-composed music. Is there a limit to artificial intelligence? Are there tasks that only a human mind can do? Is our mind just a computer or is our intelligence more than just a functioning machine? A computer can be very intelligent (e.g., in playing chess), but can it develop consciousness and cognition? With all our modern understanding of the structure and operations of human brains, we do not have a conclusive answer to these questions.

If we have learned anything in the history of science, it is that the search for the truth is elusive. Ptolemy's theory of epicycles which successfully explained the movements of planets was replaced by Newton's action-at-a-distance gravitational force, which has since been overtaken by the wrapped space-time field theory of Einstein. With each step, we have been able to achieve better accuracy in matching the observations and expand the scope of the application of the theory. We have certainly made progress, but there is no estimate of how far away the end is, or whether we will ever reach the end.

Every generation believed that they had a good understanding of Nature, and they have been proven wrong time and time again. Less than 300 years ago, scientists believed the Earth was a few thousand years old and all life was created at the same time. We did not know that microorganisms are a major part of the living world and play a major role in causing human diseases. Our knowledge of the Universe was limited to stars within the Milky Way, with no awareness of the billions of galaxies beyond. It would be arrogant of us to believe that we have all the answers now.

What will drive science in the future? Our progress in science in the last three hundred years has been driven by two main factors: advances in technology and the adoption of the scientific method as a way of reasoning. Will these two pillars of science stand forever?

Are discoveries in science inevitable? Is it only a matter of time until all the secrets of the working of Nature will be revealed by the process of science? Is there a limit to what we can explore with science? Science has had a successful run. To what extent are we confident that our current view of the Universe and our place in it is the truth? How long will the new dogma of materialism and reductionism last? Will another way of thinking replace the scientific method? Will the success of science eventually turn out to be an illusion?

## Questions to Think About

1. Is it necessary for modern science to be so expensive? In your opinion, what is the most effective way of supporting scientific research?
2. What role does publicity play in science? Can you think of examples of people who have made great contributions to science but are virtually unknown to the public? Are there any examples of scientific celebrities whose fame is not justified by their work?
3. Will social sciences ever enter the study of the Universe? Could disciplines such as astro-ethnics, astro-sociology, or astro-anthropology become future subjects of study?
4. We have listed some examples of how technological development has helped drive scientific discoveries in the past. What future technological jumps may lead to new discoveries in science?
5. In the times of Copernicus, Tycho, and Kepler, scientific work was sponsored by the Church, kings, or aristocrats, whereas from the mid-twentieth century on, it has been supported by the state. Can different forms of sponsorship of science affect the outcome of scientific research?
6. Is a "grand unified theory" achievable? Why do physicists believe that all forces of nature should be unified into a single theory? When this is achieved, does it mean that it is the end of physics?
7. Religions were once believed to hold the key to truth, but we now believe it is science that leads us to the truth. Will science ever replace the need for religion?
8. What are the limitations of the scientific method in the search for truth?

# Appendix I. Brightest Stars in the Sky

| Name | Bayer | Apparent visual magnitude | Distance (light-years) | Luminosity (times solar) |
|------|-------|--------------------------|------------------------|--------------------------|
| Sirius | α CMa | −1.46 | 8.6 | 25.4 |
| Canopus | α Car | −0.74 | 310 | 10,700 |
| Alpha Centauri | α Cen | −0.27 | 4.4 | 1.5 |
| Arcturus | α Boo | −0.05 | 37 | 170 |
| Vega | α Lyr | 0.03 | 25 | 40.1 |
| Capella | α Aur | 0.08 | 42 | 78.7 |
| Rigel | β Ori | 0.13 | 860 | 120,000 |
| Procyon | α CMi | 0.34 | 11.5 | 6.9 |
| Achemar | α Eri | 0.46 | 140 | 3150 |
| Hadar | B Cen | 0.61 | 392 | 41,700 |
| Betelgeuse | α Ori | 0.7 | 570 | |
| Altair | α Aql | 0.76 | 16.7 | 10.6 |
| Aldebaran | α Tau | 0.85 | 67 | 439 |
| Antares | α Sco | 0.96 | 550 | 75,900 |
| Spica | α Vir | 1.04 | 250 | 20,500 |
| Pollux | β Gem | 1.14 | 34 | 33 |
| Fomalhaut | α PsA | 1.16 | 25 | 17 |
| Mimosa | β Cru | 1.25 | 280 | 34,000 |
| Deneb | α Cyg | 1.25 | 1425 | 196,000 |
| Regulus | α Leo | 1.35 | 79 | 288 |

Table adapted from Kaler, J. The 172 brightest stars through magnitude 3.00 http://stars.astro.illinois.edu/sow/bright.html

© Springer Nature Switzerland AG 2021
S. Kwok, *Our Place in the Universe - II*,
https://doi.org/10.1007/978-3-030-80260-8

# Appendix II. Units of Measurement

## 1. Length

In this book, we have used the international system (SI) of units. The basic unit for length is meter (m), and the commonly used unit for terrestrial distance is kilometer (km, $10^3$ m). For the length of smaller entities, we have centimeter (cm, $10^{-2}$ m), millimeter (mm, $10^{-3}$ m), micrometer (μm, $10^{-6}$ m), and nanometer (nm, $10^{-9}$ m). The size of bacteria is of the order of a micrometer, whereas the sizes of atoms are measured in nanometers.

## 2. Distance

Because of the large astronomical distances, the terrestrial distance unit of km is not practical. For distances to objects in the Solar System, the astronomical unit (AU) is commonly used. One AU is the mean distance between the Earth and the Sun, or about 150,000,000 km. Distances to nearby stars in our Galaxy are often quoted in units of light-years, or the distance traveled by light in one year. In professional astronomical literature, the unit of parsec is used. A parsec (pc) is the distance to a star that will have a parallax of one arc second. The nearest star Proxima Centauri is located at 1.3 pc (or 4.2 light-years). The distance between the Sun and the center of the Milky Way Galaxy is about 8000 pc, or 26,000 light-years. The Andromeda Galaxy, the nearest galaxy similar to our Milky Way galaxy, is at 770,000 pc (or 2.5 million light-years) away. There are 63,240 AU in a light year, so the size of planetary systems is much smaller than the distances between stars.

**Table A1** Distance conversion

|           | km | Radius of the Earth ($R_E$) | Astronomical unit (AU) | Light year | parsec |
|-----------|----|-----------------------------|------------------------|------------|--------|
| km        | 1  | $6.4\ 10^3$                 | $1.5\ 10^8$            | $9.5\ 10^{12}$ | $3.1\ 10^{13}$ |
| $R_E$     |    | 1                           | $2.4\ 10^4$            | $1.5\ 10^9$ | $4.8\ 10^9$ |
| AU        |    |                             | 1                      | $6.3\ 10^4$ | $2.1\ 10^5$ |
| Light year|    |                             |                        | 1          | 3.3    |
| parsec    |    |                             |                        |            | 1      |

© Springer Nature Switzerland AG 2021
S. Kwok, *Our Place in the Universe - II*,
https://doi.org/10.1007/978-3-030-80260-8

Astronomers also use red shift ($z$) as a proxy of distance. For velocities that are small fractions of the speed of light, $z \sim v/c$. By measuring the red shift of a galaxy and assuming a value for the Hubble constant ($H_0$), distance to the galaxy ($D$) is given by $zc/H_0$. For example, a red shift of 0.001 and $H_0 = 70$ km s$^{-1}$/Mpc, implies a distance of 4.3 Mpc.

3. Time

The basic unit of time for astronomy is a day (or more precisely, a mean solar day. For definition see Appendix IV, Vol. 1). A day is divided into 24 hours, each hour into 60 minutes, and each minute 60 seconds. A year (or more precisely, a tropical year. For definition see Vol. 1) is approximately 365.2422 mean solar days. For long time intervals such as geological time or age of the Universe, units of Myr (million, $10^6$ years) or Gyr (billion $10^9$ years) are also used.

4. Angle

Angular separations in the sky are measured in units of degrees ($°$). A full circle is defined as 360 degrees. This number has its origin in the number of days in the year. The number 360 is used because it is close to the number 365¼ and can be divided by 2, 3, 4, 5, 6, 8, 9, 10, 12, 15, etc. Again, since 60 is also an easily divisible number, we divide a degree into 60 arc minutes ('), and an arc minute into 60 arc seconds ("). A one-centimeter coin placed at a distance of 1 km will have an angular size of 2 arc seconds, so one arc second is a very small separation.

Celestial sources are far away and stars appear point-like because of their large distances. The apparent angular size of an object (in unit of radians) is given by the ratio of its physical size to its distance. The Sun has an angular size of about half a degree (0.533 degrees to be exact, or 0.093 radians). At the distance of 1 AU, its physical diameter is $1.4 \cdot 10^{11}$ cm. A star exactly like the Sun at a distance of 10 light-years, has an angular size of $1.5 \ 10^{-8}$ radians, or 0.003 arc seconds. The red supergiant star Betelgeuse in the constellation of Orion, one of the largest stars in the sky, has a radius 887 times the radius of the Sun. At a distance of 642 light-years, it extends an angular radius of $10^{-7}$ radians, or 0.02 arc seconds.

Nebulae are extended objects in the sky. The Orion Nebula extends over about one degree in the sky. At a distance of 1344 light-years, its physical diameter is about 12 light-years. The Andromeda Nebula is a spiral galaxy with an elongated shape and extends over 3 degrees by 1 degree in the sky. At a distance of 2.54 million light-years, its size is over 100,000 light-years.

The Virgo Cluster extends over 15 degrees in the sky. At a distance of 54 million light-years, its physical dimension is 15 million light-years across.

# Appendix III. Astronomical Measurements

Astronomical measurements can be divided into three basic categories: photometry, imaging, and spectroscopy. Photometry measures the brightness (flux) of unresolved (point-like) objects such as stars. Traditionally, stellar fluxes are expressed in units of magnitudes, which is an inverse logarithmic scale of flux. A brighter star has a smaller magnitude. A star that is 100 times brighter has a magnitude value of 5 times smaller. Specifically, the brightness ratio of two stars with a magnitude difference of $m$ is $10^{(-0.4m)}$. For example, if star A has a magnitude of 2, and star B has a magnitude of 4, then star A is brighter than star B by a factor of $10^{-0.4(2-4)} = 6.3$.

For objects that can be spatially resolved, one can take a picture (imaging). Examples of objects with finite angular sizes are galaxies, gaseous nebulae, and star clusters. Imaging can be performed on a telescope using photographic plates or CCD cameras. The total flux of the object is distributed over an angular area in a picture. The technical term for the brightness of an object with the finite angular size is intensity, which is proportional to the amount of energy flow per second per unit area per unit angle. In our everyday life, most of the objects we see with our eyes have finite angular sizes and intensity is what our eyes sense.

Our eyes can also distinguish colors. Sunlight can be spread out by a prism to show a continuous distribution of colors from red to violet. Because stars have different temperatures, their brightness is color dependent. Astronomers use photometric filters to limit light with a narrow range of colors to enter the telescope and measure the flux received within this color range. For example, the visible (V) filter is centered on 0.55 μm and has a width of 0.09 μm. The visual magnitudes of some of the brightest stars are listed in Appendix I. A cool star (e.g., the Sun is cooler than Vega) will be relatively brighter (smaller magnitude) in the red than in the visual range. Using a set of color filters allows astronomers to estimate the temperature of a star.

With a spectrograph, light can be spread out in much finer details and one can see discrete dark lines in the solar spectrum. The use of spectrographs on telescopes allows us to identify atomic and molecular lines in the spectra of stars and nebulae.

© Springer Nature Switzerland AG 2021
S. Kwok, *Our Place in the Universe - II*,
https://doi.org/10.1007/978-3-030-80260-8

# Appendix IV. Photometric Method to Estimate the Distances of Stars

Let $L_O$ be the luminosity (energy output per unit time) of the Sun. The solar flux received by a planet at distance $D_p$ is

$$F_p = \frac{L_O}{4\pi D_p^2}$$

Assuming that the planet has a radius of $Rp$, the rate of solar energy received by the planet is

$$\gamma = \pi R_p^2 F_p$$

Assuming a fraction $\alpha$ (albedo) of the incident energy is reflected, the planetary flux received by Earth at the distance $d$ from the planet is

$$F_{pE} = \frac{\alpha\gamma}{4\pi d^2} = \frac{\alpha R_p^2 L_O}{16\pi D_p^2 d^2} = \frac{\alpha \theta_p^2 L_O}{16\pi D_p^2}$$

where $\theta_p = R_p/d$ is the angular radius of the planet. For a star of the same luminosity as the Sun to be as bright as the planet, the distance of the star ($D_*$) is

$$D_* = \sqrt{\frac{L_O}{4\pi F_{pE}}} = \frac{2D_p}{\theta_p}\sqrt{\frac{1}{\alpha}}$$

For example, Copernicus determined that the planet Saturn has an orbital radius of 9.5 AU from the Sun (Table 16.2, Vol. 1). In opposition, the angular radius of Saturn was observed to be approximately 10 arc sec. Assuming an albedo of 1, a star of the same luminosity of the Sun will have the same brightness as Saturn if it is located at a distance of 400,000 AU. Using the modern value of AU, it would be at a

© Springer Nature Switzerland AG 2021
S. Kwok, *Our Place in the Universe - II*,
https://doi.org/10.1007/978-3-030-80260-8

distance of 6.2 light-years. If the albedo of Saturn is 0.5, the star will be at a distance of 8.8 light-years.

For a planet in the outer Solar System such as Uranus and Neptune, or objects in the Kuiper Belt, $D_p \approx d$, so the brightness of the planet due to reflected sunlight decreases as $D_p^{-4}$.

# Appendix V. Mass of the Milky Way

From the rotational speed of the Sun around the Milky Way and its distance to the center of the Milky Way, the total mass of the Milky Way ($M_{MW}$) can be estimated. From Newton's law of gravitation, the total mass of the Milky Way interior to the orbit of the Sun can be assumed to be concentrated at a point at the center of the Milky Way. If the Sun is situated at a distance of $D_\Theta$ from the center, the gravitational acceleration felt by the Sun is $GM_{MW}/D_\Theta^2$, where $G$ is the gravitational constant. This acceleration must be balanced by the centrifugal acceleration of $V^2/D_\Theta$, where $V$ is the rotational speed of the Sun around the Milky Way. We have therefore

$$M_{MW} = \frac{V^2 D_\Theta}{G}$$

Assuming values of $D_\Theta = 25{,}000$ light-years and $V = 250$ km/s, we have $M_{MW} = 7 \cdot 10^{10}$ mass of the Sun ($M_\Theta$). This mass is likely to be an underestimate because there are stars beyond the orbit of the Sun. Modern estimates of the total mass of the Milky Way make use of the orbital velocities of stars near the outer edge of the Milky Way, giving values as high as a trillion solar masses. It is currently believed that most of this mass is not in the form of stars, but in an unknown form called "dark matter."

© Springer Nature Switzerland AG 2021
S. Kwok, *Our Place in the Universe - II*,
https://doi.org/10.1007/978-3-030-80260-8

# Appendix VI. Examples of Inertial Forces

In our everyday lives, we experience many examples of inertial forces. When we are in a car that goes at a uniform speed and in a straight line, we can easily fall asleep and do not feel anything related to the car's movement. But as soon as the car brakes, we will feel the deceleration and fall forward. A car braking from 80 km/hour to zero in 1 second will cause a deceleration of $(80 \text{ km/hour}–0 \text{ km/hour})/1 \text{ sec} = 22 \text{ m sec}^{-2}$, which is more than twice the surface gravity of $9.8 \text{ m sec}^{-2}$.

When we are on a rotating platform, the centrifugal acceleration is given by $\omega^2 R$, where $\omega$ is the angular speed of the rotation and $R$ is the radius of the platform. When we stand at the edge of a merry-go-round of 1 m radius which turns with a period $T$ of 5 seconds, the angular speed is given by $\omega = 2\pi/T$ and the centrifugal acceleration is $1.6 \text{ m sec}^{-2}$. Although this value is small compared to surface gravity of $9.8 \text{ m sec}^{-2}$, we can still nevertheless feel it.

A human body can tolerate accelerations up to several times of surface gravity. A pilot in a fighter jet traveling at 1000 km/hour turning with a radius of 1 km will suffer a centrifugal acceleration close to 8 times gravity, which is close to the human tolerance limit.

The Earth rotates once every 24 hours and the angular speed of rotation is $\omega = 2\pi/24 \text{ hour}$. The rotational speed at the equator is $\omega R_E = 1668 \text{ km/hour}$, where $R_E$ is the radius of the Earth. The centrifugal acceleration is maximum at the equator and has a value of $\omega^2 R_E = 0.034 \text{ m sec}^{-2}$, which is 0.34% of the value of gravity. The Earth revolves around the Sun for a period of one year at a distance of 1 AU. The rotational speed of the Earth is 10,723 km/hour. The centrifugal force felt by the Earth is exactly balanced by the gravitational attraction by the Sun, which has a value of $GM_\odot/AU^2 = 0.006 \text{ m sec}^{-2}$.

An Earth-orbiting space station at an altitude $h$ of 400 km will suffer a centrifugal acceleration of $\omega^2(R_E + h)$, which must be exactly balanced by the gravitational attraction by the Earth of $GM_E/(R_E + h)^2$. This gives an angular velocity of $\omega = [GM_E/(R_E + h)^3]^{1/2}$. The orbital period of the space station is $2\pi/\omega$, giving a value of 92 minutes. Astronauts inside this orbiting space station will be entirely weightless as they are in total free fall and the net force on them is zero.

© Springer Nature Switzerland AG 2021
S. Kwok, *Our Place in the Universe - II*,
https://doi.org/10.1007/978-3-030-80260-8

# Appendix VII. Astronomy from Other Planetary Systems

Although there may be billions of planets in our Galaxy, their physical conditions and surroundings under which they exist are likely to be very different from Earth. It would be chauvinistic for us to assume that the science of astronomy will develop the same way everywhere in the cosmos.

1. Planetary rotation period and axis orientation

The Earth rotates around its north-south axis once a day, resulting in the apparent diurnal motions of the Sun and the stars. Let us consider the scenario for a planet that does not undergo rotation and faces the same direction in space as it revolves around its parent star. Inhabitants on this planet will have half of a planetary year of daylight and half a year of darkness, with no stars visible for half a year and the other half a year with the sky filled with stars. They will see the same stellar constellations during half a year of darkness and will be unaware of the constellations in the other half of the celestial sphere. In order to gain knowledge of the complete sky, they have to travel to the other hemisphere (Fig. A1).

The Earth's rotational axis is inclined 23.5° w.r.t. to its plane of revolution around the Sun, resulting in seasons. An exo-planet's axis of rotation can be along any direction. For example, the rotational axis of Uranus lies almost exactly in its plane of revolution around the Sun. A person living near the north or south poles of a planet like Uranus will experience continuous sunlight for half a year with no observations of other stars possible and for the other half a year of total darkness and non-stop astronomical observations.

2. Planets in a binary star system

The Earth has one parent star, the Sun. However, over half of all the stars in the Galaxy are in binary or multiple-star systems. Depending on the configurations of the binary system, the motions of planets in a binary stellar system can be extremely complicated, and so will the experience of the inhabitants. For example, as the planet turns away from one of its parent stars, it could be facing the other parent star. Its "day" could still be defined by one rotation period of the planet, but the association

© Springer Nature Switzerland AG 2021
S. Kwok, *Our Place in the Universe - II*,
https://doi.org/10.1007/978-3-030-80260-8

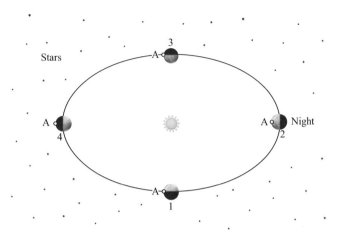

**Fig. A1**  A non-rotating Earth. If the Earth has no rotation, an observer at point A of the Earth will experience daylight for half of the year (from 1 to 3) and darkness (shaded region) for the other half of the year (from 3 to 1). Only stars on the left side of the celestial sphere can be seen by observer A

of bright day and dark night with its rotation would be completely lost. The exact timing of light and dark as the planet rotates and revolves around its parent stars would be determined by the stars' orbits and their respective orientations to the planet. At times there will be two suns in the sky, and other times one during the "day" and the other during the "night."

How would astronomy have developed under these circumstances? Without the regularity of nightly observations, the task would be much more difficult. No doubt with diligence and mathematical analysis, an intelligent population would eventually figure out the motions of the two parent stars and the other planets in this binary system. If the two stars were close enough, their gravitational pull would disturb the planets' orbits, but, hopefully, this perturbation would not be large enough to disrupt completely the emergence of life. Still, the irregularity of the planetary orbits in such a system would make the Copernican view of the solar system, the Keplerian laws of planetary motion, and the Newtonian theory of gravitation much more difficult to formulate.

Part of the reason it took so long for us to figure out the planetary motions in our Solar System is that planets move in elliptical, rather than circular, orbits. Just imagine how difficult a mathematical problem it would be if the planets moved in erratic orbits in a three-body gravitational system, which has no analytical mathematical solution. It would be extremely difficult for Kepler to determine the erratic orbit of Mars by observing it from an erratically moving Earth.

3. An obstructive atmosphere

Although the Earth's atmosphere is mostly opaque in the X-ray, ultraviolet, infrared, and submillimeter wavelengths, it is largely transparent in the visible and radio wavelengths. Except for the occasional presence of clouds, we have an unobstructed view of stars, which mostly radiate in visible light. Although Venus also has a solid surface like the Earth, it has a thick atmosphere which prevents astronomical observations from its surface. It is quite conceivable that life could develop in an exo-planet with a similarly thick atmosphere. Such a civilization would likely develop sciences such as biology, chemistry, and physics, but not astronomy, at least not until they develop the technology of space flight. They must be amazed by the discovery of all the stars and galaxies in the Universe upon emerging from the confines of their planet's atmosphere.

# Review Exercises

1. Galileo's telescopic observations helped heliocentrism gain wider support amongst astronomers and scholars of his time. Do any of his observations disprove geocentrism? If not, in what ways do they lend support to a heliocentric model?
2. What are the implications of Galileo's discovery of the changing positions of sunspots with time?
3. There is a common perception that Copernicus' heliocentric model is superior to Ptolemy's geocentric model because it is simpler. In what ways that it is simpler?
4. How did ancient people determine the length of a year?
5. Tycho believed that a 3rd magnitude star has an apparent size of 1 arc min. If this is true, what would be the physical size of the star if it is located 1 light year away?
6. Light travels at a very high speed. How could scientists measure the speed of light with such a high degree of accuracy in the seventeenth century?
7. If John weighs 70 kg at the north pole, what would be his weight at the equator?
8. If the Earth rotates with a period of 5 hours instead of 24 hours, what would be the effects on our lives?
9. Assuming that Mars is revolving around the Sun in a circular orbit with a radius of 1.52 AU, what are the minimum and maximum amount of time for a radio signal from Mars to reach the Earth? (use the modern value of AU and speed of light).
10. Assuming that Mars is revolving around the Sun in a circular orbit with a radius of 1.52 AU, what is the expected ratio of maximum to minimum brightness of Mars as observed from Earth?
11. What methods did ancient sailors use to navigate at sea?
12. Describe two different methods to determine the direction north.
13. A sea captain uses a sextant to determine the altitude of the star Polaris to be 45°. What is the latitude of the location of his ship?

© Springer Nature Switzerland AG 2021
S. Kwok, *Our Place in the Universe - II*,
https://doi.org/10.1007/978-3-030-80260-8

14. A sea caption uses a sextant to determine the altitude of the Sun at noon to be 60°
    on the date of the summer solstice. What is the latitude of the location of his
    ship?

15. A sea caption uses a sextant to determine the altitude of the Sun at noon to be 60°
    on the date of the winter solstice. What is the latitude of the location of his ship?

16. A sea caption who carries with him at sea a watch that keeps good Greenwich
    time and finds that noon occurs at 17:00 Greenwich time. What is the longitude
    of his ship?

17. From the island of St. Helena (latitude 16° south), what is the declination range
    of stars that can be observed from there?

18. Give some evidence for the hypothesis that Moon shines by reflected sunlight.

19. Why was the number seven considered special in many ancient cultures?

20. Unlike planets, comets brighten and dim over time intervals of weeks or months.
    What is the modern explanation for this behavior of comets?

21. What are the astronomical methods used to search for planet Vulcan?

22. Why was Ceres once considered as the 5th planet from the Sun?

23. List a number of reasons why it is difficult to measure distances to celestial
    objects. Give some examples of how astronomers overcome these difficulties.

24. Assuming the heliocentric model of Copernicus, what are the observational
    evidence that suggests stars are very far away?

25. Give several pieces of evidence to support the statement that "the Sun is a star."
    What are the philosophical and religious implications of this statement?

26. Give an example of an event in your everyday life that makes use of the
    scientific method.

27. What is the significance of the work of Kirchhoff and Bunsen?

28. Which experiment did Fraunhofer perform that demonstrated the Moon and
    planets shine by reflected sunlight?

29. What evidence supports the premise that all celestial objects (planets, stars,
    nebulae, galaxies) are made up of the elements found in the periodic table? Is
    there any counter-example to this statement?

30. In what ways did the invention of the spectroscope contribute to the modern
    dogma of our view of the universe?

31. What are the discoveries that led to the beginning of the discipline of
    astrophysics?

32. Gaseous nebulae are far away. What tools allow astronomers to remotely
    determine the temperature and density in these nebulae?

33. All terrestrial motions have a beginning and an end whereas the planets seem to
    revolve around the Sun forever. How do we reconcile the difference between
    terrestrial and celestial motions with Newtonian physics?

34. How do we know how many stars are there in the Milky Way?

35. Describe the evolution of the cosmological models from the first century A.D. to
    the early twentieth century.

36. Cite three examples in your everyday life that you experience the inertial force.

37. The effect of centrifugal force on the surface of our rotation Earth is small compared to gravitational force and we do not feel it in our everyday life. If the Earth rotates once every hour instead of 24 hours, will we feel this force?
38. Air is transparent and cannot be seen. How did we know that there is something in the air? We cannot see, touch, or smell air, how did scientists determine the composition of air?
39. What are the modern interpretations of the four fundamental elements (earth, water, air, and fire) of Aristotle?
40. What are the modern interpretations of these processes: (i) transformation of ice to water; (ii) burning of wood?
41. How did scientists find out sunlight extends beyond the visible colors?
42. What is the modern interpretation of the "caloric rays" discovered by Herschel?
43. What are the theoretical and experimental steps that led scientists to realize that light is a form of electromagnetic waves?
44. What is the philosophical significance of the unification of electricity, magnetism, and optics?
45. Assuming that stars radiate like a blackbody, what is the peak wavelength of radiation for a star of 10,000 K?
46. Explain why the Sun is the ultimate source of chemical and gravitational energy on Earth.
47. We cannot send a probe to directly observe or measure the deep interior of the Sun. How can we be sure that there are nuclear reactions going on inside the Sun?
48. What is the significance of detecting the element technetium in a star?
49. What are some of the observational evidence that suggests spiral nebulae are external to the Milky Way?
50. List some of the methods that astronomers use to estimate the age of the Universe.
51. Give some examples of how spectroscopy contributes to our modern understanding of the Universe.
52. What is the observational evidence that shows the Universe has been evolving?
53. Why cannot astronomers use telescopes to observe the Universe during the first one hundred years after the Big Bang?
54. Why are the oceans salty?
55. What are the methods geologists have used to estimate the age of the Earth?
56. What is the difference between meteors and meteorites?
57. Ancient astronomers believed that both comets and meteors are atmospheric phenomena. What are the observational facts that led us to abandon this view?
58. What is the evidence that meteors are caused by celestial events?
59. What is the evidence that shows the Moon was formed from materials of the Earth?
60. What is the order sequence of the formation of the Sun, the Moon, the planets, and the stars? Provide evidence in support of your answer.
61. What is the difference between the theories of the evolution of Darwin and Lamarck?

62. Describe the basic mechanism of natural selection.
63. We often hear the statement that "humans evolved from apes." Is this correct? If not, what is the correct evolutionary relationship between humans and apes?
64. What is the significance of the discovery of microorganisms?
65. List some of the major events in the history of the Earth.
66. What is the modern explanation of the unequal lengths of the seasons?
67. What is the observational evidence that the length of a day was shorter in the past?
68. What are some of the effects of human utilization of fire?
69. Astronomers used to believe that stars evolve from hot to cold along the main sequence. What are the observational/theoretical developments that led to the abandonment of this view?
70. Can life develop on a planet around the star Vega?
71. What are the signatures of extraterrestrial intelligent life that we can search for?
72. What are some of the conditions that make a planet habitable?
73. Compare the impact of the inventions of the telescope and the microscope on the understanding of our place in the Universe.
74. Give some examples of historical events or discoveries that led to major changes in our perception of our place in the Universe. Discuss the philosophical significance of each case.

# Glossary

**Absorption** The destruction of light as the result of light encountering matter. During the process of absorption, energy of light is converted to the internal energy of the absorbing material.

**Aliphatics** The term "aliphatics" is derived from the Greek word "aleiphas," meaning "fat." Common animal fats and oils are made of long chains of hydrocarbons.

**Altitude** The angle measured along the great circle perpendicular to the horizon. It is measured from the horizon, positive to the zenith (90 degrees), and negative to the nadir ($-90$ degrees). It is sometimes called the elevation.

**Aphelion** A point in a planet's orbit which is farthest from the Sun

**Aromatics** The class of aromatic molecules has its origin in chemical extracts from pleasant-smelling plants. These are hydrocarbons with a ring-like structure.

**Astrobiology** The study of the possibility of life beyond the terrestrial environment. The field encompasses several sub-disciplines, including the search for life in the Solar System, the search for signals from extraterrestrial intelligence by remote sensing, and the study of the origin of life.

**Astrochemistry** The discipline of studying chemical processes in space.

**Astronomical Unit (AU)** Mean Sun–Earth distance

**Atmosphere** A layer of gas that covers the surface of a planet. Although there is no definite outer boundary of the atmosphere, the altitude of 100 km is commonly used as the line separating the Earth's atmosphere and outer space.

**Atomic number** Number of protons in the atomic nucleus. The atomic number uniquely identifies an element

**Atomic weight** Total number of protons and neutrons in an atomic nucleus. Because of a mix of isotopes, an element's atomic weight represents a weighted average of the atomic weights of all the isotopes.

**Autumnal equinox** Spatial definition: The intersection of the ecliptic and the celestial equator where the Sun goes from positive to negative declination. Temporal definition: date on which the Sun crosses the celestial equator moving southward, occurring on or near September 22.

© Springer Nature Switzerland AG 2021
S. Kwok, *Our Place in the Universe - II*,
https://doi.org/10.1007/978-3-030-80260-8

**Avogadro's law**  Equal volumes of all gases under the same conditions contain the same number of molecules

**Bacteria**  Single-cell microorganisms without a membrane-bound nucleus. Bacteria are found everywhere on Earth, including our digestive tracts and in deep-sea vents.

**Biosphere**  Parts of the Earth where life exists. These include the surface of the Earth, the oceans, and the lower atmosphere.

**Carbon dating**  Use of the ratios of isotopes of carbon to determine the age of an object.

**Catastrophism**  The idea that the Earth was shaped by abrupt, cataclysmic events

**CCD**  Charged coupled device

**Celestial equator**  The projection of Earth's equator onto the celestial sphere.

**Celestial pole**  Projection of Earth's north or south pole onto the celestial sphere.

**Celestial sphere**  An imaginary sphere surrounding Earth to which all stars were once considered to be attached.

**Cepheid variables**  A class of variable stars named after $\delta$ Cephei, the first one discovered. Cepheid variables are giant stars with luminosities thousands of times that of the Sun. Their variability in brightness is the result of pulsation. They possess a unique period-luminosity relationship, allowing them to be used as standard candles for distance determination.

**Conjunction**  Orbital configuration in which a planet lies in the same direction as the Sun, as seen from Earth. The value of elongation at conjunction is 0 degrees for superior planets. For inferior planets, there can be two conjunctions, inferior or superior conjunction.

**Day**  1. The time when the Sun is above the horizon. Opposite of night. 2. The length of time from noon to noon, see Solar day and Mean solar day.

**Deferent**  The large circle upon which the center of the epicycle moves

**Declination**  The elevation of the point on the celestial sphere from the plane of the celestial equator. The celestial equator, therefore, has declination $0°$, and the north and south celestial poles have declinations $90°$ and $-90°$, respectively. It is the equivalent of latitude on the celestial sphere.

**Degree** (°)  The unit of angular measure defined such that an entire rotation is 360 degrees. This unit dates back to the Babylonians, who used a base 60 number system. The number 360 likely arose from the Babylonian year, which was composed of 360 days (12 months of 30 days each). The degree is subdivided into 60 (arc) minutes per degree, and 60 (arc) seconds per (arc) minute. $1° = 60'$ and $1' = 60''$.

**Dendrochronology**  Tree ring dating

**Diffraction**  The pattern of light that results from light passing through a small aperture.

**Diurnal motion**  The daily motion of the Sun and the stars

**East**  One of the two intersection points of the celestial equator and the horizon. If we face north, east is 90 degrees to the right.

**Ecliptic**  The path of the Sun on the celestial sphere over the course of a year.

**Elements**  A term introduced in the late eighteenth century to refer to entities that cannot be broken down into a simpler substances by a chemical reaction.

**Elongation**  Angular separation between a planet and the Sun. The angle between the line joining the planet and Earth and the line joining the Sun and Earth.

**Emission**  The creation of light by matter. The light emitted can be in a continuous or discrete (quantized) forms.

**Enzymes**  Proteins that act as catalysts to accelerate biochemical reactions

**Epicycle**  A circle whose center is on the boundary of another circle.

**Ether (or aether)**  1. A hypothetical substance that acts as medium for light propagation. 2. An element that makes up celestial bodies.

**Eukaryotes**  An organism with a clearly defined membrane-bound nucleus. The earliest eukaryotes appeared about 1.8 billion years ago.

**Evolution**  In general, it means a process of gradual change, for example as in geological or cosmic evolution. More specifically, in biology, it refers to the process which living organisms change from their ancestors, including the appearance of new species.

**Exogenous delivery**  the delivery of materials to Earth from the outside

**Extrasolar planets**  Planets orbiting stars outside the Solar System

**Extremophiles**  Organisms that survive in extreme environmental conditions such as high temperature, high acidity, high salinity, or high ionizing radiation backgrounds.

**Fission**  The process that splits an atomic nucleus into smaller components

**Fictitious force**  A force that does not originate from matter

**Fireball**  A meteor that is brighter than any of the planets

**Flux**  The amount of radiative energy flow per unit time per unit area. Our perception of brightness of a star (a point-like object in the sky) is proportional to the flux we receive from the star.

**Frequency**  A measure of the color of light. Frequencies are measured in units of Hertz (cycles per second).

**Fusion**  The process of combining several smaller atomic nuclei into a larger one.

**Geocentric model**  A cosmological worldview where the Earth is located at the center of the Universe.

**Geochronology**  The science of determining the age of rocks, fossils, and sediments.

**Globular clusters**  Hundreds of thousands of stars bound together by gravity into a spherical group. They are generally found in the haloes of galaxies and contain stars that are older than stars in the disc of the galaxies.

**Great circle**  A great circle on the surface of a sphere is the circle that divides the sphere into two equal hemispheres. The plane that defines the great circle also passes through the center of the sphere. The radius of the great circle is therefore equal to the radius of the sphere.

**Half-life**  The length of time for a population of radioactive isotopes to decay into half of its original number.

**Heavy bombardment** The collision of large number of asteroids with the Earth, Moon, Mercury, Venus, and Mars during the early history of the Solar System.

**Heliocentric** Sun-centered

**Horizon** The maximum visible extent of the horizontal plane on which an observer stands. On a spherical Earth, the horizon of an observer is the tangential plane perpendicular to the line to the center of the Earth.

**Hour** 1. The unit of time measure defined such that the period is one twenty-fourth of a mean solar day. The hour is subdivided into 60 minutes per hour, and 60 seconds per minute. $1^h = 60^m$ and $1^m = 60^s$. 2. The unit of measure of right ascension representing 15 degrees, or one twenty-fourth of a great circle. $1^h = 15°$, $1^m = 15'$, and $1^s = 15''$. Hence, $1'' = 0.0667^s$.

**Ice ages** Periods of the Earth when one or both of its polar regions are covered with ice

**Inferior conjunction** An inferior conjunction occurs when an inferior planet (Mercury or Venus) lies along a straight line between the Earth and the Sun. At this point, the elongation is zero degrees.

**Inferior planet** Planets whose orbits lie within the orbit of the Earth. Mercury and Venus are the two inferior planets.

**Inertial force** A fictitious force that is felt by someone in a non-inertial frame of reference.

**Inertial frame of reference** A frame of reference which is either at rest of moving uniformly in a straight line.

**Ions** Atoms having lost one or more electrons and therefore left with a net positive charge.

**Island universes** Stellar systems outside of the Milky Way. The modern term for island universe is galaxy.

**Isotopes** Variation of an element with the same number of protons but different number of neutrons in the nuclei. Isotopes of an element will have the same atomic number but different atomic weights

**Isotropic** Same in all directions

**Interference** When two waves interact with each other, it can result if an increase in amplitude (constructive interference) or decrease in amplitude (destructive interference) depending on their relative phases.

**Interplanetary dust particles** micrometer-size particles from asteroids or comets that enter the Earth's atmosphere.

**Insoluble organic matter** the insoluble component of organic matter in carbonaceous meteorites.

**Interstellar medium** The general space between stars. The Milky Way galaxy contains about 100 billion stars, but the stars are widely separated from each other. The space in between, however, is not totally empty. There are low concentrations of gas and dust in the vast volumes between stars. These materials are referred to as the interstellar medium.

**Kelvin** A temperature scale with the same degree magnitude as the Celsius scale but the starting point ($0°$ Kelvin) is set at absolute zero, 273 degrees below the freezing point of water (zero degree Celsius).

**Law** An empirical relationship between two or more observable quantities.

**Light year** A unit of distance equal to the distance traveled by light in a year. It has a value of $9.5 \ 10^{12}$ km or 63,240 AU.

**Liquid** A state of matter that can transform its shape to that of its container while maintaining a nearly constant volume.

**Luminosity** The total amount of energy radiated by a star per unit time. This is equivalent to the concept of power used for light bulbs.

**Magnitude** A logarithmic scale unit to measure stellar brightness. See Appendix III.

**Main sequence** The band on the luminosity-temperature plot that many stars occupy. It corresponds to the period of a star's life that relies on nuclear fusion of hydrogen into helium in the core for power.

**Mantle of the Earth** A rocky shell of about 2900 km thick which separates the core and the crust of the Earth.

**Materialism** The belief that everything that exists is matter or results from matter (including life and consciousness).

**Mean solar day** Average length of time from one noon to the next, taken over the course of a year.

**Meridian** An imaginary line on the celestial sphere through the north and south celestial poles, passing directly overhead at a given location. A meridian is a line of constant longitude.

**Meteor** A streak of light caused by the entry of an extraterrestrial solid into the Earth's atmosphere.

**Meteorites** Remnants of meteoroids on the ground having survived their trips through the atmosphere

**Meteoroids** Small interplanetary solids orbiting the Sun. Meteoroids are similar to asteroids except smaller ($<10$ m). When a meteoroid enters the Earth's atmosphere, it creates the phenomenon of meteor.

**Minerals** A naturally occurring solid element or chemical compound, usually an inorganic substance with a crystalline structure.

**Minor planets** Astronomical objects that orbit the Sun and are not planets or comets. They include objects such as asteroids and Kuiper Belt Objects.

**Minute** 1. The unit of time, one-sixtieth of an hour. See Hour definition 1. 2. The unit of angle, one-sixtieth of a degree. See Degree. 3. The unit of angle, one-sixtieth of an hour. See Hour definition 2.

**Non-inertial frame of reference** A frame of reference which is either accelerating or rotating

**Noon** The instance when the Sun is the highest in the sky in one day. It is also the time when the Sun crosses the local (celestial) meridian.

**North** Found by locating the north celestial pole and dropping a vertical line from it to the horizon.

**Novae** A form of exploding stars that brightens on time scale of days. Novae are the result of binary evolution where matter is accreted to the surface of a white dwarf from its companion. When sufficient materials accumulate, nuclear reaction converting hydrogen into helium is initiated, resulting in a sudden increase in the stellar luminosity.

**Obliquity of the ecliptic** The angle between the plane of the ecliptic relative to the celestial equator. A modern interpretation of this term is the tilt of the earth's rotation axis relative to the orbital plane of the Earth around the Sun.

**Oparin-Haldane hypothesis** The origin of life on Earth can be understood through the laws of chemistry and physics.

**Opposition** Orbital configuration in which a planet lies in the opposite direction from the Sun, as seen from Earth. The value of elongation at opposition is 180 degrees.

**Panspermia** The theory that life is common in the Universe and the seeds of life are transported across the Universe

**Parallax** The change in the relative positions of stars as the result of the changing position of the observer.

**Perihelion** The point in a planet's orbit which is closest to the Sun

**Photometry** The astronomical technique to measure the brightness of celestial objects. Because stars and distant galaxies have small angular sizes, all their light is essentially concentrated in one point. The measurement of the brightness of small light sources (called "flux" in technical terms) is different from our usual concept of brightness in everyday life, which is distributed over an extended area (called "intensity" in technical terms).

**Planetary conjunction** Planets coming together sharing similar apparent positions in the sky. Planetary conjunctions were believed to have astrological significance.

**Planetary transit** Orbital configuration in which an inferior planet is observed to pass directly in front of the Sun.

**Planetesimals** Small-sized (1 to 10 km) bodies formed out of the solar nebula. Planets are believed to have formed from aggregation as the result of collisions between planetesimals.

**Pole star** The bright star closest to the celestial pole. Currently, the Pole Star in the north is Polaris. There is no pole star in the south.

**Polyhedron** Three-dimensional solids with flat polygonal faces and straight edges.

**Prebiotic molecules** A subset of organic molecules which are the structural basis of biomolecules, which are molecules directly responsible for the functions of life. Prebiotic molecules can be considered as precursors to biomolecules in the pathway to the origin of life.

**Precession** The slow gyration of the rotation axis of the Earth relative to the ecliptic polar axis as the result of external gravitational influence. It makes the vernal equinox drift slowly around the zodiac. This drift is in a clockwise direction as seen over the north ecliptic pole.

**Primordial soup** The liquid environment consisting of organic molecules from which life arose on the primeval Earth.

**Principle** An idea that is assumed to be true without a need for proof.

**Principle of Equivalence** There are no experimental means that we can distinguish an inertial force from a gravitational force.

**Principle of relativity** One cannot tell by performing any experiment whether one is at rest or moving uniformly in a straight line

**Prokaryotes** A unicellular organism without an internal membrane-bound structure. Both bacteria and archaea are prokaryotes.

**Proteins** Long polymers of amino acids

**Protista** Simple eukaryotic organisms

**Radian** A unit of measure of angles. 180 degrees of angle equal to $\pi$ radians. One degree is 0.0175 radians and one arc second is $2.42407 \ 10^{-6}$ radians.

**Reddening** Change in the color of a star as the result of scattering by foreground interstellar solids along our line of sight.

**Red giants** The evolutionary stage of stars after the Main Sequence. When hydrogen in a stellar core is exhausted by nuclear burning, a star begins to burn hydrogen in an envelope surrounding the helium core. The stellar envelope expands and the star becomes more luminous.

**Reflection** The change in direction of light back to the incoming direction upon hitting a surface of matter.

**Refraction** The change in direction of light as the result of light passing from one medium to another. The most common examples of refraction are light entering water or passing through a piece of glass

**Rotational transitions** A molecule can rotate around a common axis. The simplest example one can have is a two-atom molecule such as CO. The C and O atoms rotate around an axis perpendicular to the line connecting the two atoms. Quantum principles dictate that the molecule can have only discrete rotational speeds. When they change from a higher to lower rotation rate, radio waves (usually in the millimeter wavelength range) are emitted.

**Satellites** Natural bodies that revolve around planets. For example, Titan is a satellite of Saturn. It is also referred to as a moon of Saturn as the Moon is a satellite of the Earth. Artificial satellites are man-made objects that are sent into orbits around the Earth. An artificial satellite can in principle stay in orbit forever, but low-orbit satellites suffer from atmospheric drag and will eventually fall back on Earth.

**Scattering** Change in the direction of light upon encountering matter. Scattering is a general form of reflection. The change in direction can be either forward or backward. The term reflection is used only in backward scattering.

**Scientific notation** The expression of a number in the form $a \times 10^p$, where $p$ is an integer called the "order of magnitude". For example, the scientific notation of 101,325 is $1.01325 \times 10^5$. The order of magnitude is 5.

**Second** 1. The unit of time, one-sixtieth of a minute. See Hour definition 1. 2. The unit of angle, one-sixtieth of an arc minute. See Degree and Hour definition 2.

**Sidereal day** The time between successive risings of a given star, or the time for a star to pass the celestial meridian on successive nights. One sidereal day $= 23^{\text{h}}56^{\text{m}}4.091^{\text{s}}$ which is roughly 4 minutes shorter than a solar day.

**Sidereal month** Time required for the Moon to complete one trip around the celestial sphere (27.32166 days, $27^{\text{d}}7^{\text{h}}43^{\text{m}}11.5^{\text{s}}$).

**Sidereal period (of a planet)** Time required for a planet to complete one cycle around the Sun and return to the starting position among the stars. It is also called the sidereal orbital period.

**Sidereal year** Time required for the constellations to complete one cycle around the sky and return to their starting points, as seen from a given point on Earth. Earth's orbital period around the Sun is 1 sidereal year (365.256 mean solar days), or 20 minutes longer than the tropical year because of precession.

**Silicates** Silicate is a mineral that is commonly found on the surface of the Earth. Silicate is primarily made up of elements silicon and oxygen but can contain other metals such as iron and magnesium. An amorphous form of silicates is commonly found in old, oxygen-rich, red giant stars.

**Smoke** Solid particles generated as the result of burning. Examples include the burning of wood, candles, oil lamps, etc. When the burning is not complete, the surplus carbon atoms congregate themselves into particles that make up smoke.

**Soot** Carbon particles produced by the incomplete combustion of hydrocarbons. The most commonly observed example is automobile exhaust generated by an internal combustion engine (in particular diesel engines).

**Solar day** The period of time between one noon and the next.

**Solar flux** The rate of solar radiation reaching the Earth per unit area. Current value is 340 W m$^{-2}$.

**South** The opposite direction (180 degrees) from north.

**Spectroscopy** A technique of splitting light into small intervals of color. Because atoms and molecules emit light of unique frequencies, spectroscopy is the technique that allows us to detect atoms and molecules in space.

**Spontaneous creation** The hypothesis that life can arise from nonliving matter

**Stellar atmosphere** Because stars are gaseous objects, the term "atmosphere" has a different meaning from the atmosphere of the Earth. The Earth has a solid surface, and the gaseous atmosphere is easily distinguishable. In a star, there is no clear physical separation between the atmosphere and the star itself. Instead, astronomers use the concept of "stellar atmosphere" to refer to the top layer of a star which we can see through. Beyond a certain depth, the atmosphere is no longer transparent. This point is called the "photosphere" which refers to the surface of the star. It is important to realize that the photosphere is not a physical surface.

**Summer solstice** Spatial definition: point on the ecliptic where the Sun is at its northernmost point above the celestial equator. Temporal definition: the date on which the Sun reverses direction from going north to going south, occurring on or near June 21.

**Superior conjunction** A superior conjunction occurs when an inferior planet lies along a straight line joining the Earth and the Sun but is on the opposite side of the Sun from the Earth.

**Superior planet** Planets whose orbits lie outside that of the orbit of the Earth, specifically, Mars, Jupiter, Saturn, Neptune, and Uranus.

**Synodic month** Time required for the Moon to complete a full cycle of phases (29.53059 days).

**Synodic period** Time required for a planet to return to the same apparent position relative to the Sun, e.g., from opposition to opposition, or from inferior conjunction to inferior conjunction.

**Time zone** Region on Earth in which all clocks keep the same time, regardless of the precise position of the Sun in the sky, for consistency in travel and communications. The standard time zones have been adopted around the world since 1884. The time of each zone is defined to be the local mean solar time of the central longitude of the zone except some curving of the time zone boundaries is introduced to cater for the non-straight boundaries of some countries.

**Tropic of Cancer** The northernmost latitude that the Sun can be seen directly overhead. For the year 2013 it is at the latitude of 23° 26' 14" N.

**Tropic of Capricorn** The southernmost latitude that the Sun can be seen directly overhead. For the year 2013, it is at latitude 23° 26' 15" S.

**Tropics** The geographical region between latitudes of 23.5 N and 23.5 S, between the Tropic of Cancer and Tropic of Capricorn.

**Tropical year** The time interval between one vernal equinox and the next. It is approximately 365.2422 mean solar days. In terms of hours and minutes, it is 365 days, 5 hours, 48 minutes, 45.19 seconds.

**Tropical period** The amount of time for a planet to go once around the ecliptic

**Ultraviolet** Light beyond the bluest color that human eye can see.

**Uniformitarianism** The idea that changes on Earth are gradual and happen over long time scales

**Vernal equinox** Spatial definition: one of the two intersections of the ecliptic and the celestial equator when the Sun passes from negative to positive declination. Temporal definition: date on which the Sun crosses the celestial equator moving northward, occurring on or near March 21.

**Vibration** A molecule can undergo vibrational motion such as stretching and bending. For a two-atom molecule, the link between the two atoms can stretch into different lengths. A 3 or more atom molecule can also bend (change of the angles between atoms). A jump from a higher rate of stretch (or bend) to a lower rate of stretch (or bend) gives off infrared light of characteristic wavelengths, therefore allowing us to identify the motion.

**Vitalism** The belief that a living organism is more than the sum of its chemical parts.

**Vitality** A hypothesized element that separates living from non-living.

**Wavelength** A quantitative measure of color of light. The human eye response range extends from red (corresponding to a wavelength of approximately 0.7 micrometers) to violet (a wavelength of approximately 0.3 micrometers).

**West** One of the intersections of the celestial equator and the horizon. If we face north, west is 90 degrees to the left.

**Winter solstice** Spatial definition: point on the ecliptic where the Sun is at its southernmost point below the celestial equator. Temporal definition: the date on which the Sun reverses direction from going south to going north, occurring on or near December 21.

**Year** See tropical year

**Zodiac** The 12 constellations on the ecliptic. With modern constellation boundaries defined by the International Astronomical Union (IAU) in 1930, the ecliptic also goes through the modern constellation of Ophiuchus.

**Zodiac signs** The ecliptic is divided into 12 equal zones, each is assigned a sign, in the order of Aires, Pisces, Aquarius, Capricornus, Sagittarius, Scorpius, Libra, Virgo, Leo, Cancer, Gemini, and Taurus.

# Further Reading

## Chapter 1

Evans, J. (1998). *The history and practice of ancient astronomy.* Oxford University Press.
Kuhn, T. S. (1957). *The Copernican revolution.* Harvard University Press.
Koestler, A. (1959). *The sleepwalkers.* Hutchinson.
Walker, C. (1996). *Astronomy before the telescope.* British Museum Press.

## Chapter 2

Danielson, D. R. (2000). *The book of the Cosmos.* Perseus.
Hirshfeld, A. (2001). *Parallax: The race to measure the Cosmos.* W.H. Freeman and Company.

## Chapter 4

Van Helden, A. (2004). *Huygen's ring, Cassini's division, and Saturn's children.* Smithsonian Institution.

## Chapter 5

Sobel, D. (1995). *Longitude: The true story of a lone genius who solved the greatest scientific problem of his time.* Penguin.
Wilson, D. (2003). *A brief history of the circumnavigators.* Robinson.

© Springer Nature Switzerland AG 2021
S. Kwok, *Our Place in the Universe - II*,
https://doi.org/10.1007/978-3-030-80260-8

# Chapter 6

Dick, S. J. (2013). *Discovery and classification in astronomy: Controversy and consensus.* Cambridge University Press.
Levenson, T. (2015). *The hunt for Vulcan.* Random House.

# Chapter 9

Kwok, S. (2007). *Physics and chemistry of the interstellar medium.* University Science Books.

# Chapter 10

Fernandez, J. A. (2005). *Comets: Nature, dynamics, origin, and their cosmogonical relevance.* Springer.
Goldsmith, D. (1887). *Worlds unnumbered: The search for extrasolar planets.* University Science Books.

# Chapter 11

Dick, S. J. (1982). *Plurality of the worlds: The origins of the extraterrestrial life debate from Democritus to Kant.* Cambridge University Press.

# Chapter 13

Bernstein, J. (1973). *Einstein.* Penguin.
Collins, H. M., Pinch, T., & The Golem. (1993). *What everyone should know about science.* Cambridge University Press.
Mermin, N. D. (1968). *Space and time in special relativity.* McGraw Hill.
Reichenbach, H. (1958). *The philosophy of space and time.* Dover.
Sciama, D. W. (1969). *The physical foundations of general relativity.* Doubleday.

# Chapter 14

Goldstein, M., & Goldstein, I. F. (1978). *How we know.* Plenum Press.

# Chapter 15

Hoyle, F. (1994). *Home is where the wind blows: Chapters from a cosmologist's life*. University Science Books.

# Chapter 16

Bartusiak, M. (2010). *The day we found the universe*. Panthon Books.
Gamow, G. (1947). *One two three ... infinity*. Viking.
Mitton, S. (2005). *Conflict in the Cosmos: Fred Hoyle's life in science*. Gardners Books.

# Chapter 18

Porter, R. (1996). *Cambridge illustrated history of medicine*. Cambridge University Press.
Kennedy, M. (2004). *A brief history of disease, science, and medicine*. Asklepiad Press.

# Chapter 19

Brack, A. (1998). *The molecular origins of life*. Cambridge University Press.
Gargaurd, M., Barbier, B., Martin, H., & Reisse, J. (2005). *Lectures in astrobiology* (Vol. 1). Springer.
Garguard, M., et al. (2015). *Encyclopedia of astrobiology*. Springer.
Horneck, G., & Rettberg, P. (2007). *Complete course in astrobiology*. Wiley.
Kolb, V. (2019). *Handbook of astrobiology*. CRC.
Smith, E., & Morowitz, H. (2016). *The origin and nature of life on earth: The emergence of the fourth geosphere*. Cambridge University Press.

# Chapter 20

Kwok, S. (2011). *Organic matter in the universe*. Wiley.

# Chapter 21

Christian, D., Brown, C. S., & Benjamin, C. (2014). *Big history: Between nothing and everything.* McGraw Hill.

Kwok, S. (2013). *Stardust: The cosmic seeds of life.* Springer.

Lunine, J. I. (2005). *Astrobiology.* Addison Wesley.

# Chapter 23

Hoyle, F. (1955). *Frontiers of astronomy.* Harper and Row.

Kwok, S. (2000). *Origin and evolution of planetary nebulae.* Cambridge University Press.

Kwok, S. (2001). *Cosmic butterflies.* Cambridge University Press.

# Chapter 24

Delsemme, A. (1998). *Our cosmic origins: From the big bang to the emergence of life and intelligence.* Cambridge University Press.

Dick, S. J. (2018). *Astrobiology, discovery, and societal impact.* Cambridge University Press.

Goldsmith, D., & Owen, T. (2001). *The search for life in the universe.* University Science Books.

Shklovskii, I. S., & Sagan, C. (1966). *Intelligent life in the universe.* Holden-Day.

Ward, P. D., & Borwnlee, D. (2000). *Rare earth: Why complex life is uncommon in the universe.* Copernicus Books.

# Chapter 26

Bernal, J. D. (1954). *Science in history.* MIT Press.

Dick, S. J. (2013). *Discovery and classification in astronomy: Controversy and consensus.* Cambridge University Press.

Kuhn, T. S. (1996). *The structure of scientific revolutions* (3rd ed.). U of Chicago Press.

Rothman, T. (2004). *Everything's relative and other fables from science and technology.* Wiley.

Zimmermann, L. (2011). *Bad science: A brief history of bizarre misconceptions, totally wrong conclusions and incredibly stupid theories.* Eagle Press.

# Index

© Springer Nature Switzerland AG 2021
S. Kwok, *Our Place in the Universe - II*,
https://doi.org/10.1007/978-3-030-80260-8